James Ripp

Engineering field theory with applications

To Lynn, Ann, Richard, David and Peter and to Sona

Engineering field theory with applications

LEO SETIAN

Professor of Electrical Engineering, John Brown University, Arkansas

CAMBRIDGE
UNIVERSITY PRESS

Published by the Press Syndicate of the University of Cambridge
The Pitt Building, Trumpington Street, Cambridge CB2 1RP
40 West 20th Street, New York, NY 10011–4211, USA
10 Stamford Road, Oakleigh, Victoria 3166, Australia

First published 1992

Printed in Great Britain at the University Press, Cambridge

British Library cataloguing in publication data

Setian, Leo
Engineering field theory with applications.
1. Engineering. Mathematics
I. Title
620.4015123

Library of Congress cataloguing in publication data

Setian, Leo
Engineering field theory with applications/Leo Setian.
p. cm.
Includes bibliographical references and index.
ISBN 0 521 37554 1 ISBN 0 521 37697 1 (pbk.)
1. Field theory (Physics) 2. Engineering mathematics. I. Title.
QC173.7.S47 1992
530.1′4—dc20 91-9212 CIP

ISBN 0 521 37554 1 hardback
ISBN 0 521 37697 1 paperback

CONTENTS

PREFACE

Electrons flowing through a conductor; fluid flowing through a pipe; heat moving along a cooling fin – what do these engineering fields have in common? It's the mathematics. They all use the same mathematical models to describe the physical behavior.

Engineering field theory with applications is a text which presents an interdisciplinary approach to the physical fields of interest to the engineer, mathematician or physicist. In view of the demand for greater versatility in engineering and the sciences, the text develops the general framework or core material for numerous engineering disciplines. The analogous patterns in fields such as electrical, magnetic, gravitational, fluid and thermal are noted; at the same time their distinctives are revealed. Included in the text is an applications chapter using the theory to develop numerical and experimental solutions to real world problems.

The book is the result of lectures given for several years in the Division of Engineering at John Brown University. It is intended as a core course for electrical and mechanical engineering students who typically follow with courses such as electromagnetics, fluid mechanics, or thermodynamics.

What justifies a text which departs from the traditional approach to teaching engineering in separate compartments? Most engineering students begin to specialize in the freshman or sophomore years. Most courses taught to *all* engineers are either taught by the physics or mathematics departments or are taught freshman and sophomore as engineering courses. It is unusual to have an upper level course taught to all engineering students covering many engineering disciplines.

The justification for a text which teaches several fields of engineering in one semester is that there is more material to be covered compared with 10 or 20 years ago. Today, all engineering students have more information available which needs to be assimilated in four years. Something must be done. It appears to be unrealistic to expect the problem of "too much information/too little time" to disappear. Either the information must be cut back or the education must be extended. By combining the material in a number of engineering fields we can teach more efficiently. *Engineering field theory with applications* may provide a solution.

Chapter 1 is an introduction and includes the rationale for the text as well as a listing of the equations common to engineering, such as Laplace's (Poisson's) equation, Gauss' law and Stokes' theorem.

Chapter 2 is a study of the mathematical basis needed for the text. This includes the three most common coordinate systems – Cartesian, cylindrical and spherical – and transformations between systems. Also included in this chapter are the dot, and cross products, the gradient, divergence, curl, Gauss' law, and Stokes' theorem.

The first engineering field studied is electrostatics, covered in Chapter 3. Electric force, electric field, and electric flux density are defined and explained. Gauss' law is applied. Included are the analysis of such configurations as the point charge, the line charge and plane surfaces and combinations of these. Streamlines and potential lines are introduced. Maxwell's first and second equations (the divergence equation) are introduced. Boundary conditions for the electric field are derived and applied.

Chapter 4 is a study of magnetostatics. Magnetic force, magnetic field, and magnetic flux are defined and explained. The similarities and differences between electrostatics and magnetostatics are shown. Some of the derivations in Chapter 4 are simplified through the use of the duality principle. Maxwell's third and fourth equations (the curl equation) are introduced. A magnetic Gauss' law is introduced, as is Ampere's law, the Biot–Savart equation, and the Lorentz force equation. Boundary conditions for the magnetic field are derived and applied.

Chapter 5 is a study of gravitational fields. Gauss' law is applied to various physical configurations. Comparisons are made between gravitational and electrostatic and magnetic fields.

Chapter 6 is a study of fluid fields. Gauss' law and Stokes' theorem are explained as they apply to fluid fields. Compressibility and vortices are discussed. In this chapter there is a brief study of fluids in motion; most of the text is concerned with statics. Bernoulli's equation is derived and applied.

Chapter 7 is a study of thermal fields. Heat is transferred from one body to another via three methods – conduction, convection, and radiation. The first two are considered static fields; radiation is a high frequency phenomenon.

Chapter 8 is a miscellaneous chapter which includes four fields – electric current, flow through permeable media, diffusive fields and acoustic fields. Analogous equations and parameters are introduced and defined.

Chapter 9 is a wrap-up chapter and a highlight of the text. This chapter on applications includes solutions to Laplace's equation (and one on Poisson's equation) – finite differences, the relaxation method, separation of variables, a resistive analog, measurements in a conductive solution, Teledeltos paper measurements, orthogonality, curvilinear squares, the magnetic vector potential, stress lines, heat transfer, permeability, fluid flow modeling, the method of moments, and the finite element method. Chapter 9 concludes with 15 assignments for the 'laboratory'. These are assignments to be completed using numerical methods and experimental work to be completed in the laboratory.

This text is used in a single semester, four hour course at John Brown University. (Prerequisites for the course are the typical calculus sequence and differential equations.) The four hours consist of three lecture hours and one three hour laboratory each week for a 15 week semester. It is possible that some chapters could be omitted to fit a three hour course. However, the spirit of the book is the unification of many fields of engineering. Ideally the book should be treated as a whole.

Many centuries ago King Lemuel wrote.

> An excellent wife, who can find?
> For her worth is far above jewels

My appreciation goes to my 'anoosh' who typed the entire manuscript for this book. Her patience and love are amazing. She may be the only registered nurse in the world with a minor in engineering field theory. Also my thanks to my son Peter Setian for doing all of the graphics in the book.

Leo Setian

1

Introduction

1.1 Introduction

Engineering Field Theory with Applications is intended to give a unified mathematical treatment to several disciplines of engineering. This unified field theory is common to electric, magnetic and gravitational fields to name a few; this commonality aids our understanding. First, however, what is a field? Most of us experience fields regularly and may not realize it. For example, as you read this page it is in an electric field and you are sitting (or lying down) in a thermal field. Then, all of us are in a gravitational field. These fields are characterized by the fact that each point in space has its own (field) value. So we say a field is a spatial arrangement – imaginary or real – of any parameter. The space may be one-, two-, or three-dimensional depending on the quantity. A temperature field is a one-, two-, or three-dimensional arrangement of points at which we determine a temperature. All the points together constitute a field. The kitchen oven is an example of a three-dimensional temperature field. A gravitational field is a three-dimensional arrangement of forces of gravity. A spacecraft is subjected to these forces as it flies through the field. The gravitational force is exerted by the planet on the craft. Each spatial point can be represented by a different gravitational constant; there are an infinite number of values. These fields exist individually or coexist. In addition, there are fields which interact. For example, in electromagnetics, electric and magnetic fields interact. The electric field gives rise to the magnetic field and the magnetic field gives rise to the electric field. From this we get wave propagation. In fluids, temperature and velocity fields interact. An increase in the temperature of a fluid causes atoms and electrons to be agitated at a higher velocity. The fluid's velocity increases. The inverse is true: an increase (or decrease) in the velocity field of a fluid will cause an increase (or decrease) in the temperature field.

Since there is this interaction and coexistence between engineering fields, it makes sense to look at their bases to see what they have in common. The common language of engineering is mathematics. Are the equations used in the fields the same, similar, or completely different? It turns out that the answer is *yes*. Some equations are the same, some equations are similar and some equations are different. We will point these out as we proceed on our odyssey through the book. This technique of examining sameness, similarities and differences will hopefully strengthen our study. This trend or evolution of pedagogy is actually a broadening of what is already done in most curricula. For example, electricity and magnetism are taught together in electrical engineering courses today when at first they were not even connected by any unifying laws.

Today the engineering graduate faces the demands of society in a variety of fields. The engineer designing an amplifier must be concerned with more than the electronic design. In addition, he must be concerned with heat transfer and space optimization. Gone are the days when the electrical engineers, mechanical engineers and industrial engineers would operate in isolated cloisters. As another example take the automotive engineer. What once was the domain of the mechanical engineer now includes complex problems such as air pollution, fuel efficiency, electronic ignitions and on-board computers. More could be said about areas such as space exploration, microelectronics and electrical power generation. Suffice it to say that today's engineer must be versed in more than one field. Let's now look at the common language of engineering.

1.2 Mathematical basis

Mathematics is the language of engineering. Mathematical equations or expressions describe certain behavior in engineering. There are several mathematical expressions common to the engineering disciplines we study in this book. The first is Poisson's equation.

$$\nabla^2 f = \text{constant} \tag{1.1}$$

where f is some spatial function.

If the right side of equation (1.1) is zero, we obtain Laplace's equation, or

$$\nabla^2 f = 0 \tag{1.2}$$

The solutions of equations (1.1) and (1.2) are the potential functions. From these we can obtain the field intensity vectors. This is obtained by taking the derivative of the potential. Once we have the field intensity, we are able to determine the flux density.

Another important mathematical principle we will utilize in this book is Gauss' law. Gauss' law states, 'the flux emanating from a closed surface is the result of the charge enclosed by that surface'.

$$\Phi \equiv Q = \oint \mathbf{D} \cdot \mathrm{d}\mathbf{A} \tag{1.3}$$

where Q is some enclosed source, **D** is a flux density and **A** is the area which encloses the source. It is a most useful law when the problem has symmetry.

That is, for symmetric sources the flux density, **D**, in equation (1.3) is moved outside the integral. This simplifies the integration. Once we obtain the density, we can find the field intensity and potential. Equations (1.1) (or (1.2)) and (1.3) are two different methods of solving field problems. Equation (1.1) gives us the potential function while equation (1.3) gives us the density function. The potential and density functions are related and their relationship will be shown in later chapters. Naturally both methods should give the same results.

A third important equation we will utilize is Stokes' theorem. Stokes' theorem relates a surface integral to a line integral. The surface in this case is an open surface as opposed to the Gaussian surface in equation (1.3) above, which is a closed surface. Stokes' theorem is written

$$\oint \mathbf{F} \cdot \mathrm{d}\mathbf{l} = \iint \nabla \times \mathbf{F} \cdot \mathrm{d}\mathbf{A} \tag{1.4}$$

for any vector **F**. The **dl** path in equation (1.4) encloses the open surface indicated on the right-hand side.

A fourth important equation is the divergence theorem. While Stokes' theorem relates a closed line integral to an open area integral, the divergence theorem relates a closed surface to a volume. This is written

$$\oint \mathbf{D} \cdot \mathrm{d}\mathbf{A} = \iiint \nabla \cdot \mathbf{D}\, \mathrm{d}v \tag{1.5}$$

where v is volume. Thus, for any vector density, **D**, the divergence theorem says, if we add all the densities which are at an enclosing surface, the result will equal the divergence of **D** throughout the volume. The surface on the left of equation (1.5) encloses the volume on the right. This is the integral form of Maxwell's first equation. More will be said about that later.

1.3 Gradient

The last mathematical operation we mention in this introduction is the gradient. Gradient is the maximum rate of change of potential with position. The gradient is a vector quantity. For conservative fields the gradient of the potential equals minus the field intensity.

In electrostatics we write

$$\mathbf{E} = -\nabla V \quad \text{volts/meter}$$

where **E** is the electric field intensity.

In magnetostatics we write

$$\mathbf{H} = -\nabla V_{\mathrm{m}} \quad \text{amps/meter}$$

where **H** is the magnetic field intensity and V_{m} is magnetic scalar potential.

In gravitational fields we write

$$\mathbf{g} = -\nabla V_g \quad \text{Newtons/km or meters/second}^2$$

where **g** is the gravitational field intensity and V_{g} is gravitational potential.

In fluids we write

$$\mathbf{u} = -\nabla V_{\mathrm{f}} \quad \text{meters/second}$$

where **u** is velocity and V_{f} is fluid potential.

In heat transfer by conduction we write

$$\mathbf{q} = -k\nabla T \quad \text{watts/meter}^2$$

where **q** is heat flux density, k is thermal conductivity and T is temperature.

In electrical conduction we write

$$\mathbf{J} = -\sigma\nabla V \quad \text{amps/meter}^2$$

where **J** is a current density vector and σ is conductivity.

In fluid flow through permeable media we write

$$\mathbf{u} = -k\nabla p \quad \text{meters/second}$$

where **u** is velocity, k is soil permeability and p is pressure.

In diffusion we write

$$\mathbf{D}_{\mathrm{d}} = -k_{\mathrm{d}}\nabla N \quad \text{particles/meter}^2\ \text{second}$$

where $\mathbf{D}_d$ is particle flux density, k_d is the diffusion coefficient and N is particles per unit volume.

Finally, in acoustics we write

$$\mathbf{u} = -\left(\frac{1}{j\omega\rho_0}\right)\nabla p \quad \text{meters/second}$$

where $\mathbf{u}$ is velocity of the sound wave, ω is radian frequency, ρ_0 is density and p is pressure.

We have attempted to show a pattern in this introduction of concepts involving fields and fluxes. It is obvious that the fields and fluxes are different quantities with different units. The mathematical exercises are the same.

Problems

1. List five engineering occupations which require more than one engineering field. For each occupation list the different engineering fields required.

2. List three countries other than North American countries which can utilize engineers with a broad background – preferable to highly specialized engineers – and specify the desirable engineering backgrounds.

2

Mathematical basis

2.1 Introduction

In Chapter 1 we showed some of the equations common to various fields in engineering. In this chapter, we will attempt to show how these equations are manipulated to work for us. Any attempt to solve problems more quickly is welcomed with open arms. Vector analysis aids us in that we are generally able to spend less time in solving complex problems with its usage.

2.2 Vector/scalar operations

The use of vector analysis facilitates the understanding of 'charges' and their effects in the study of physical fields. Our first need is to define what we mean when we say vector. A vector is a quantity which has magnitude and direction. It will be noted in this text by bold face letters, e.g., **A**, **F**, or **G**. A scalar is a quantity which has magnitude only. Examples of vectors are forces, velocities, and electric fields. Examples of scalars are temperature, mass, and height.

Let us review the laws of addition and multiplication for scalars and vectors.

From Table 2.1, we can observe the operations which are similar for scalar and vector algebra. Also note the operations which are dissimilar and not allowed under vector algebra. (The products $\mathbf{A} \cdot \mathbf{B}$ and $\mathbf{A} \times \mathbf{B}$ will be defined in sections 2.4 and 2.5 respectively.) In addition, we note that the product of a scalar a and a vector $\mathbf{A}$ is $a\mathbf{A}$, but the addition of a scalar and a vector is not allowed.

A vector can be described in terms of unit vectors; for example if $\mathbf{A} = 3\hat{\mathbf{a}}_x - 2\hat{\mathbf{a}}_y + 4\hat{\mathbf{a}}_z$, then the meaning is three units in the $+x$ direction, two units in the $-y$ direction, and four units in the $+z$ direction. These unit vectors are shown as lower case letters with a cap, that is $\hat{\mathbf{a}}_x$, $\hat{\mathbf{a}}_y$, and so forth.

Let us look at some examples:

Example 2.1 Let

$$\mathbf{A} = 3\hat{\mathbf{a}}_x - 2\hat{\mathbf{a}}_y + 4\hat{\mathbf{a}}_z$$

$$\mathbf{B} = \hat{\mathbf{a}}_x + 4\hat{\mathbf{a}}_y + 5\hat{\mathbf{a}}_z$$

Determine $\mathbf{C}$ where $\mathbf{C} = \mathbf{A} + \mathbf{B}$

$$\mathbf{C} = (3+1)\hat{\mathbf{a}}_x + (-2+4)\hat{\mathbf{a}}_y + (4+5)\hat{\mathbf{a}}_z$$

This is illustrated in Figure 2.1 where $\mathbf{C}$ is the resultant vector $\mathbf{A} + \mathbf{B}$.

In general vectors are represented in rectangular (Cartesian) coordinates as

$$\mathbf{F} = F_x\hat{\mathbf{a}}_x + F_y\hat{\mathbf{a}}_y + F_z\hat{\mathbf{a}}_z$$

where F_x, F_y and F_z are the components of $\mathbf{F}$ in the given directions.

It should be clear that these vectors could be forces, velocities, or any such entity. The representation is the same in any case, that is the magnitude of the vector is illustrated by the *arbitrary* length of the arrow. Later we will suggest another way. Consider another example:

Example 2.2 Determine $a(\mathbf{A} - \mathbf{B})$ where

$$a = 2.2,$$

$$\mathbf{A} = \hat{\mathbf{a}}_x + \hat{\mathbf{a}}_y + 3\hat{\mathbf{a}}_z$$

Table 2.1. *Laws for addition and multiplication for scalars and vectors*

Scalars	Vectors
Addition	
$a + b = b + a$	$\mathbf{A} + \mathbf{B} = \mathbf{B} + \mathbf{A}$
$a + (b + c) = (a + b) + c$	$\mathbf{A} + (\mathbf{B} + \mathbf{C}) = (\mathbf{A} + \mathbf{B}) + \mathbf{C}$
Multiplication	
$ab = ba$	$\mathbf{AB}$ not allowed
$(ab)c = a(bc)$	$\mathbf{A} \cdot \mathbf{B} = \mathbf{B} \cdot \mathbf{A}$
	$\mathbf{A} \times \mathbf{B} \neq \mathbf{B} \times \mathbf{A}$
	$\mathbf{A} \cdot \mathbf{B} \cdot \mathbf{C}$ not allowed
	$(\mathbf{A} \times \mathbf{B}) \times \mathbf{C} \neq \mathbf{A} \times (\mathbf{B} \times \mathbf{C})$

and

$$\mathbf{B} = 2\hat{\mathbf{a}}_x + \hat{\mathbf{a}}_y - 2\hat{\mathbf{a}}_z$$

$$a(\mathbf{A} - \mathbf{B}) = a\mathbf{A} - a\mathbf{B}$$

$$= -2.2\hat{\mathbf{a}}_x + 11\hat{\mathbf{a}}_z$$

D2.1 If $\mathbf{A} = 3\hat{\mathbf{a}}_x + (\frac{3}{2})\hat{\mathbf{a}}_y + (\frac{10}{3})\hat{\mathbf{a}}_z$ and $\mathbf{B} = \hat{\mathbf{a}}_x - (\frac{1}{2})\hat{\mathbf{a}}_y - (\frac{1}{3})\hat{\mathbf{a}}_z$ find (*a*) $\mathbf{B} - \mathbf{A}$ (*b*) $\mathbf{A} - \mathbf{B}$ (*c*) $\mathbf{A} - 2\mathbf{B}$.
Answer. (*a*) $-2\hat{\mathbf{a}}_x - 2\hat{\mathbf{a}}_y - \frac{11}{3}\hat{\mathbf{a}}_z$ (*b*) $2\hat{\mathbf{a}}_x + 2\hat{\mathbf{a}}_y + \frac{11}{3}\hat{\mathbf{a}}_z$ (*c*) $\hat{\mathbf{a}}_x + \frac{5}{2}\hat{\mathbf{a}}_y + 4\hat{\mathbf{a}}_z$

2.3 Fields – scalar/vector

A field is a spatial arrangement of some parameter. There are scalar fields and vector fields. An example of a scalar field is a temperature field. An example of a vector field is a gravitational field.

Figure 2.1. Three vectors **A**, **B**, and **C**, shown in the Cartesian coordinate system. The vector components are shown as dashed lines.

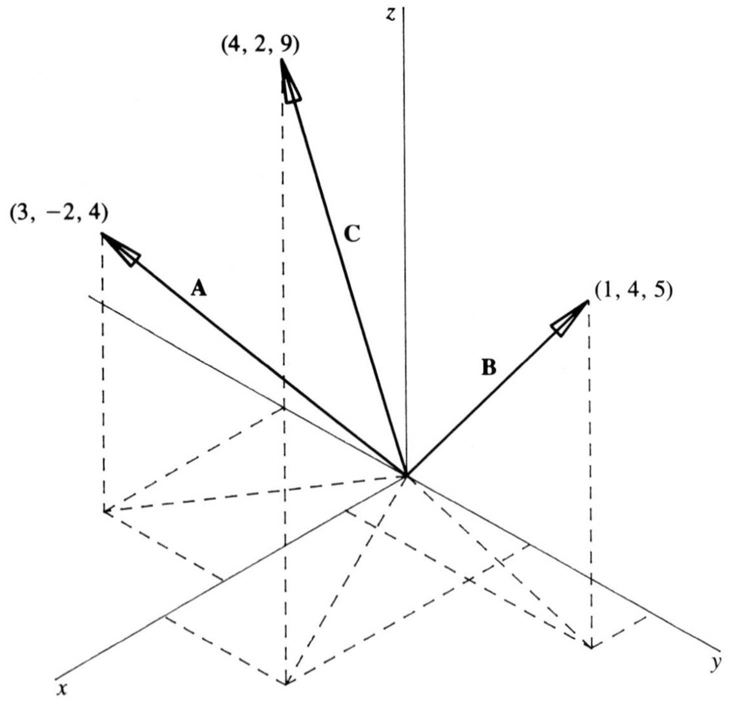

D2.2 Represent graphically, in the xy plane several values of $f(x, y) = x^2y$.

We also define a vector field as one in which the vector varies with position. For example, if $\mathbf{A} = \frac{1}{2}x\hat{\mathbf{a}}_x + 3y\hat{\mathbf{a}}_y + z\hat{\mathbf{a}}_z$ then there are an infinite number of vectors and $\mathbf{A}$ represents a vector field. For every point (x, y, z) in space we can find a vector $\mathbf{A}$. If we attempt to draw a vector field for $\mathbf{A}$ in three dimensions, we realize how difficult it would be. We illustrate with one- and two-dimensional fields.

Example 2.3 Draw several vectors for the one-dimensional vector field

$$\mathbf{B} = (x - \tfrac{1}{2})\hat{\mathbf{a}}_x$$

How shall we graphically represent the field? There is only one way to show direction, but there are many ways to show magnitude. One has been shown above: see Figure 2.1.

Figure 2.2 shows two ways: magnitude represented by (a) the length of the vector and (b) the width of the vector. In both, the unit length of the vector is arbitrarily chosen to be $\frac{1}{2}$. In both, vectors are drawn at $x = 0$, 1, 2, and 4. We can see advantages and disadvantages to both. We will use the method shown in Figure 2.2(a); that is, the length will indicate the magnitude. Let us consider next a two-dimensional vector.

Example 2.4 Draw several vectors for the two-dimensional vector field

$$\mathbf{C} = (2x - y + 3)\hat{\mathbf{a}}_x + y\hat{\mathbf{a}}_y$$

In the example shown in Figure 2.3, the choice of the vector unit length is $\frac{1}{4}$. Again, there are an infinite number of vectors which can be illustrated. Clarity and legibility demand that we limit the number of vectors. Several are drawn at integer values of x and y.

Figure 2.2. Two methods of representing magnitude from example 2.3. (a) length (b) width.

D2.3 Represent graphically, in the xy plane, several values of $\mathbf{A} = (x+y)\hat{\mathbf{a}}_x + (x-y)\hat{\mathbf{a}}_y$.

There are two important multiplicative rules for vectors which we will be using, the dot and cross products.

2.4 Dot product

Given two vectors **A** and **B** the dot product is defined as

$$\mathbf{A} \cdot \mathbf{B} \triangleq |\mathbf{A}|\,|\mathbf{B}| \cos \alpha \tag{2.1}$$

In words we say '**A** dot **B** equals the magnitude of vector **A** times the magnitude of vector **B** times the cosine of the angle between them'. The resultant is a scalar.

D2.4 If $\mathbf{A} = 3\hat{\mathbf{a}}_x + 4\hat{\mathbf{a}}_y$ and $\mathbf{B} = 4\hat{\mathbf{a}}_x + 3\hat{\mathbf{a}}_y$ find (a) $\mathbf{A} \cdot \mathbf{B}$ (b) $\mathbf{B} \cdot \mathbf{A}$.
Answer. (a) 24 (b) 24

Figure 2.3. Two-dimensional representation of the vector field in example 2.4.

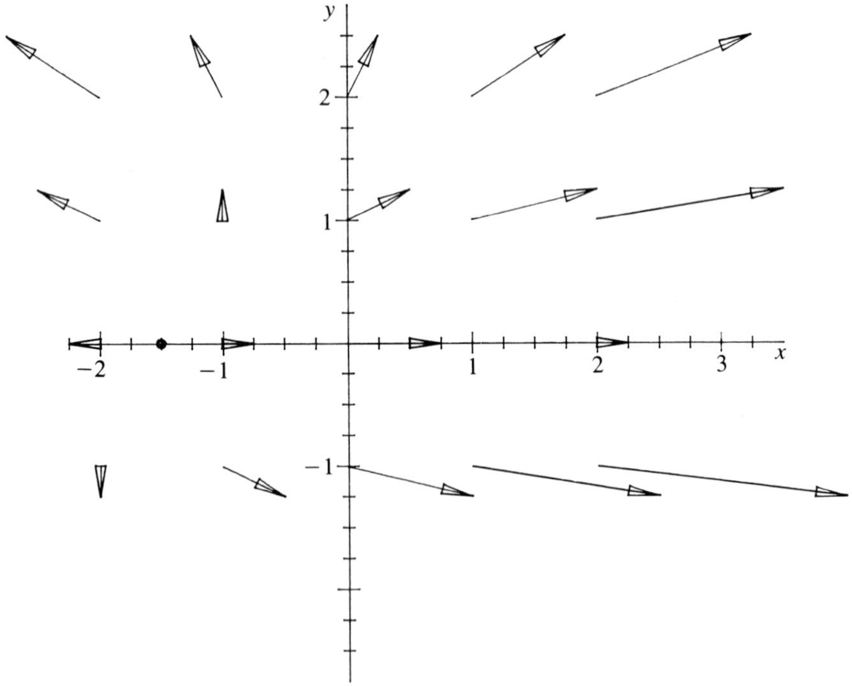

2.5 Cross product

Given two vectors **A** and **B** the cross product is defined as

$$\mathbf{A} \times \mathbf{B} \triangleq |\mathbf{A}|\,|\mathbf{B}| \sin \alpha \, \hat{\mathbf{a}}_{\mathrm{N}} \tag{2.2}$$

In words we say 'the magnitude of **A** cross **B** equals the product of the magnitude of vectors **A** and **B** times the sine of the angle between them'. The resultant is a vector. The direction is normal to the plane which includes both vectors. This is shown in Figure 2.4.

D2.5 If $\mathbf{A} = 3\hat{\mathbf{a}}_x + 4\hat{\mathbf{a}}_y$ and $\mathbf{B} = 4\hat{\mathbf{a}}_x + 3\hat{\mathbf{a}}_y$ find (*a*) $\mathbf{A} \times \mathbf{B}$ (*b*) $\mathbf{B} \times \mathbf{A}$.
Answer. (*a*) $-7\hat{\mathbf{a}}_z$ (*b*) $7\hat{\mathbf{a}}_z$

Note that the dot product is a scalar while the cross product is a vector. Also, the dot product is commutative; that is, $\mathbf{A} \cdot \mathbf{B} = \mathbf{B} \cdot \mathbf{A}$. The cross product is not commutative; that is, $\mathbf{A} \times \mathbf{B} \neq \mathbf{B} \times \mathbf{A}$ but rather $\mathbf{A} \times \mathbf{B} = -\mathbf{B} \times \mathbf{A}$.

Figure 2.4. Representation of the cross product **A** × **B**.

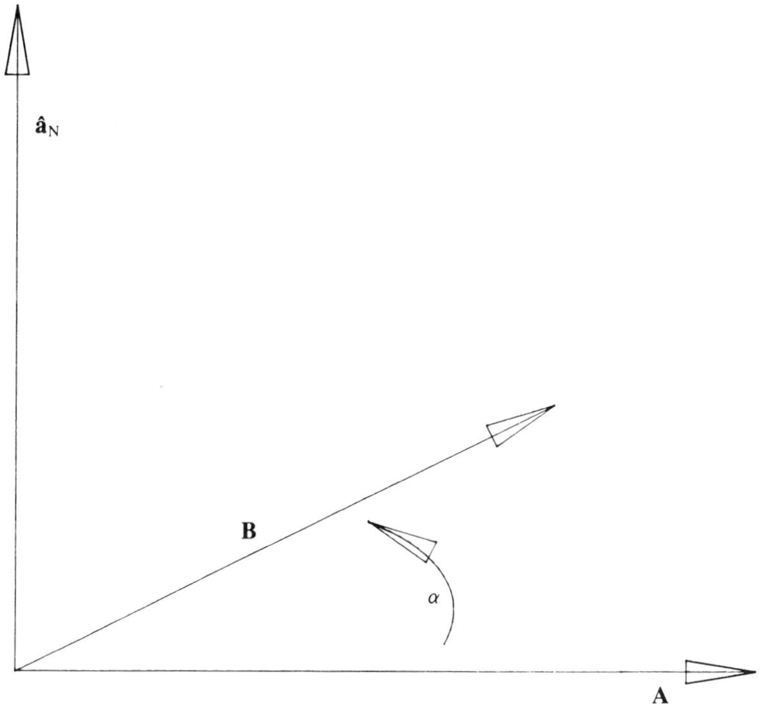

2.6 Unit vector

The unit vector is defined as the vector divided by its magnitude, or

$$\hat{\mathbf{a}}_A \triangleq \frac{\mathbf{A}}{|\mathbf{A}|} \tag{2.3}$$

This may be difficult to picture, but think of the unit vector as a 'small arrow' in the $\mathbf{A}$ direction.

D2.6 Let $\mathbf{A} = 5\hat{\mathbf{a}}_x - 2\hat{\mathbf{a}}_y + \sqrt{7}\hat{\mathbf{a}}_z$. Determine $\hat{\mathbf{a}}_A$.
Answer. $\frac{5}{6}\hat{\mathbf{a}}_x - \frac{1}{3}\hat{\mathbf{a}}_y + \frac{\sqrt{7}}{6}\hat{\mathbf{a}}_z$

2.7 Coordinate-systems – Cartesian

What coordinate system should we use for solving our problems? It turns out that no one coordinate system will suit all our needs. The choice of the correct system can simplify a problem. For example, given heat flow in a cylindrical rod, to find the temperature field we would be something less than wise to not choose the cylindrical coordinate system. We are most familiar with the Cartesian coordinate system which is shown in Figure 2.5. The Cartesian coordinate system is one in which the three axes are normal to each other and the distances are parallel to the axes. A right-handed system is one where the x-axis rotated into the y-axis, by the shorter route, would produce a normal we call the z-axis. Again, if we think in terms of the cross product, 'rotating x into y gives z'.

A differential length is

$$\mathrm{d}\mathbf{l} = \mathrm{d}x\hat{\mathbf{a}}_x + \mathrm{d}y\hat{\mathbf{a}}_y + \mathrm{d}z\hat{\mathbf{a}}_z \tag{2.4}$$

where the magnitude is

$$|\mathrm{d}\mathbf{l}| = \sqrt{(\mathrm{d}x)^2 + (\mathrm{d}y)^2 + (\mathrm{d}z)^2} \tag{2.5}$$

and the '$\hat{\mathbf{a}}_x$' in equation (2.4) indicates a unit length in the x-direction and similarly for the other two unit vectors. A differential area is a vector. The vector is *normal* to the area described. From Figure 2.5 we have

$$\mathrm{d}\mathbf{A}_x = \mathrm{d}y\ \mathrm{d}z\hat{\mathbf{a}}_x$$

$$\mathrm{d}\mathbf{A}_y = \mathrm{d}x\ \mathrm{d}z\hat{\mathbf{a}}_y$$

and so forth for the remaining sides. Four of the six areas are shown in Figure 2.5. A differential volume is a scalar. The differential volume enclosed by the six areas is simply the product of the sides, that is

$$\mathrm{d}v = \mathrm{d}x\ \mathrm{d}y\ \mathrm{d}z \tag{2.6}$$

Now let us determine two products of vectors in the Cartesian system. First, we write the dot product, $\mathbf{A} \cdot \mathbf{B}$.

$$\begin{aligned}\mathbf{A} \cdot \mathbf{B} = {} & (A_x\hat{\mathbf{a}}_x + A_y\hat{\mathbf{a}}_y + A_z\hat{\mathbf{a}}_z) \\ & \cdot (B_x\hat{\mathbf{a}}_x + B_y\hat{\mathbf{a}}_y + B_z\hat{\mathbf{a}}_z) \\ = {} & A_xB_x\hat{\mathbf{a}}_x \cdot \hat{\mathbf{a}}_x + A_xB_y\hat{\mathbf{a}}_x \cdot \hat{\mathbf{a}}_y + A_xB_z\hat{\mathbf{a}}_x \cdot \hat{\mathbf{a}}_z \\ & + A_yB_x\hat{\mathbf{a}}_y \cdot \hat{\mathbf{a}}_x + A_yB_y\hat{\mathbf{a}}_y \cdot \hat{\mathbf{a}}_y + A_yB_z\hat{\mathbf{a}}_y \cdot \hat{\mathbf{a}}_z \\ & + A_zB_x\hat{\mathbf{a}}_z \cdot \hat{\mathbf{a}}_x + A_zB_y\hat{\mathbf{a}}_z \cdot \hat{\mathbf{a}}_y + A_zB_z\hat{\mathbf{a}}_z \cdot \hat{\mathbf{a}}_z\end{aligned}$$

Under the dot product rule the product of two orthogonal vectors is zero, since $\cos(\pi/2) = 0$. Note that in the Cartesian system $\hat{\mathbf{a}}_x$, $\hat{\mathbf{a}}_y$ and $\hat{\mathbf{a}}_z$

Figure 2.5. An elemental cube shown in the Cartesian coordinate system.

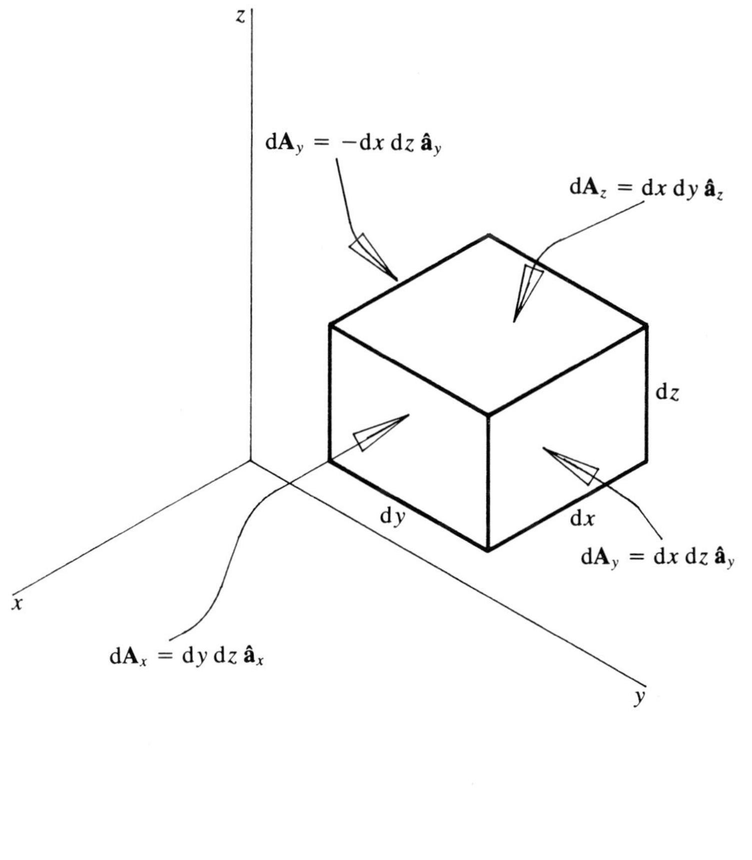

are orthogonal. Thus from $\mathbf{A} \cdot \mathbf{B}$

$$\hat{\mathbf{a}}_x \cdot \hat{\mathbf{a}}_y = \hat{\mathbf{a}}_x \cdot \hat{\mathbf{a}}_z = \hat{\mathbf{a}}_y \cdot \hat{\mathbf{a}}_x = \hat{\mathbf{a}}_y \cdot \hat{\mathbf{a}}_z$$
$$= \hat{\mathbf{a}}_z \cdot \hat{\mathbf{a}}_x = \hat{\mathbf{a}}_z \cdot \hat{\mathbf{a}}_y = 0$$

and

$$\hat{\mathbf{a}}_x \cdot \hat{\mathbf{a}}_x = \hat{\mathbf{a}}_y \cdot \hat{\mathbf{a}}_y = \hat{\mathbf{a}}_z \cdot \hat{\mathbf{a}}_z = 1$$

so

$$\mathbf{A} \cdot \mathbf{B} = A_x B_x + A_y B_y + A_z B_z$$

Example 2.5 If

$$\mathbf{A} = -3\hat{\mathbf{a}}_x + \hat{\mathbf{a}}_y + 2\hat{\mathbf{a}}_z$$

and

$$\mathbf{B} = 2\hat{\mathbf{a}}_x + \hat{\mathbf{a}}_y + 2\hat{\mathbf{a}}_z$$

then

$$\mathbf{A} \cdot \mathbf{B} = (-3\hat{\mathbf{a}}_x + \hat{\mathbf{a}}_y + 2\hat{\mathbf{a}}_z) \cdot (2\hat{\mathbf{a}}_x + \hat{\mathbf{a}}_y + 2\hat{\mathbf{a}}_z)$$
$$= -6 + 1 + 4$$
$$= -1$$

D2.7 If $\mathbf{A} = -\hat{\mathbf{a}}_x + x\hat{\mathbf{a}}_y + \hat{\mathbf{a}}_z$ and $\mathbf{B} = \hat{\mathbf{a}}_x + 2\hat{\mathbf{a}}_y - 5\hat{\mathbf{a}}_z$ find $\mathbf{A} \cdot \mathbf{B}$.
Answer. $2x - 6$

Next we write the cross product, $\mathbf{A} \times \mathbf{B}$.

$$\mathbf{A} \times \mathbf{B} = (A_x\hat{\mathbf{a}}_x + A_y\hat{\mathbf{a}}_y + A_z\hat{\mathbf{a}}_z)$$
$$\times (B_x\hat{\mathbf{a}}_x + B_y\hat{\mathbf{a}}_y + B_z\hat{\mathbf{a}}_z)$$
$$= A_x B_x \hat{\mathbf{a}}_x \times \hat{\mathbf{a}}_x + A_x B_y \hat{\mathbf{a}}_x \times \hat{\mathbf{a}}_y$$
$$+ A_x B_z \hat{\mathbf{a}}_x \times \hat{\mathbf{a}}_z + A_y B_x \hat{\mathbf{a}}_y \times \hat{\mathbf{a}}_x$$
$$+ A_y B_y \hat{\mathbf{a}}_y \times \hat{\mathbf{a}}_y + A_y B_z \hat{\mathbf{a}}_y \times \hat{\mathbf{a}}_z$$
$$+ A_z B_x \hat{\mathbf{a}}_z \times \hat{\mathbf{a}}_x + A_z B_y \hat{\mathbf{a}}_z \times \hat{\mathbf{a}}_y$$
$$+ A_z B_z \hat{\mathbf{a}}_z \times \hat{\mathbf{a}}_z$$

From the definition of cross product

$$\hat{\mathbf{a}}_x \times \hat{\mathbf{a}}_x = \hat{\mathbf{a}}_y \times \hat{\mathbf{a}}_y = \hat{\mathbf{a}}_z \times \hat{\mathbf{a}}_z = 0$$

and

$$\hat{\mathbf{a}}_x \times \hat{\mathbf{a}}_y = \hat{\mathbf{a}}_z$$

$$\hat{\mathbf{a}}_x \times \hat{\mathbf{a}}_z = -\hat{\mathbf{a}}_y$$

$$\hat{\mathbf{a}}_y \times \hat{\mathbf{a}}_x = -\hat{\mathbf{a}}_z$$

$$\hat{\mathbf{a}}_y \times \hat{\mathbf{a}}_z = \hat{\mathbf{a}}_x$$

$$\hat{\mathbf{a}}_z \times \hat{\mathbf{a}}_x = \hat{\mathbf{a}}_y$$

$$\hat{\mathbf{a}}_z \times \hat{\mathbf{a}}_y = -\hat{\mathbf{a}}_x$$

In matrix form

$$\mathbf{A} \times \mathbf{B} = \begin{vmatrix} \hat{\mathbf{a}}_x & \hat{\mathbf{a}}_y & \hat{\mathbf{a}}_z \\ A_x & A_y & A_z \\ B_x & B_y & B_z \end{vmatrix} \tag{2.7}$$

Expanding along the top row, we obtain

$$\begin{aligned} \mathbf{A} \times \mathbf{B} &= \hat{\mathbf{a}}_x \begin{vmatrix} A_y & A_z \\ B_y & B_z \end{vmatrix} - \hat{\mathbf{a}}_y \begin{vmatrix} A_x & A_z \\ B_x & B_z \end{vmatrix} + \hat{\mathbf{a}}_z \begin{vmatrix} A_x & A_y \\ B_x & B_y \end{vmatrix} \\ &= \hat{\mathbf{a}}_x(A_yB_z - A_zB_y) - \hat{\mathbf{a}}_y(A_xB_z - A_zB_x) \\ &\quad + \hat{\mathbf{a}}_z(A_xB_y - A_yB_x) \end{aligned}$$

Example 2.6 If $\mathbf{A} = -3\hat{\mathbf{a}}_x + \hat{\mathbf{a}}_y + 2\hat{\mathbf{a}}_z$ and $\mathbf{B} = 2\hat{\mathbf{a}}_x + \hat{\mathbf{a}}_y + 2\hat{\mathbf{a}}_z$ then $\mathbf{A} \times \mathbf{B} = 10\hat{\mathbf{a}}_y - 5\hat{\mathbf{a}}_z$.

D2.8 Given $\mathbf{A} = x\hat{\mathbf{a}}_x + 2\hat{\mathbf{a}}_y$ and $\mathbf{B} = 2\hat{\mathbf{a}}_x + y\hat{\mathbf{a}}_y$ determine $\mathbf{A} \times \mathbf{B}$.
Answer. $(xy - 4)\hat{\mathbf{a}}_z$.

In addition, we may determine the angle between the vectors from the definitions of dot and cross products. That is, from equation (2.1)

$$\alpha = \cos^{-1} \frac{\mathbf{A} \cdot \mathbf{B}}{|\mathbf{A}|\,|\mathbf{B}|} \tag{2.8}$$

where α is the angle between **A** and **B**. And from the cross product, equation (2.2)

$$\alpha = \sin^{-1} \frac{|\mathbf{A} \times \mathbf{B}|}{|\mathbf{A}|\,|\mathbf{B}|} \tag{2.9}$$

where α is the angle from **A** to **B**.

2.8 Coordinate systems – cylindrical

The cylindrical coordinate system, like the Cartesian, is an orthogonal system. It has three unit vectors labeled $\hat{\mathbf{a}}_r$, $\hat{\mathbf{a}}_\phi$, and $\hat{\mathbf{a}}_z$. An elemental cube in the cylindrical coordinate system is shown in Figure 2.6. The three unit vectors are shown. For a right-handed system, we rotate $\hat{\mathbf{a}}_r$ into $\hat{\mathbf{a}}_\phi$ to give $\hat{\mathbf{a}}_z$, and so forth. A differential length is

$$d\mathbf{l} = dr\hat{\mathbf{a}}_r + r\,d\phi\hat{\mathbf{a}}_\phi + dz\hat{\mathbf{a}}_z$$

The magnitude of the differential length is

$$|d\mathbf{l}| = \sqrt{(dr)^2 + (r\,d\phi)^2 + (dz)^2}$$

From Figure 2.6 the various surface areas are (in vector form)

$$d\mathbf{A}_r = r\,d\phi\,dz\hat{\mathbf{a}}_r$$

$$d\mathbf{A}_\phi = dr\,dz\hat{\mathbf{a}}_\phi$$

$$d\mathbf{A}_z = r\,dr\,d\phi\hat{\mathbf{a}}_z$$

$$d\mathbf{A}_r = -r\,d\phi\,dz\hat{\mathbf{a}}_r$$

Figure 2.6. An elemental cube shown in the cylindrical coordinate system.

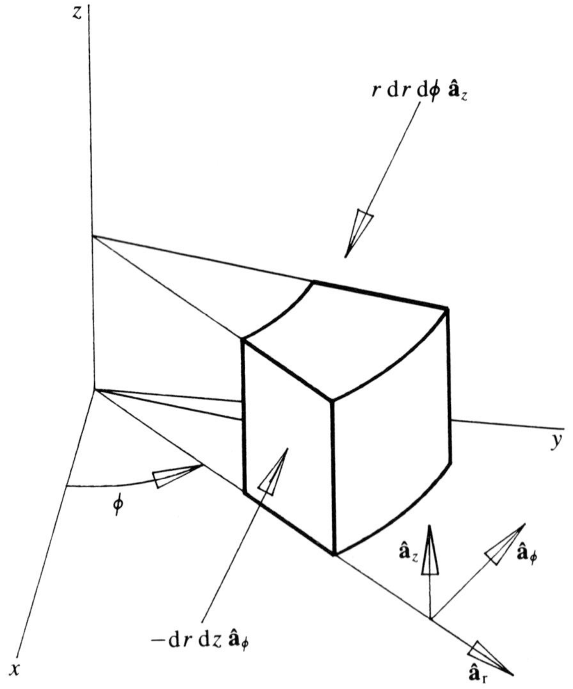

$$d\mathbf{A}_\phi = -dr\, dz\hat{\mathbf{a}}_\phi$$

$$d\mathbf{A}_z = -r\, dr\, d\phi\hat{\mathbf{a}}_z$$

Two of the differential areas $d\mathbf{A}_z$ and $d\mathbf{A}_\phi$ are marked in Figure 2.6, and the differential volume is

$$dv = r\, dr\, d\phi\, dz$$

By convention, the area vectors point away from the enclosed volume. In general, the unit vector in the **A** direction is written

$$\hat{\mathbf{a}}_A = \frac{A_r\hat{\mathbf{a}}_r + A_\phi\hat{\mathbf{a}}_\phi + A_z\hat{\mathbf{a}}_z}{\sqrt{A_r^2 + A_\phi^2 + A_z^2}} \tag{2.10}$$

Example 2.7 Given $\mathbf{A} = 3\hat{\mathbf{a}}_r + 2\hat{\mathbf{a}}_\phi - 4\hat{\mathbf{a}}_z$; determine $\hat{\mathbf{a}}_A$.

$$\hat{\mathbf{a}}_A = \frac{1}{\sqrt{29}}[3\hat{\mathbf{a}}_r + 2\hat{\mathbf{a}}_\phi - 4\hat{\mathbf{a}}_z]$$

Example 2.8 Let

$$\mathbf{A} = 3\hat{\mathbf{a}}_r + 2\hat{\mathbf{a}}_\phi - 4\hat{\mathbf{a}}_z$$

and

$$\mathbf{B} = \hat{\mathbf{a}}_r + \hat{\mathbf{a}}_\phi + 2\hat{\mathbf{a}}_z$$

For the dot product

$$\mathbf{A} \cdot \mathbf{B} = -3$$

and for the cross product

$$\mathbf{A} \times \mathbf{B} = 8\hat{\mathbf{a}}_r - 10\hat{\mathbf{a}}_\phi + \hat{\mathbf{a}}_z$$

We obtain the angle between **A** and **B** by recalling the definitions of the dot and cross products from which we obtained equations (2.8) and (2.9).

$$\alpha = \cos^{-1}\frac{\mathbf{A} \cdot \mathbf{B}}{|\mathbf{A}|\,|\mathbf{B}|} \tag{2.8}$$

and

$$\alpha = \sin^{-1}\frac{|\mathbf{A} \times \mathbf{B}|}{|\mathbf{A}|\,|\mathbf{B}|} \tag{2.9}$$

Example 2.9 From example 2.8 determine α using equation (2.8).

$$\alpha = \cos^{-1}\frac{-3}{\sqrt{174}}$$

$$= \pm 103.15°$$

and using equation (2.9)

$$\alpha = \sin^{-1}\frac{\sqrt{165}}{\sqrt{174}}$$

$$= 76.85°, 103.15°$$

Note that 103.15° is common to both $\mathbf{A} \cdot \mathbf{B}$ and $\mathbf{A} \times \mathbf{B}$; it is chosen. The ambiguity results from the fact that the vectors in cylindrical coordinates, **A** and **B** are not fully defined.

D2.9 Given $\mathbf{A} = 2\hat{\mathbf{a}}_r - 3\hat{\mathbf{a}}_z$ and $\mathbf{B} = \hat{\mathbf{a}}_r + \hat{\mathbf{a}}_z$, calculate the angle between vectors **A** and **B**.
Answer. $\alpha = 101.3°$.

These last two examples of multiplication in the cylindrical coordinate system show that there is no difference in the *mechanics* of taking the dot and cross products regardless of the system. Let us proceed to the spherical coordinate system.

2.9 Coordinate systems – spherical

The spherical coordinate system, like the first two systems, is an orthogonal system with the three unit vectors being $\hat{\mathbf{a}}_r$, $\hat{\mathbf{a}}_\theta$, and $\hat{\mathbf{a}}_\phi$. It is the most complex of the three. An elemental cube is shown in Figure 2.7. A differential length is

$$\mathbf{dl} = \mathrm{d}r\hat{\mathbf{a}}_r + r\,\mathrm{d}\theta\hat{\mathbf{a}}_\theta + r\sin\theta\,\mathrm{d}\phi\hat{\mathbf{a}}_\phi$$

and the magnitude is

$$|\mathbf{dl}| = \sqrt{(\mathrm{d}r)^2 + (r\,\mathrm{d}\theta)^2 + (r\sin\theta\,\mathrm{d}\phi)^2}$$

Three of the surface areas are, from Figure 2.7,

$$d\mathbf{A}_r = r^2 \sin\theta \, d\theta \, d\phi \hat{\mathbf{a}}_r$$

$$d\mathbf{A}_\theta = r \sin\theta \, dr \, d\phi \hat{\mathbf{a}}_\theta$$

$$d\mathbf{A}_\phi = r \, dr \, d\theta \hat{\mathbf{a}}_\phi$$

and the differential volume is

$$dv = r^2 \sin\theta \, dr \, d\theta \, d\phi$$

Example 2.10 Find the volume of one octant (1/8 of a sphere) of a sphere of radius r_0 by integrating the differential volume.

We choose the octant where x, y, and z are positive as shown in

Figure 2.7. An elemental cube shown in the spherical coordinate system.

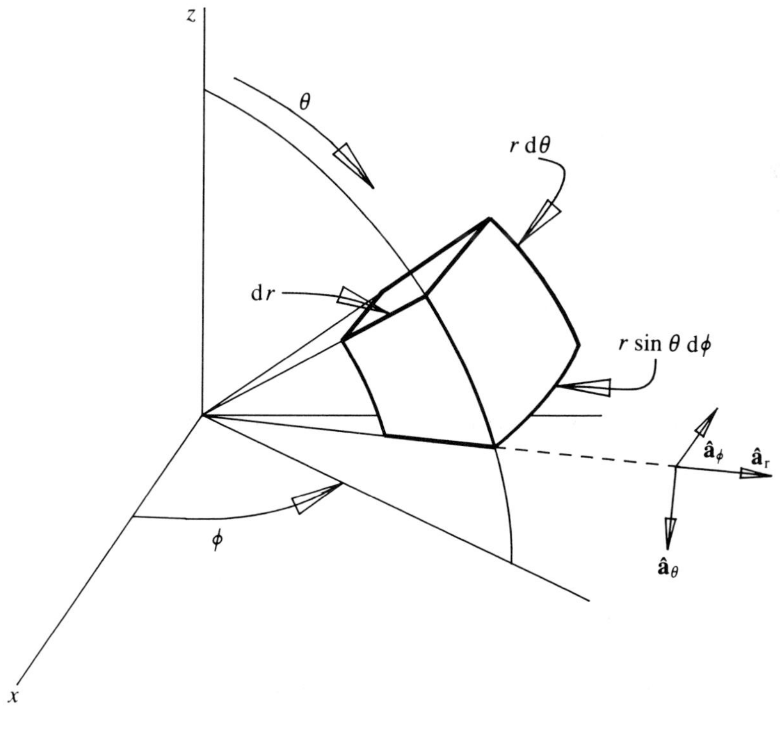

Figure 2.7 and write

$$\text{Volume} = \iiint dv$$

$$= \iiint r^2 \sin\theta \, dr \, d\theta \, d\phi$$

$$= \tfrac{1}{6}\pi r_0^3$$

which is what we expect since the volume of a cube is $\frac{4}{3}\pi r_0^3$.

Finally, the unit vector in spherical coordinates is written

$$\hat{\mathbf{a}}_A = \frac{A_r\hat{\mathbf{a}}_r + A_\theta\hat{\mathbf{a}}_\theta + A_\phi\hat{\mathbf{a}}_\phi}{\sqrt{(A_r^2 + A_\theta^2 + A_\phi^2)}} \tag{2.11}$$

D2.10 Find the length of the arc in spherical coordinates from $\phi = 30°$ to $\phi = 60°$ where $r = 2$ m and $\theta = 60°$.
Answer. 0.907 m.

2.10 Transformations

At times it is useful to change from one coordinate system to another. For example, if a cylindrical problem were given in Cartesian coordinates, it would be more convenient to change to cylindrical coordinates before proceeding with the solution.

Example 2.11 As an illustration let us choose a two-dimensional problem. Let us find the area of a circle shown in Figure 2.8 with radius 3 m in two ways: using 1. the rectangular system and 2. the polar system. The equation of the circle centered at the origin is $x^2 + y^2 = 9$.

1. Using the rectangular system, we obtain

$$\text{Area} = 4\int_0^3 \int_0^{\sqrt{9-y^2}} dx \, dy$$

$$= 9\pi$$

2. Using the polar system we obtain

$$\text{Area} = 4\int_0^{\pi/2} \int_0^3 r \, dr \, d\phi$$

$$= 9\pi$$

In this example we see that the integration associated with the polar system is easier. Thus a problem given in one coordinate system may be solved more easily in another system, and we need to be able to transform between coordinates. Now let us consider the rectangular and cylindrical coordinate systems as shown in Figure 2.9. From Figure 2.9 we can see both Cartesian and cylindrical coordinates displayed.

We first pursue a transformation in two dimensions and then proceed to three dimensions. Let us change the differential length, **dl**, in Figure 2.10 from rectangular (Cartesian) to polar coordinates.

We want to change

$$\mathrm{d}\mathbf{l} = \mathrm{d}x\hat{\mathbf{a}}_x + \mathrm{d}y\hat{\mathbf{a}}_y$$
$$= \mathrm{d}l_x\hat{\mathbf{a}}_x + \mathrm{d}l_y\hat{\mathbf{a}}_y$$

in rectangular to

$$\mathrm{d}\mathbf{l} = \mathrm{d}r\hat{\mathbf{a}}_r + r\,\mathrm{d}\phi\hat{\mathbf{a}}_\phi$$
$$= \mathrm{d}l_r\hat{\mathbf{a}}_r + \mathrm{d}l_\phi\hat{\mathbf{a}}_\phi$$

Figure 2.8. Circle area calculation in example 2.11 using two different coordinate systems.

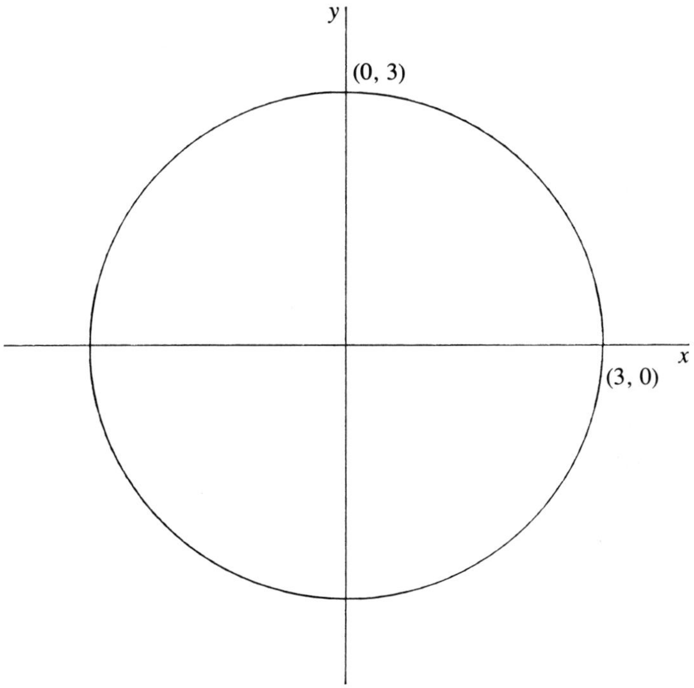

Figure 2.9. Coordinate transformation between Cartesian and cylindrical coordinates.

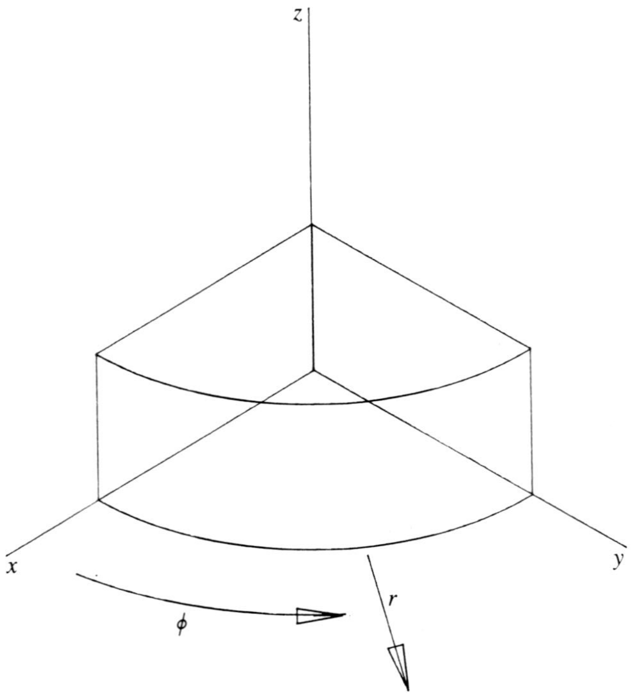

Figure 2.10. A two-dimensional coordinate transformation between Cartesian and cylindrical coordinates (xy plane).

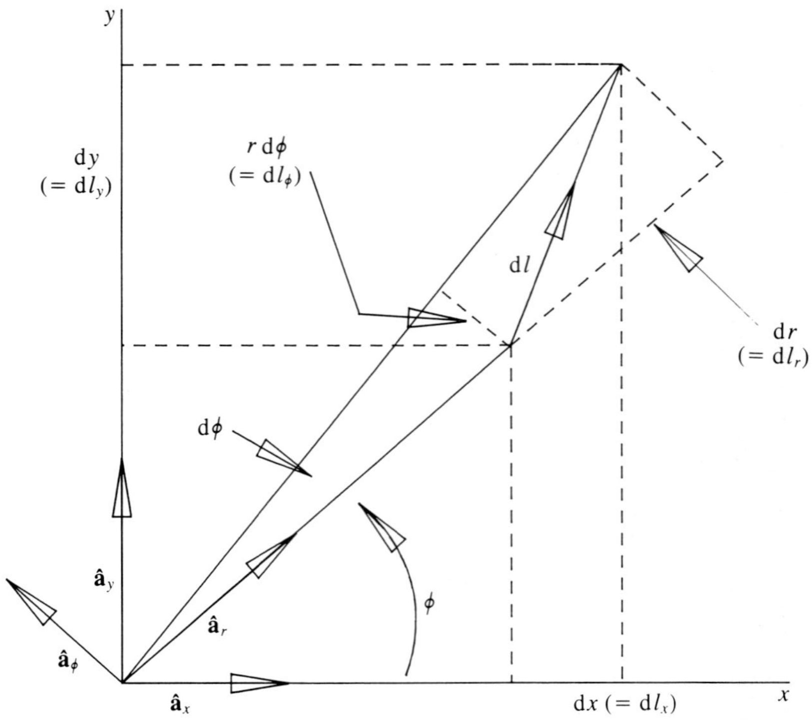

in polar. We see that

$$x = r\cos\phi \quad \text{and} \quad y = r\sin\phi$$

from which

$$\mathrm{d}x = \mathrm{d}r\cos\phi - r\,\mathrm{d}\phi\sin\phi$$

and

$$\mathrm{d}y = \mathrm{d}r\sin\phi + r\,\mathrm{d}\phi\cos\phi$$

We observe that

$$\begin{aligned}
\mathrm{d}l_r &= \mathbf{dl}_{\mathrm{rect}} \cdot \hat{\mathbf{a}}_r \\
&= \mathrm{d}x\hat{\mathbf{a}}_x \cdot \hat{\mathbf{a}}_r + \mathrm{d}y\hat{\mathbf{a}}_y \cdot \hat{\mathbf{a}}_r \\
&= \mathrm{d}r\cos^2\phi - r\,\mathrm{d}\phi\sin\phi\cos\phi + \mathrm{d}r\sin^2\phi \\
&\quad + r\,\mathrm{d}\phi\sin\phi\cos\phi \\
&= \mathrm{d}r
\end{aligned}$$

Also

$$\begin{aligned}
\mathrm{d}l_\phi &= \mathbf{dl}_{\mathrm{rect}} \cdot \hat{\mathbf{a}}_\phi \\
&= \mathrm{d}x\hat{\mathbf{a}}_x \cdot \hat{\mathbf{a}}_\phi + \mathrm{d}y\hat{\mathbf{a}}_y \cdot \hat{\mathbf{a}}_\phi \\
&= \mathrm{d}x(-\sin\phi) + \mathrm{d}y\cos\phi \\
&= \mathrm{d}r\cos\phi(-\sin\phi) + r\,\mathrm{d}\phi\sin^2\phi \\
&\quad + \mathrm{d}r\sin\phi\cos\phi + r\,\mathrm{d}\phi\cos^2\phi \\
&= r\,\mathrm{d}\phi
\end{aligned}$$

Finally

$$\mathbf{dl}_{\mathrm{polar}} = \mathrm{d}r\hat{\mathbf{a}}_r + r\,\mathrm{d}\phi\hat{\mathbf{a}}_\phi$$

In transformations any component of either vector may be found by taking the dot product of the appropriate vectors.

Example 2.12 Determine the r, ϕ, and z components in cylindrical coordinates of any vector $\mathbf{A}$ in Cartesian.

$$A_r = \mathbf{A} \cdot \hat{\mathbf{a}}_r = (A_x\hat{\mathbf{a}}_x + A_y\hat{\mathbf{a}}_y + A_z\hat{\mathbf{a}}_z) \cdot \hat{\mathbf{a}}_r$$

$$A_\phi = \mathbf{A} \cdot \hat{\mathbf{a}}_\phi = (A_x\hat{\mathbf{a}}_x + A_y\hat{\mathbf{a}}_y + A_z\hat{\mathbf{a}}_z) \cdot \hat{\mathbf{a}}_\phi$$

and

$$A_z = A_z$$

From Figure 2.10

$$\hat{\mathbf{a}}_x \cdot \hat{\mathbf{a}}_r = (1)(1)\cos\phi$$

$$\hat{\mathbf{a}}_y \cdot \hat{\mathbf{a}}_r = \sin\phi$$

$$\hat{\mathbf{a}}_x \cdot \hat{\mathbf{a}}_\phi = -\sin\phi$$

$$\hat{\mathbf{a}}_y \cdot \hat{\mathbf{a}}_\phi = \cos\phi$$

Example 2.13 Determine the x, y, and z components in Cartesian coordinates of any vector **A** in cylindrical.

$$A_x = \mathbf{A} \cdot \hat{\mathbf{a}}_x = (A_r\hat{\mathbf{a}}_r + A_\phi\hat{\mathbf{a}}_\phi + A_z\hat{\mathbf{a}}_z) \cdot \hat{\mathbf{a}}_x$$

$$A_y = \mathbf{A} \cdot \hat{\mathbf{a}}_y = (A_r\hat{\mathbf{a}}_r + A_\phi\hat{\mathbf{a}}_\phi + A_z\hat{\mathbf{a}}_z) \cdot \hat{\mathbf{a}}_y$$

and

$$A_z = A_z$$

Example 2.14 Transform the vector in rectangular coordinates,

$$\mathbf{A} = 2\hat{\mathbf{a}}_x + 3\hat{\mathbf{a}}_y + 4\hat{\mathbf{a}}_z$$

into cylindrical coordinates.

$$A_r = (2\hat{\mathbf{a}}_x + 3\hat{\mathbf{a}}_y + 4\hat{\mathbf{a}}_z) \cdot \hat{\mathbf{a}}_r$$

$$= 2\cos\phi + 3\sin\phi$$

$$A_\phi = -2\sin\phi + 3\cos\phi$$

$$A_z = 4$$

therefore, we write in cylindrical coordinates

$$\mathbf{A} = (2\cos\phi + 3\sin\phi)\hat{\mathbf{a}}_r + (-2\sin\phi + 3\cos\phi)\hat{\mathbf{a}}_\phi + 4\hat{\mathbf{a}}_z$$

Example 2.15 Transform the vector in cylindrical coordinates,

$$\mathbf{B} = \hat{\mathbf{a}}_r + 3\hat{\mathbf{a}}_\phi - \hat{\mathbf{a}}_z$$

into rectangular coordinates.

$$B_x = (\hat{\mathbf{a}}_r + 3\hat{\mathbf{a}}_\phi - \hat{\mathbf{a}}_z) \cdot \hat{\mathbf{a}}_x$$

$$= \cos\phi - 3\sin\phi$$

$$B_y = (\hat{\mathbf{a}}_r + 3\hat{\mathbf{a}}_\phi - \hat{\mathbf{a}}_z) \cdot \hat{\mathbf{a}}_y$$

$$= \sin\phi + 3\cos\phi$$

$$B_z = -1$$

which leads us to write

$$\mathbf{B} = (\cos\phi - 3\sin\phi)\hat{\mathbf{a}}_x + (3\cos\phi + \sin\phi)\hat{\mathbf{a}}_y - \hat{\mathbf{a}}_z \tag{2.12}$$

However, we are not finished yet; this equation still contains 'ϕ'. From Figure 2.9 we note that

$$\cos\phi = \frac{x}{r}$$

$$= \frac{x}{\sqrt{(x^2 + y^2)}}$$

and

$$\sin\phi = \frac{y}{\sqrt{(x^2 + y^2)}}$$

Substituting these into equation (2.12) we obtain the vector, **B**, in rectangular coordinates.

$$\mathbf{B} = \frac{x - 3y}{\sqrt{(x^2 + y^2)}}\hat{\mathbf{a}}_x + \frac{3x + y}{\sqrt{(x^2 + y^2)}}\hat{\mathbf{a}}_y - \hat{\mathbf{a}}_z$$

Figure 2.11. Coordinate transformation between Cartesian and spherical coordinates.

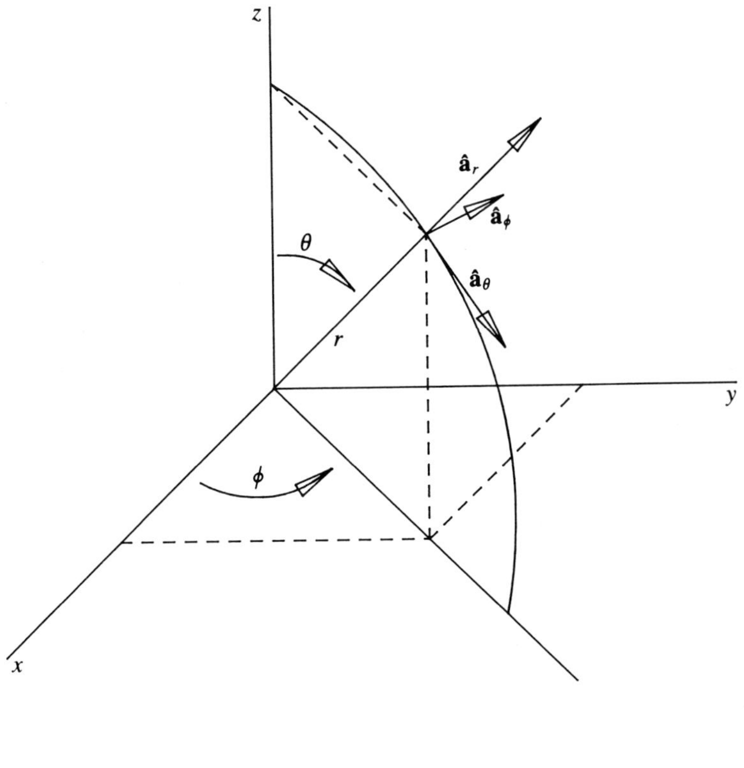

D2.11 Transform the vector in rectangular coordinates $\mathbf{A} = -\hat{\mathbf{a}}_x + \hat{\mathbf{a}}_y + 4\hat{\mathbf{a}}_z$ into cylindrical coordinates.
Answer. $(\sin\phi - \cos\phi)\hat{\mathbf{a}}_r + (\sin\phi + \cos\phi)\hat{\mathbf{a}}_\phi + 4\hat{\mathbf{a}}_z$.

D2.12 Transform the vector in cylindrical coordinates $\mathbf{B} = 3\hat{\mathbf{a}}_r + \hat{\mathbf{a}}_\phi$ into rectangular coordinates.
Answer.

$$\frac{3x - y}{\sqrt{(x^2 + y^2)}}\hat{\mathbf{a}}_x + \frac{3y + x}{\sqrt{(x^2 + y^2)}}\hat{\mathbf{a}}_y.$$

Next let us examine the relationship between rectangular and spherical coordinates. Figure 2.11 shows the relationship between rectangular and spherical coordinates. They are listed below:

$$x = r \sin\theta \cos\phi$$

$$y = r \sin\theta \sin\phi$$

$$z = r \cos\theta$$

$$r = (x^2 + y^2 + z^2)^{1/2}$$

$$\theta = \tan^{-1}\frac{(x^2 + y^2)^{1/2}}{z}$$

$$= \cos^{-1}\frac{z}{(x^2 + y^2 + z^2)^{1/2}}$$

$$\phi = \tan^{-1}\left(\frac{y}{x}\right)$$

$$\hat{\mathbf{a}}_x \cdot \hat{\mathbf{a}}_r = (1)(1) \sin\theta \cos\phi$$

$$\hat{\mathbf{a}}_y \cdot \hat{\mathbf{a}}_r = \sin\theta \sin\phi$$

$$\hat{\mathbf{a}}_z \cdot \hat{\mathbf{a}}_r = \cos\theta$$

$$\hat{\mathbf{a}}_x \cdot \hat{\mathbf{a}}_\theta = \cos\theta \cos\phi$$

$$\hat{\mathbf{a}}_y \cdot \hat{\mathbf{a}}_\theta = \cos\theta \sin\phi$$

$$\hat{\mathbf{a}}_z \cdot \hat{\mathbf{a}}_\theta = -\sin\theta$$

$$\hat{\mathbf{a}}_x \cdot \hat{\mathbf{a}}_\phi = -\sin\phi$$

$$\hat{\mathbf{a}}_y \cdot \hat{\mathbf{a}}_\phi = \cos\phi$$

$$\hat{\mathbf{a}}_z \cdot \hat{\mathbf{a}}_\phi = 0$$

Let us illustrate their usage with an example.

Example 2.16 Transform the vector **D** in rectangular (Cartesian) coordinates into spherical coordinates.

$$\mathbf{D} = -2\hat{\mathbf{a}}_x + 3\hat{\mathbf{a}}_y + \hat{\mathbf{a}}_z$$

$$D_r = \mathbf{D} \cdot \hat{\mathbf{a}}_r = -2 \sin\theta \cos\phi + 3 \sin\theta \sin\phi + \cos\theta$$

$$D_\theta = \mathbf{D} \cdot \hat{\mathbf{a}}_\theta = -2 \cos\theta \cos\phi + 3 \cos\theta \sin\phi - \sin\theta$$

$$D_\phi = \mathbf{D} \cdot \hat{\mathbf{a}}_\phi = 2 \sin\phi + 3 \cos\phi$$

Thus we obtain

$$\begin{aligned}\mathbf{D} = {} & (-2 \sin\theta \cos\phi + 3 \sin\theta \sin\phi + \cos\theta)\hat{\mathbf{a}}_r \\ & + (-2 \cos\theta \cos\phi + 3 \cos\theta \sin\phi - \sin\theta)\hat{\mathbf{a}}_\theta \\ & + (2 \sin\phi + 3 \cos\phi)\hat{\mathbf{a}}_\phi\end{aligned}$$

D2.13 Transform the vector in rectangular coordinates $\mathbf{A} = -2\hat{\mathbf{a}}_x - 3\hat{\mathbf{a}}_y - \hat{\mathbf{a}}_z$ into spherical coordinates.
Answer. $(-2 \sin\theta \cos\phi - 3 \sin\theta \sin\phi - \cos\theta)\hat{\mathbf{a}}_r + (-2 \cos\theta \cos\phi - 3 \cos\theta \sin\phi + \sin\theta)\hat{\mathbf{a}}_\theta + (2 \sin\phi - 3 \cos\phi)\hat{\mathbf{a}}_\phi$.

Transforming a vector from spherical to rectangular will be left as an exercise for the ambitious student (who else?).

2.11 The gradient

Now let us utilize the above in defining equations we will need in the remainder of the text. The first is the gradient of a difference. Imagine a room as shown in Figure 2.12 with a fireplace at one end used for heating the room. The temperature will vary through the room. The different temperatures establish a scalar field; they also establish a gradient, where the gradient is a vector. The temperature varies in the direction of the vector. We place the source at the origin and note that heat is transferred radially. We choose cylindrical coordinates; the constant surfaces are also constant temperature surfaces. Assume no z variation in temperature so that we have a two-dimensional problem. We write for the gradient

$$\text{Temperature gradient} = \nabla T = \frac{\partial T}{\partial r}\hat{\mathbf{a}}_r \tag{2.13}$$

since

$$\frac{\partial T}{\partial \phi} = \frac{\partial T}{\partial z} = 0$$

We define a heat field intensity

$$\text{Heat Field Intensity} \triangleq -\nabla \text{T} \tag{2.14}$$

that is, the heat field intensity is defined as the negative gradient of the temperature. The minus sign indicates heat flows from high to low temperature.

In general, the gradient is written in Cartesian coordinates

$$\nabla T = \frac{\partial T}{\partial x}\hat{\mathbf{a}}_x + \frac{\partial T}{\partial y}\hat{\mathbf{a}}_y + \frac{\partial T}{\partial z}\hat{\mathbf{a}}_z \tag{2.15}$$

in cylindrical coordinates

$$\nabla T = \frac{\partial T}{\partial r}\hat{\mathbf{a}}_r + \frac{\partial T}{r\,\partial \phi}\hat{\mathbf{a}}_\phi + \frac{\partial T}{\partial z}\hat{\mathbf{a}}_z \tag{2.16}$$

and in spherical coordinates

$$\nabla T = \frac{\partial T}{\partial r}\hat{\mathbf{a}}_r + \frac{\partial T}{r\,\partial \theta}\hat{\mathbf{a}}_\theta + \frac{1}{r \sin\theta}\frac{\partial T}{\partial \phi}\hat{\mathbf{a}}_\phi \tag{2.17}$$

Figure 2.12. Graphical illustration of the gradient. (An isothermal surface is a surface at constant temperature.)

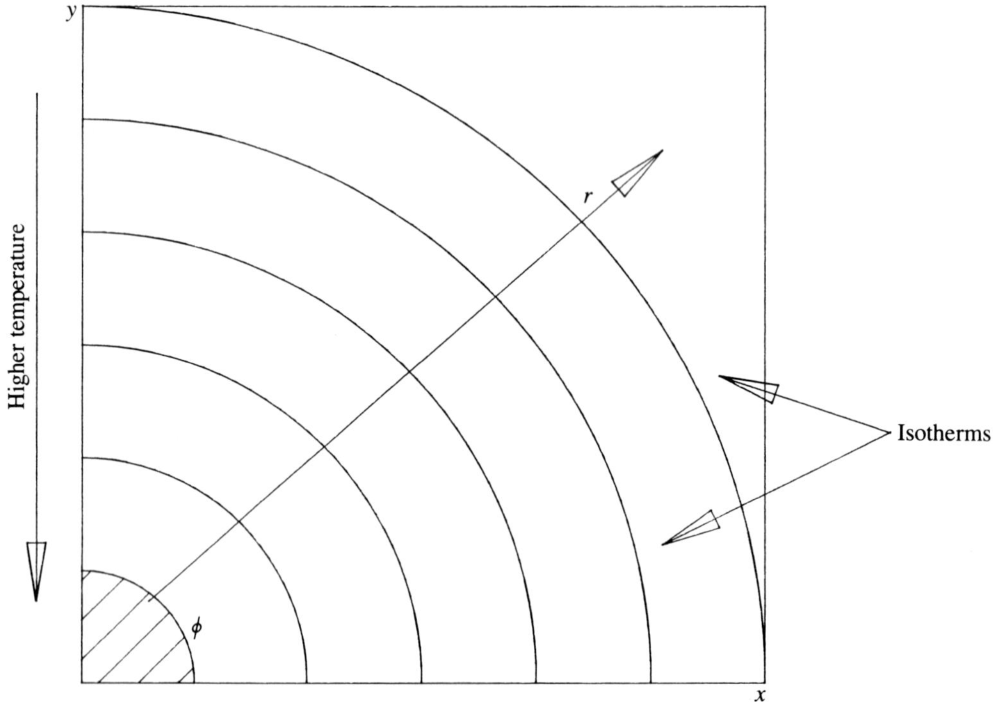

The equations are derived in numerous other texts and are presented here without further explanation.

D2.14 If (a) $T = x^2 + y$, find ∇T in Cartesian coordinates (b) $T = rz \cos \phi$, find ∇T in cylindrical coordinates, (c) $T = 2r \cos \theta \sin \phi$, find ∇T in spherical coordinates.
Answer. (a) $2x\hat{\mathbf{a}}_x + \hat{\mathbf{a}}_y$. (b) $z \cos \phi \hat{\mathbf{a}}_r - z \sin \phi \hat{\mathbf{a}}_\phi + r \cos \phi \hat{\mathbf{a}}_z$. (c) $2 \cos \theta \sin \phi \hat{\mathbf{a}}_r - 2 \sin \theta \sin \phi \hat{\mathbf{a}}_\theta + 2 \cot \theta \cos \phi \hat{\mathbf{a}}_\phi$.

We observe that the gradient is a vector. What is the direction? Let us take a closer look at the heat transfer problem above. This expanded view is shown in Figure 2.13. The (heat) field intensity vector is drawn in general. We represent the vector by components tangential and normal to the isotherms. The tangential component is in the direction of zero change of temperature; the normal component is in the direction of maximum (magnitude) change. Clearly there can be no change in the direction of no change – an isotherm. By convention the flow from high concentration to low concentration is defined as

Figure 2.13. Expanded view of a two-dimensional heat transfer.

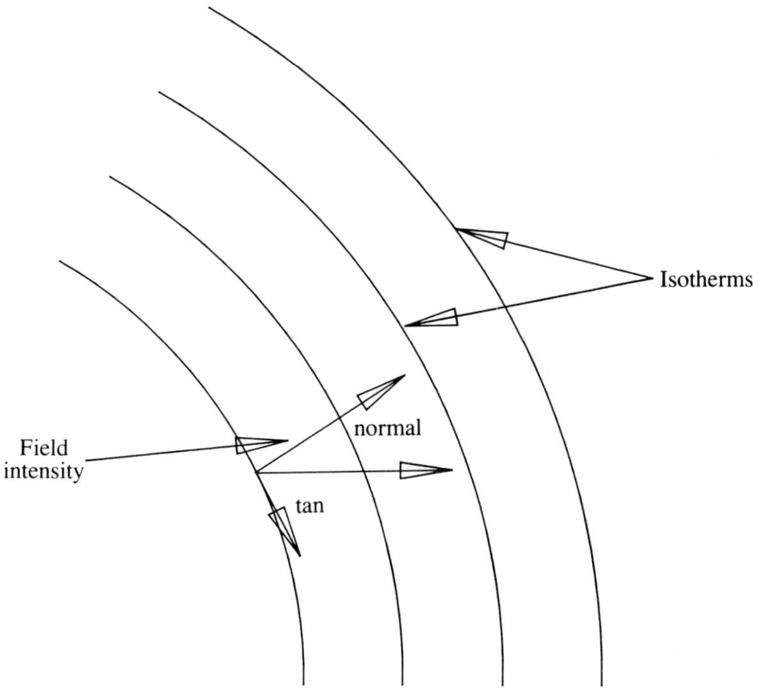

negative. Conclusion: the gradient must point in the direction of maximum rate of change of field.

Now equations (2.15), (2.16) and (2.17) contain the scalar field T. The gradient is an operation that determines the spatial rate of change of the temperature. Note, however, that the temperature could be replaced by any scalar field. The gradient is simply an operator. Thus, the gradient operating on a scalar field produces a vector indicating the greatest rate of change of the field. Other operations in which del (∇) is used are

$$\text{the divergence of } F \quad \nabla \cdot \mathbf{F}$$

and

$$\text{the curl of } F \quad \nabla \times \mathbf{F}$$

where $\mathbf{F}$ is a vector field instead of a scalar field such as we have been discussing. These will be covered later.

We now examine the integral relationship between the field intensity and the potential surfaces (isotherms in the previous example). Recall equation (2.14) in which

$$\text{Heat Field Intensity} \quad \mathbf{q} \propto \nabla T$$

or

$$\mathbf{q} = -k\,\nabla T \tag{2.18}$$

where k is a constant of proportionality called the thermal conductivity.

Example 2.17 Calculate the approximate heat field intensity along a 3 m long, narrow rod. One end is in boiling water and the other end is at room temperature.

Let $k = 1$. Assume the rod is vertical (z direction).

$$\mathbf{q} = -k\,\nabla T$$
$$= -1\left(\frac{27-100}{3}\right)\hat{\mathbf{a}}_z$$

In one dimension for rectangular coordinates, we write

$$\mathbf{q} = -k\frac{\mathrm{d}T}{\mathrm{d}x}\hat{\mathbf{a}}_x$$

or

$$\approx -k\frac{\Delta T}{\Delta x}\hat{\mathbf{a}}_x$$

and

$$\Delta\mathbf{x}\cdot\mathbf{q} = -k\frac{\Delta T}{\Delta x}\hat{\mathbf{a}}_x\cdot\Delta\mathbf{x}$$

$$\mathbf{q}\cdot\Delta\mathbf{x} = -k\,\Delta T$$

Finally

$$T_1 - T_2 = -\frac{1}{k}\int \mathbf{q}\cdot \mathrm{d}\mathbf{x}$$

In general

$$V_1 - V_2 = -\int \mathbf{F}\cdot\mathrm{d}\mathbf{l} \tag{2.19}$$

where V is potential and **F** is any vector field.

For a conservative field

$$\oint \mathbf{F}\cdot\mathrm{d}\mathbf{l} = 0$$

where $\oint$ indicates a closed path in the field. If

$$\mathbf{F} = \nabla f$$

where f is any scalar field then

$$\oint \nabla f\cdot\mathrm{d}\mathbf{l} = 0 \tag{2.20}$$

for any contour.

The change in any function f is written

$$\mathrm{d}f = \frac{\partial f}{\partial x}\mathrm{d}x + \frac{\partial f}{\partial y}\mathrm{d}y + \frac{\partial f}{\partial z}\mathrm{d}z$$

The differential element of distance is written

$$\mathrm{d}\mathbf{l} = \mathrm{d}x\hat{\mathbf{a}}_x + \mathrm{d}y\hat{\mathbf{a}}_y + \mathrm{d}z\hat{\mathbf{a}}_z$$

So we write

$$\mathrm{d}f = \left(\frac{\partial f}{\partial x}\hat{\mathbf{a}}_x + \frac{\partial f}{\partial y}\hat{\mathbf{a}}_y + \frac{\partial f}{\partial z}\hat{\mathbf{a}}_z\right) \cdot \mathrm{d}\mathbf{l} \tag{2.21}$$

so that

$$\begin{aligned}\mathrm{d}f &= \nabla f \cdot \mathrm{d}\mathbf{l} \\ &= |\nabla f|\, \mathrm{d}l \cos\alpha\end{aligned}$$

This is a maximum when $\alpha = 0$ and a minimum when $\alpha = \pi$. That is

$$\left.\frac{\mathrm{d}f}{\mathrm{d}l}\right|_{\max} = |\nabla f|$$

In other words, the gradient is zero along lines of constant potential.

D2.15 If $f = xy^2z$, (*a*) find ∇f and (*b*) evaluate ∇f at $(1, 2, -4)$.
Answer. (*a*) $y^2z\hat{\mathbf{a}}_x + 2xyz\hat{\mathbf{a}}_y + xy^2\hat{\mathbf{a}}_z$ (*b*) $-16\hat{\mathbf{a}}_x - 16\hat{\mathbf{a}}_y + 4\hat{\mathbf{a}}_z$.

2.12 Divergence

Flux is a term used to represent imaginary lines in a field along which forces occur. One of the more important theorems which we will use in virtually every discipline is the divergence theorem. It relates volume and surface integrals and involves flux. We will derive the theorem beginning with Gauss' law. It is stated

$$\text{Flux} = \Phi \equiv Q = \oint \mathbf{D} \cdot \mathrm{d}\mathbf{A} \tag{2.22}$$

where $\mathbf{D}$ is flux density, and Q is a source.

Equation (2.22) tells us that the total flux passing through a closed area is the accumulation of that passing through small elements of the total areas. In addition the flux is the result of the source enclosed. Picture a red hot ball bearing in the center of a larger sphere. Heat will flow through the larger sphere like an imaginary fluid. In equation (2.22) the hot ball bearing is represented by Q, the imaginary fluid is the flux, and heat flow per unit area over each element of the spherical surface is the flux density. Now consider a small cube.

In Figure 2.14 we have a cube with flux flowing through it. At the surface $y = 0$ the flux is

$$\mathbf{D} \cdot \mathrm{d}\mathbf{A}|_{y=0} \approx -D_y\, \Delta x\, \Delta z$$

and at $y = \Delta y$

$$\mathbf{D} \cdot \mathrm{d}\mathbf{A}|_{y=\Delta y} \approx D_y \, \Delta x \, \Delta z + \frac{\partial D_y}{\partial y} \Delta x \, \Delta y \, \Delta z$$

The net flux leaving in the y-direction is

$$\mathbf{D} \cdot \mathrm{d}\mathbf{A}|_{y=0} + \mathbf{D} \cdot \mathrm{d}\mathbf{A}|_{y=\Delta y} \approx \frac{\partial D_y}{\partial y} \Delta x \, \Delta y \, \Delta z$$

Also, the net flux leaving in the x-direction is

$$\mathbf{D} \cdot \mathrm{d}\mathbf{A}|_{x=0} + \mathbf{D} \cdot \mathrm{d}\mathbf{A}|_{x=\Delta x} \approx \frac{\partial D_x}{\partial x} \Delta x \, \Delta y \, \Delta z$$

and in the z-direction

$$\mathbf{D} \cdot \mathrm{d}\mathbf{A}|_{z=0} + \mathbf{D} \cdot \mathrm{d}\mathbf{A}|_{z=\Delta z} \approx \frac{\partial D_z}{\partial z} \Delta x \, \Delta y \, \Delta z$$

Figure 2.14. Flux flowing through an elemental cube.

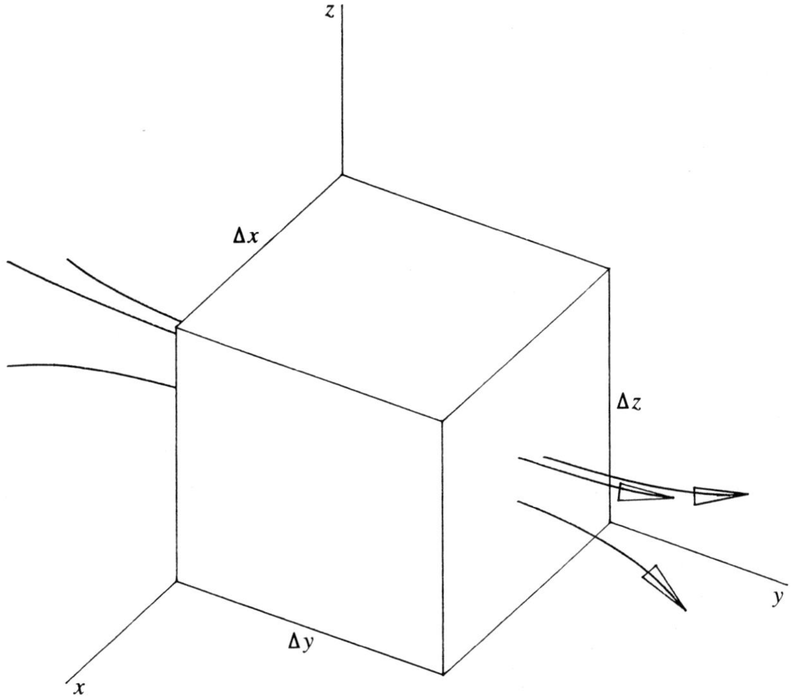

Now if we sum the flux in all three directions

$$\Phi = \oint \mathbf{D} \cdot \mathrm{d}\mathbf{A}$$

$$\approx \left(\frac{\partial D_x}{\partial x} + \frac{\partial D_y}{\partial y} + \frac{\partial D_z}{\partial z} \right) \Delta x \, \Delta y \, \Delta z \tag{2.23}$$

In the limit

$$\lim_{\Delta v \to 0} \frac{\oint \mathbf{D} \cdot \mathrm{d}\mathbf{A}}{\Delta v} = \frac{\partial D_x}{\partial x} + \frac{\partial D_y}{\partial y} + \frac{\partial D_z}{\partial z} \tag{2.24}$$

Definition:

The divergence of **D** is

$$\nabla \cdot \mathbf{D} \triangleq \lim_{\Delta v \to 0} \frac{\oint \mathbf{D} \cdot \mathrm{d}\mathbf{A}}{\Delta v}$$

$$= \frac{\partial D_x}{\partial x} + \frac{\partial D_y}{\partial y} + \frac{\partial D_z}{\partial z} \tag{2.25}$$

in Cartesian coordinates. Similar expressions may be obtained in the other coordinate systems by using the appropriate Δv in equation (2.24). In cylindrical coordinates this becomes

$$\nabla \cdot \mathbf{D} = \frac{1}{r} \frac{\partial (r D_r)}{\partial r} + \frac{1}{r} \frac{\partial D_\phi}{\partial \phi} + \frac{\partial D_z}{\partial z} \tag{2.26}$$

and in spherical coordinates this becomes

$$\nabla \cdot \mathbf{D} = \frac{1}{r^2} \frac{\partial (r^2 D_r)}{\partial r} + \frac{1}{r \sin \theta} \frac{\partial (\sin \theta D_\theta)}{\partial \theta} + \frac{1}{r \sin \theta} \frac{\partial D_\phi}{\partial \phi} \tag{2.27}$$

D2.16 (*a*) $\mathbf{D} = x\hat{\mathbf{a}}_x + 2\hat{\mathbf{a}}_y + z^2\hat{\mathbf{a}}_z$ in rectangular, (*b*) $\mathbf{D} = r\hat{\mathbf{a}}_r + r \sin \phi \hat{\mathbf{a}}_\phi$ in cylindrical, (*c*) $\mathbf{D} = r\hat{\mathbf{a}}_r + \hat{\mathbf{a}}_\theta + \sin \phi \hat{\mathbf{a}}_\phi$ in spherical. Calculate $\nabla \cdot \mathbf{D}$.

Answer. (*a*) $2z + 1$, (*b*) $2 + \cos \phi$, (*c*) $3 + \dfrac{\cos \theta}{r \sin \theta} + \dfrac{\cos \phi}{r \sin \theta}$.

Now if we return to equation (2.24), which we repeat here

$$\lim_{\Delta v \to 0} \frac{\oint \mathbf{D} \cdot \mathrm{d}\mathbf{A}}{\Delta v} = \frac{\partial D_x}{\partial x} + \frac{\partial D_y}{\partial y} + \frac{\partial D_z}{\partial z} = \nabla \cdot \mathbf{D}$$

Multiply both sides by Δv and add all differential volumes

$$\oint \mathbf{D} \cdot \mathrm{d}\mathbf{A} = \sum \nabla \cdot \mathbf{D}\, \Delta v$$

In the limit

$$\oint \mathbf{D} \cdot \mathrm{d}\mathbf{A} = \iiint \nabla \cdot \mathbf{D}\, \mathrm{d}v \tag{2.28}$$

where the left side is a surface integral over a closed surface and the right side is a volume integral. We can understand equation (2.28) better if we consider that the left hand side is an imaginary surface which includes the real volume on the right hand side. The flux lines diverge from the volume (source) and intersect the imaginary surface.

This is the divergence theorem.

Example 2.18

$$\mathbf{D} = 2x\hat{\mathbf{a}}_x + xy^2\hat{\mathbf{a}}_y + 3yz\hat{\mathbf{a}}_z$$

Calculate the flux leaving unit cube shown in Figure 2.15 using divergence alone.

$$\oint \mathbf{D} \cdot \mathrm{d}\mathbf{A} = \iiint \nabla \cdot \mathbf{D}\, \mathrm{d}v$$

$$\int_1 + \int_2 + \int_3 + \int_4 + \int_5 + \int_6 = \int (2 + 2xy + 3y)\, \mathrm{d}v$$

$$4 = 4C$$

D2.17 Calculate the flux leaving the unit cube of Figure 2.15 if $\mathbf{D} = 2\hat{\mathbf{a}}_x + x^2y\hat{\mathbf{a}}_y + xyz\hat{\mathbf{a}}_z$ using both sides of equation (2.28).
Answer. $\frac{7}{12}$.

This example illustrates the fact that there is a *source* within the closed surface which has a value of four coulombs. If the sign were minus, it would imply that there is a *sink* within. If the integration of either side yielded zero, it would imply no source or sink; the flux lines are the result of some outside source.

2.13 Curl

One of the questions of interest in field theory is whether or not a field is conservative. From equation (2.20) the definition for conser-

Figure 2.15. The unit cube of example 2.18 and D2.17.

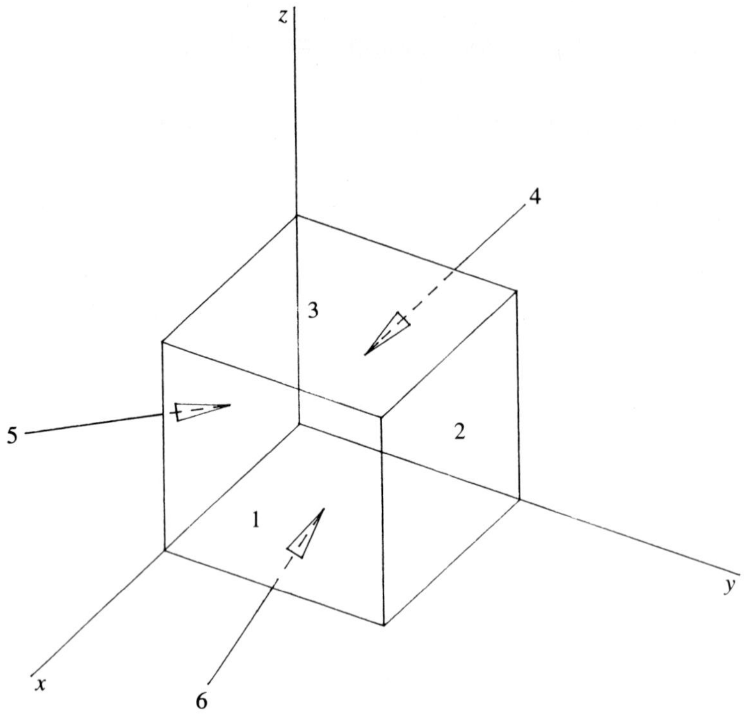

Figure 2.16. Integration over a closed path in the *xy* plane.

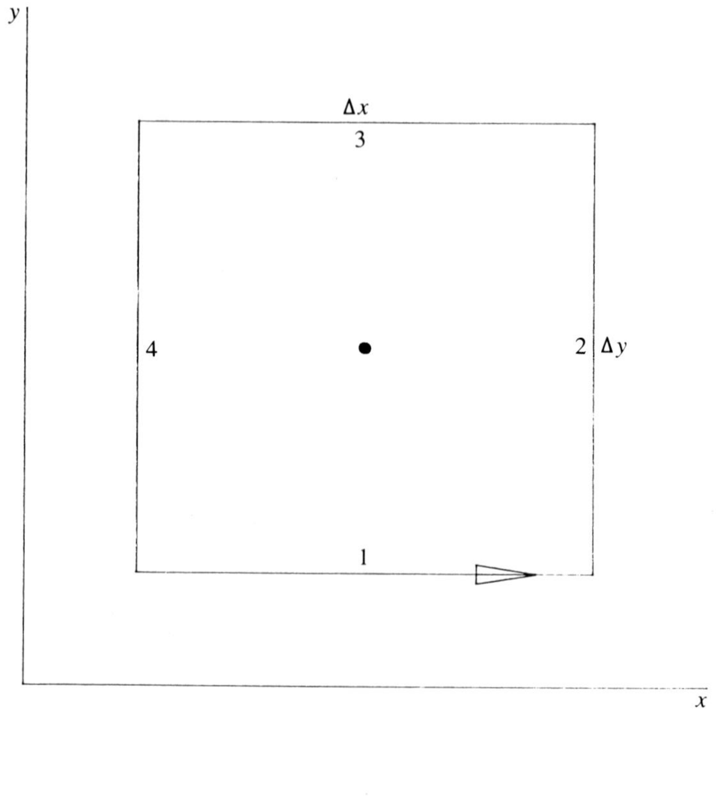

vation is $\oint \mathbf{F} \cdot d\mathbf{l} = 0$ over *any* closed path. If there is at least one path where $\oint \mathbf{F} \cdot d\mathbf{l} \neq 0$ then the field is not conservative; there is a *circulation*. Instead of trying several paths to see whether or not $\oint \mathbf{F} \cdot d\mathbf{l} \neq 0$, we look for an easier way.

There is a mathematical operation which determines the circulation of a field. It is called a curl.

Consider the circulation of **F** over a small rectangle in the xy plane with sides Δx and Δy. We carry out the integration in the counterclockwise direction as shown in Figure 2.16, about the point $(x, y, 0)$.

$$\int_1 \mathbf{F} \cdot d\mathbf{l} = \int F_x \, dx \approx F_x\left(x, y - \frac{\Delta y}{2}\right) \Delta x$$

$$\int_2 \mathbf{F} \cdot d\mathbf{l} = \int F_y \, dy \approx F_y\left(x + \frac{\Delta x}{2}, y\right) \Delta y$$

$$\int_3 \mathbf{F} \cdot d\mathbf{l} = \int F_x \, dx = -F_x\left(x, y + \frac{\Delta y}{2}\right) \Delta x$$

$$\int_4 \mathbf{F} \cdot d\mathbf{l} = \int F_y \, dy \approx -F_y\left(x - \frac{\Delta x}{2}, y\right) \Delta y$$

$$\begin{aligned}\oint \mathbf{F} \cdot d\mathbf{l} = \int_1 + \int_2 + \int_3 + \int_4 &\approx F_x\left(x, y - \frac{\Delta y}{2}\right) \frac{\Delta x \, \Delta y}{\Delta y} \\ &- F_x\left(x, y + \frac{\Delta y}{2}\right) \frac{\Delta x \, \Delta y}{\Delta y} + F_y\left(x + \frac{\Delta x}{2}, y\right) \frac{\Delta x \, \Delta y}{\Delta x} \\ &- F_y\left(x - \frac{\Delta x}{2}, y\right) \frac{\Delta x \, \Delta y}{\Delta x} \\ &= \left[\frac{F_y\left(x + \frac{\Delta x}{2}, y\right) - F_y\left(x - \frac{\Delta x}{2}, y\right)}{\Delta x} \right. \\ &\qquad \left. - \frac{F_x\left(x, y + \frac{\Delta y}{2}\right) - F_x\left(x, y - \frac{\Delta y}{2}\right)}{\Delta y} \right] \Delta x \, \Delta y\end{aligned}$$

But

$$\Delta x \, \Delta y = \Delta A_z$$

and

$$\frac{\oint \mathbf{F}\cdot d\mathbf{l}}{\Delta A} \approx \frac{F_y\left(x+\frac{\Delta x}{2}, y\right) - F_y\left(x-\frac{\Delta x}{2}, y\right)}{\Delta x} - \frac{F_x\left(x, y+\frac{\Delta y}{2}\right) - F_x\left(x, y-\frac{\Delta y}{2}\right)}{\Delta y}$$

or

$$\lim_{\Delta A\to 0} \frac{\oint \mathbf{F}\cdot d\mathbf{l}}{\Delta A} = \frac{\partial F_y}{\partial x} - \frac{\partial F_x}{\partial y}$$

In the *yz* plane we do a similar contour integration as shown in Figure 2.17. We obtain as a result

$$\oint \mathbf{F}\cdot d\mathbf{l} = \int_1 + \int_2 + \int_3 + \int_4$$

Figure 2.17. Integration over a closed path in the *yz* plane.

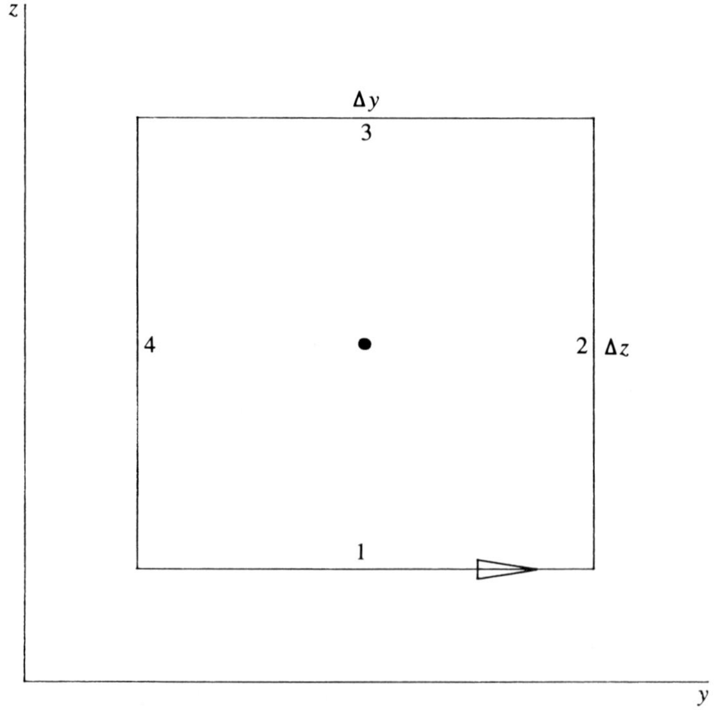

and

$$\int_1 \mathbf{F} \cdot d\mathbf{l} \approx F_y\left(y, z - \frac{\Delta z}{2}\right) \Delta y$$

$$\int_2 \mathbf{F} \cdot d\mathbf{l} \approx F_z\left(y + \frac{\Delta y}{2}, z\right) \Delta z$$

$$\int_3 \mathbf{F} \cdot d\mathbf{l} \approx -F_y\left(y, z + \frac{\Delta z}{2}\right) \Delta y$$

$$\int_4 \mathbf{F} \cdot d\mathbf{l} \approx -F_z\left(y - \frac{\Delta y}{2}, z\right) \Delta z$$

$$\oint \mathbf{F} \cdot d\mathbf{l} \approx \left\{ \frac{F_y\left(y, z - \frac{\Delta z}{2}\right) - F_y\left(y, z + \frac{\Delta z}{2}\right)}{\Delta z} + \frac{F_z\left(y + \frac{\Delta y}{2}, z\right) - F_z\left(y - \frac{\Delta y}{2}, z\right)}{\Delta y} \right\} \Delta y \, \Delta z$$

or

$$\lim_{\Delta A \to 0} \frac{\oint \mathbf{F} \cdot d\mathbf{l}}{\Delta A} = \frac{\partial F_z}{\partial y} - \frac{\partial F_y}{\partial z}$$

In the xz plane we do a similar contour integration as shown in Figure 2.18. We obtain as a result

$$\lim_{\Delta A \to 0} \frac{\oint \mathbf{F} \cdot d\mathbf{l}}{\Delta A} = \frac{\partial F_x}{\partial z} - \frac{\partial F_z}{\partial x}$$

The convention used for integration around the contour is that the area vector is normal to the area around which the contour is drawn. The right-hand rule is as the fingers of the right hand point in the direction of integration, the thumb points in the vector direction normal to the plane of integration. This is illustrated by referring to Figures 2.16, 2.17, and 2.18. In Figure 2.16 as we integrate along the path 1–2–3–4 (finger direction) the normal points toward the reader (thumb direction) in the z-direction. Similarly in Figure 2.17 the thumb points in the x-direction and in Figure 2.18 the normal is in the $-y$-direction. Mathematically

$$\lim_{\Delta A_x \to 0} \frac{\oint \mathbf{F} \cdot d\mathbf{l}}{\Delta A_x} = \frac{\partial F_z}{\partial y} - \frac{\partial F_y}{\partial z}$$

Similarly

$$\lim_{\Delta A_y \to 0} \frac{\oint \mathbf{F} \cdot d\mathbf{l}}{\Delta A_y} = \frac{\partial F_x}{\partial z} - \frac{\partial F_z}{\partial x}$$

and

$$\lim_{\Delta A_z \to 0} \frac{\oint \mathbf{F} \cdot d\mathbf{l}}{\Delta A_z} = \frac{\partial F_y}{\partial x} - \frac{\partial F_x}{\partial y}$$

Figure 2.18. Integration over a closed path in the xz plane.

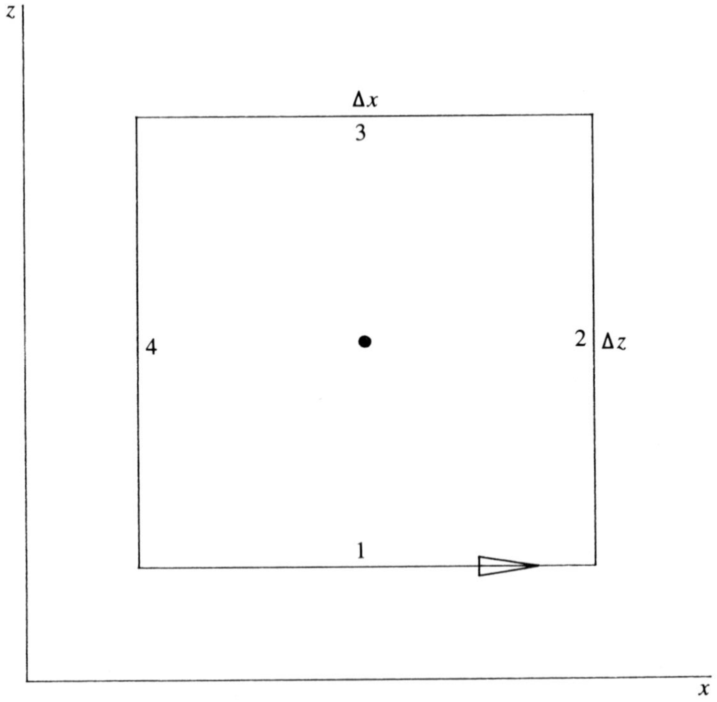

The sum of these is called the curl, an indication of the circulation of the vector field **F**.

$$\text{curl}\,\mathbf{F} \triangleq \left(\frac{\partial F_z}{\partial y} - \frac{\partial F_y}{\partial z}\right)\hat{\mathbf{a}}_x + \left(\frac{\partial F_x}{\partial z} - \frac{\partial F_z}{\partial x}\right)\hat{\mathbf{a}}_y + \left(\frac{\partial F_y}{\partial x} - \frac{\partial F_x}{\partial y}\right)\hat{\mathbf{a}}_z \tag{2.29}$$

This can be written, in rectangular coordinates,

$$\nabla \times \mathbf{F} = \begin{vmatrix} \hat{\mathbf{a}}_x & \hat{\mathbf{a}}_y & \hat{\mathbf{a}}_z \\ \dfrac{\partial}{\partial x} & \dfrac{\partial}{\partial y} & \dfrac{\partial}{\partial z} \\ F_x & F_y & F_z \end{vmatrix} \tag{2.30}$$

The curl in cylindrical and spherical coordinates is derived in most calculus and many engineering texts. The derivatives are not included here. In cylindrical coordinates, we write

$$\nabla \times \mathbf{F} = \left(\frac{1}{r}\frac{\partial F_z}{\partial \phi} - \frac{\partial F_\phi}{\partial z}\right)\hat{\mathbf{a}}_r + \left(\frac{\partial F_r}{\partial z} - \frac{\partial F_z}{\partial r}\right)\hat{\mathbf{a}}_\phi + \left(\frac{1}{r}\frac{\partial (rF_\phi)}{\partial r} - \frac{1}{r}\frac{F_r}{\partial \phi}\right)\hat{\mathbf{a}}_z \tag{2.31}$$

and in spherical coordinates, we write

$$\nabla \times \mathbf{F} = \left\{\frac{1}{r \sin\theta}\left[\frac{\partial (F_\phi \sin\theta)}{\partial \theta} - \frac{\partial F_\theta}{\partial \phi}\right]\right\}\hat{\mathbf{a}}_r + \left\{\frac{1}{r \sin\theta}\frac{\partial F_r}{\partial \phi} - \frac{\partial (rF_\phi)}{r\,\partial r}\right\}\hat{\mathbf{a}}_\theta + \left\{\frac{1}{r}\left[\frac{\partial (rF_\theta)}{\partial r} - \frac{\partial F_r}{\partial \theta}\right]\right\}\hat{\mathbf{a}}_\phi \tag{2.32}$$

D2.18 If $\mathbf{F} = xyz(\hat{\mathbf{a}}_x + \hat{\mathbf{a}}_y + \hat{\mathbf{a}}_z)$ calculate $\nabla \times \mathbf{F}$.
Answer. $(xz - xy)\hat{\mathbf{a}}_x + (xy - yz)\hat{\mathbf{a}}_y + (yz - xz)\hat{\mathbf{a}}_z$.

Example 2.19 Consider a two dimensional vector $\mathbf{F} = xy\hat{\mathbf{a}}_x + 3\hat{\mathbf{a}}_y$. Figure 2.19 shows the vector field associated with the vector **F**. To find the circulation of the vector field **F** we write

$$\nabla \times \mathbf{F} = (0-0)\hat{\mathbf{a}}_x + (0-0)\hat{\mathbf{a}}_y + (0-x)\hat{\mathbf{a}}_z = -x\hat{\mathbf{a}}_z$$

$$\lim_{\Delta A \to 0} \frac{(xy\hat{\mathbf{a}}_x + 3\hat{\mathbf{a}}_y)\cdot d\mathbf{l}}{\Delta A} = |\nabla \times (xy\hat{\mathbf{a}}_x + 3\hat{\mathbf{a}}_y)| = x$$

Thus we say there is a circulation due to the vector field **F**.

It can be seen that the curl of **F** is in the $-z$-direction for $x > 0$, $y > 0$ and in the $+z$-direction for $x < 0$, $y > 0$. Using the right-hand rule, we let our fingers lie in the general direction of flow (the arrows) and we see that, for example, in the first quadrant our thumb points into the page ($-z$-direction).

Imagine, if you will, a (dormitory) room floor covered with 3 cm of water (after a water fight). The usual outlet will be under the door. From the previous example let the $y = 0$ axis represent the bottom of the door. As the water seeps under the door toward the hall, it will follow more or less the lines shown by the vectors in the region $y > 0$ (or $y < 0$).

If we were to insert a paddlewheel with its axis pointed in the z-direction into the flowing fluid, it would turn clockwise in the first quadrant and counterclockwise in the second quadrant. The fluid flows along the arcs of circles. Thus, the flow lines show circulation.

We will discuss the result in succeeding chapters.

Figure 2.19. A model of the vector field for $\mathbf{F} = xy\hat{\mathbf{a}}_x + 3\hat{\mathbf{a}}_y$.

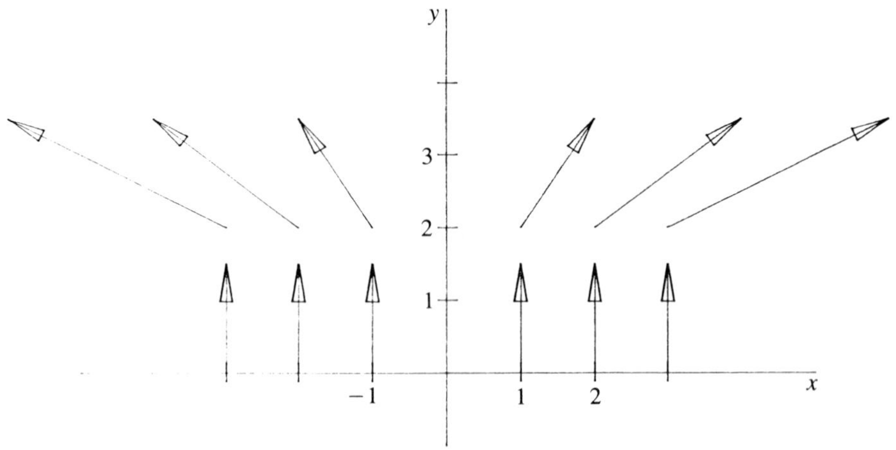

2.14 Stokes' theorem

Another useful theorem which we will cherish is Stokes' theorem. Stokes' theorem relates an open surface, say a pie plate, to the line integral – the integral around the edge of the pie plate. Consider an open surface which we divide into four squares or four of any shape as shown in Figure 2.20. We can integrate around the outside contour or integrate around the four squares; the same area answer results.

From the previous section

$$\lim_{\Delta A \to 0} \frac{\oint \mathbf{F} \cdot \mathbf{dl}}{\Delta A} \approx |\nabla \times \mathbf{F}|$$

For each of the smaller squares, we write

$$\oint \mathbf{F} \cdot \mathbf{dl} \approx \nabla \times \mathbf{F} \cdot \Delta \mathbf{A}$$

Figure 2.20. Pie plate integration.

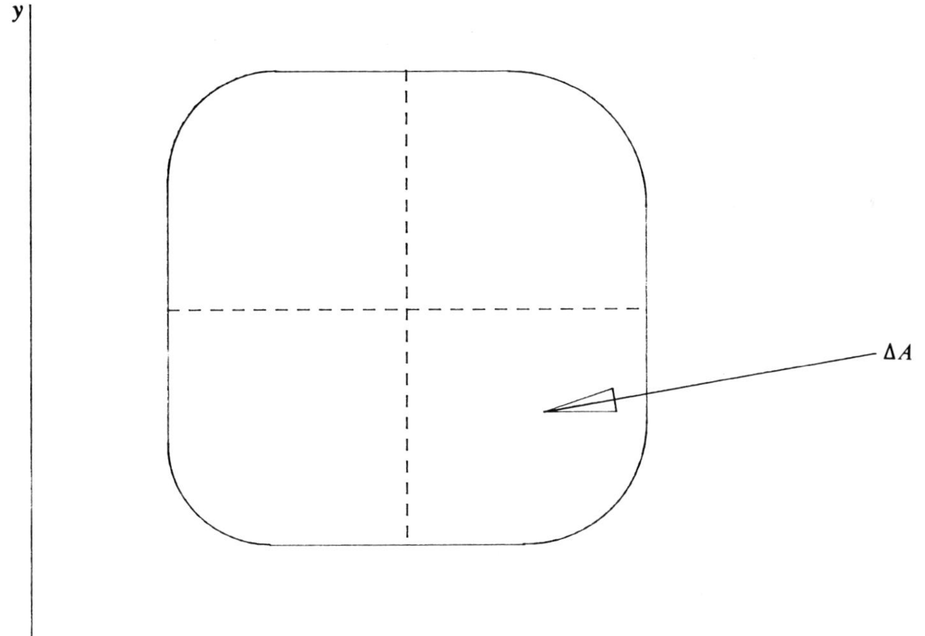

As we do this for each of the four squares, the common boundaries cancel and the contour is simply the contour around the outside. That is

$$\nabla \times \mathbf{F} = \lim_{\Delta A \to 0} \frac{\oint \boldsymbol{F} \cdot \boldsymbol{dl}}{\Delta A} \hat{\mathbf{a}}_{\mathrm{N}}$$

or

$$\iint \nabla \times \mathbf{F} \cdot \mathrm{d}\mathbf{A} = \oint \mathbf{F} \cdot \mathrm{d}\mathbf{l} \tag{2.33}$$

This is Stokes' theorem. Let us attempt an example.

Example 2.20 Evaluate Stokes' theorem for the vector field $\mathbf{F} = x\hat{\mathbf{a}}_x + 2xyz\hat{\mathbf{a}}_y + xz^2\hat{\mathbf{a}}_z$ and the area in the $z = 1$ plane bounded by the contour c. Choose a contour in the cw direction.

$$\oint \mathbf{F} \cdot \mathrm{d}\mathbf{l} = \int_1 + \int_2 + \int_3 + \int_4$$

$$= \int_2^0 2(3)y\, \mathrm{d}y + \int_3^0 x\, \mathrm{d}x + \int_0^2 z(0)\, \mathrm{d}y + \int_0^3 x\, \mathrm{d}x = -12$$

$$\iint \nabla \times \mathbf{F} \cdot \mathrm{d}\mathbf{A} = \int_0^2 \int_0^3 \left(\frac{\partial F_y}{\partial x} - \frac{\partial F_x}{\partial y} \right) \hat{\mathbf{a}}_z \cdot -\mathrm{d}x\, \mathrm{d}y\hat{\mathbf{a}}_z$$

$$= \int_0^2 \int_0^3 (2yz - 0)(-\mathrm{d}x\, \mathrm{d}y)$$

$$= -12$$

Let us apply Stokes' theorem to the illustration above where water was flowing into the hall (to the dismay of the resident advisor). From Figure 2.21 if we 'lower' the rectangle to the xy plane and then integrate the results are same; that is

$$\int_{\text{path 1-2-3-4}} (xy\hat{\mathbf{a}}_x + 3\hat{\mathbf{a}}_y) \cdot \mathrm{d}\mathbf{l} = \iint x\hat{\mathbf{a}}_z \cdot \mathrm{d}x\, \mathrm{d}y\hat{\mathbf{a}}_z$$

Just as Stokes' relates a surface to a line integral, the next law relates a volume to a surface integral.

2.15 Gauss' law

Gauss' law was introduced in section 2.12 equation (2.22). Recall that a Gaussian surface is a closed surface, e.g., a sphere. This closed surface encloses the volume involved in the integration. This time we illustrate with an example from electricity. Consider a light bulb enclosed in a clear plastic bag shown in Figure 2.22. All light rays

Figure 2.21. Volume used to illustrate Stokes' theorem.

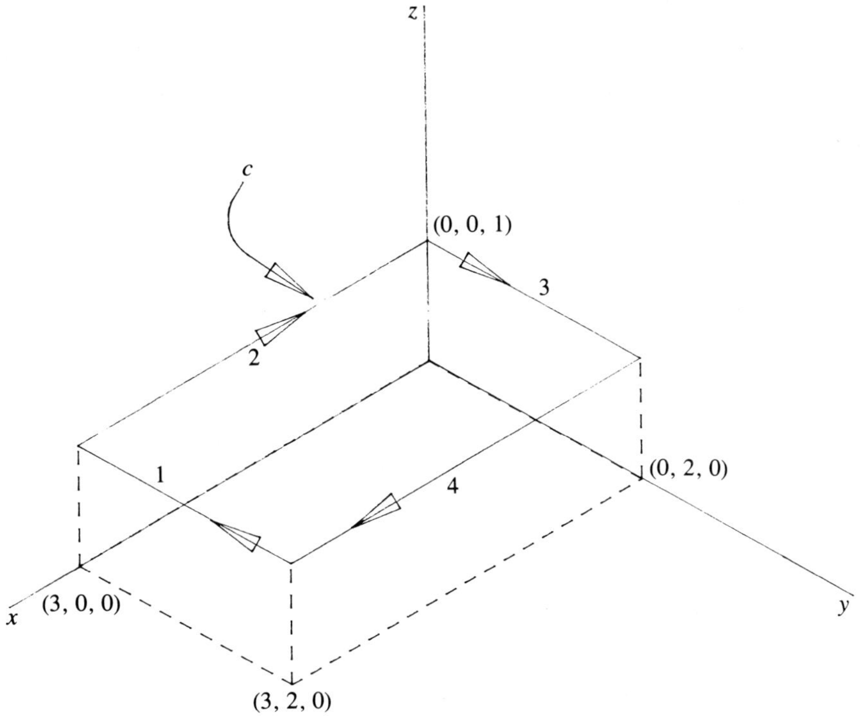

Figure 2.22. Model illustrating Gauss' law.

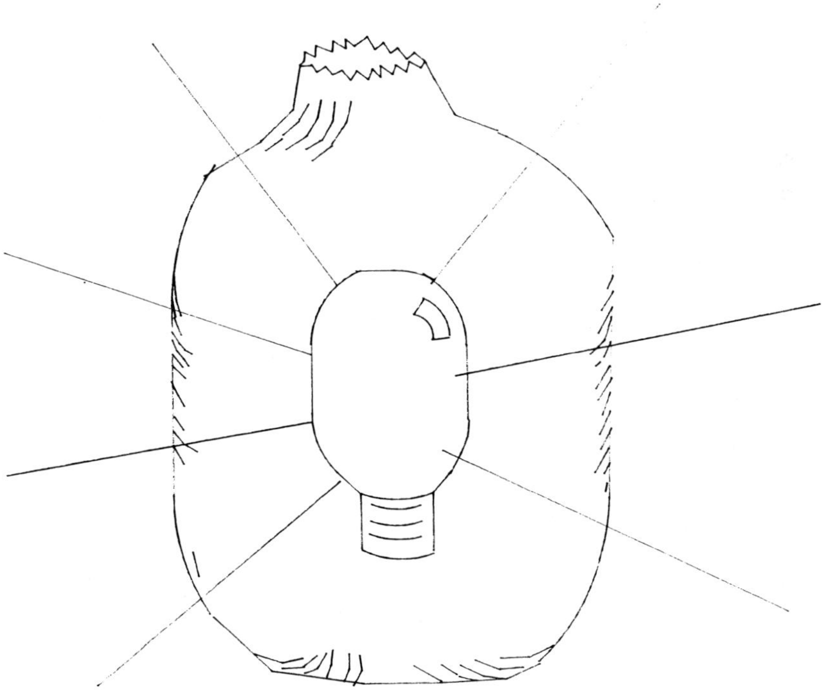

which pass through the plastic bag are the result of the light inside the bag. Simply stated: the sum of the flux lines penetrating the closed surface (plastic bag) is the result of the charge (light) enclosed.

In other words

$$\text{Flux} = \text{charge enclosed}$$

or

$$\Phi = Q$$

An integral form is

$$\Phi \equiv Q = \oint \mathbf{D} \cdot \mathrm{d}\mathbf{A} \tag{2.34}$$

where **D** is the flux density in 'charge' per meter^2 and d**A** is the area enclosing the 'charge' (after integration) with the direction normal to the surface. Also note that it is a closed integration (over our entire clear plastic bag).

D2.19 Calculate the flux leaving the cylinder centered about the z-axis, $r = 2$ m from $z = -2$ to $z = 2$ if $\mathbf{D} = r^2\hat{\mathbf{a}}_r + \sin\phi\hat{\mathbf{a}}_\phi + z^2\hat{\mathbf{a}}_z$.
Answer. 64π.

If our 'charge' consists of discrete electric charges, we write

$$Q = \sum_1^n Q_i \tag{2.35}$$

where Q_i is in coulombs (C). If our charge is a line charge, we write

$$Q = \int \rho_l \, \mathrm{d}l \tag{2.36}$$

where ρ_l is line charge density in C/m. If our charge is a surface charge, we write

$$Q = \iint \rho_s \, \mathrm{d}A \tag{2.37}$$

where ρ_s is surface charge density in C/m^2. Finally, if our charge is a volume, we write

$$Q = \iiint \rho \, \mathrm{d}v \tag{2.38}$$

where ρ is volume charge density in C/m^3.

As an illustration of Gauss' law, let us suspend in a clear liquid a sphere filled with red dye. At time $t = 0$ we puncture the sphere. At any time $t > 0$ we note that the same volume of dye is being dispersed throughout the liquid.

We can recapture all of the dye with a (closed) surface if we act quickly. However, the size of the surface must increase as $t \to \infty$. Our 'integrating' surface is a Gaussian surface which encloses all of our flux lines.

This concludes our chapter on vector analysis. Now we will put our mathematical muscles to work in various physical situations.

Problems

Let $a = 1.1$, $b = 0.5$, $\mathbf{A} = \hat{\mathbf{a}}_x + 2\hat{\mathbf{a}}_y - 3\hat{\mathbf{a}}_z$, and $\mathbf{B} = -2\hat{\mathbf{a}}_x + \hat{\mathbf{a}}_y + 2\hat{\mathbf{a}}_z$.

1. Do the following operations: (*a*) $a\mathbf{B}$, (*b*) $\mathbf{A} + \mathbf{B}$, (*c*) $\mathbf{A} - \mathbf{B}$, (*d*) $a\mathbf{A} + 2b\mathbf{B}$.

2. Find (*a*) $a\mathbf{A}$ (*b*) $|\mathbf{A}|$ (*c*) sketch $0.2a\mathbf{A}$.

3. Find the vector which extends from (*a*) the origin to $(-3, 2, 5)$ (*b*) $(1, 2, 1)$ to $(-1, 3, 2)$.

4. Temperature is a scalar. Graph several values for the temperature field $T(x, y) = x - 2y + 3$ in °C.

5. If $\mathbf{F} = (2y - x)\hat{\mathbf{a}}_x + \hat{\mathbf{a}}_y$ draw several vectors for the field – using your own notation for magnitude.

6. For the vectors **A** and **B**, find (*a*) $\mathbf{A} \cdot \mathbf{B}$ and (*b*) $\mathbf{A} \times \mathbf{B}$.

7. For the vectors **A** and **B**, find $\hat{\mathbf{a}}_A$ and $\hat{\mathbf{a}}_B$.

8. For the vectors **A** and **B**, find the angle α.

9. Find the unit vector component of (*a*) $b\mathbf{A} + \mathbf{B}$, (*b*) $a\mathbf{B}$.

10. Find (*a*) $a\mathbf{A} \cdot \mathbf{B}$ and (*b*) the angle between the two vectors, **A** and **B**.

11. Find (*a*) $\mathbf{A} \times \mathbf{B}$ and (*b*) the unit vector in the direction of $\mathbf{A} \times \mathbf{B}$.

12. If $\mathbf{C} = (x+1)\hat{\mathbf{a}}_x + 2y\hat{\mathbf{a}}_y$ find $C(1, 2, 1)$ and the unit vector at $(1, 2, 1)$ showing the direction of $\mathbf{C}$.

13. Given $\mathbf{D} = 3xy\hat{\mathbf{a}}_x + y\hat{\mathbf{a}}_y + (2x+z)\hat{\mathbf{a}}_z$, (*a*) find $\mathbf{D}$ at $(1, 2, 3)$, (*b*) plot $|\mathbf{D}|$ versus y along the line $x = 2$, $z = 0$.

14. Given $\mathbf{D} = 3xy\hat{\mathbf{a}}_x + y\hat{\mathbf{a}}_y + (2x+z)\hat{\mathbf{a}}_z$ and $\mathbf{E} = 3y\hat{\mathbf{a}}_x + 3x\hat{\mathbf{a}}_y + z\hat{\mathbf{a}}_z$. Find (*a*) $\mathbf{D} \times \mathbf{E}$ at $(3, 2, 1)$ and (*b*) the unit vector normal to $\mathbf{D}$ and $\mathbf{E}$.

15. If $\mathbf{D} = \hat{\mathbf{a}}_x + 3\hat{\mathbf{a}}_y$, $\mathbf{E} = 6\hat{\mathbf{a}}_x - 4\hat{\mathbf{a}}_y$ and $\mathbf{D} + \mathbf{C} = \mathbf{E}$ find $\mathbf{C}$.

16. (*a*) Given $(3, 2, 4)$ in rectangular coordinates, represent the point in cylindrical coordinates. (*b*) If we let $\mathbf{D} = 3\hat{\mathbf{a}}_x + 3\hat{\mathbf{a}}_y + 4\hat{\mathbf{a}}_z$ transform the vector to cylindrical coordinates.

17. Find the volume of the partial cylinder shown in Fig. 2.23.

18. If $\mathbf{C} = 2\hat{\mathbf{a}}_x + \hat{\mathbf{a}}_y - \hat{\mathbf{a}}_z$ and $\mathbf{D} = 4\hat{\mathbf{a}}_x - D_y\hat{\mathbf{a}}_y + \hat{\mathbf{a}}_z$ find D_y if $\mathbf{C}$ and $\mathbf{D}$ are perpendicular.

Figure 2.23. See problems 17 and 20.

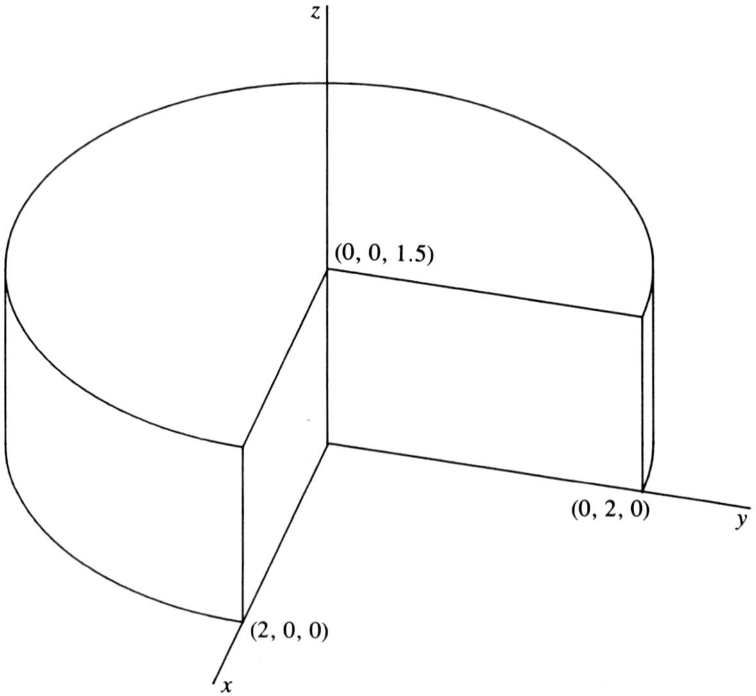

19. In cylindrical coordinates $\mathbf{D} = 31\hat{\mathbf{a}}_r - \frac{1}{2}\hat{\mathbf{a}}_\phi + \hat{\mathbf{a}}_z$ and $\mathbf{E} = 2\hat{\mathbf{a}}_r + \hat{\mathbf{a}}_\phi + \hat{\mathbf{a}}_z$. Find (*a*) $\hat{\mathbf{a}}_\mathbf{E}$ (*b*) the angle between **D** and **E** (*c*) $\mathbf{D} \cdot \mathbf{E}$, (*d*) $\mathbf{D} \times \mathbf{E}$.

20. Find the area of the surface enclosing the volume of Figure 2.23 by integration.

21. (*a*) Find the vector from $(0, 0, 0)$ to $(0, 1, 1)$, (*b*) Now rotate the point $(0, 1, 1)$ in the $\hat{\mathbf{a}}_\phi$ direction for 360°. Calculate the volume enclosed.

22. $$\mathbf{E} = \frac{x}{\sqrt{x^2 + y^2}}\hat{\mathbf{a}}_x + \frac{y}{\sqrt{x^2 + y^2}}\hat{\mathbf{a}}_y$$

Determine the unit vector at the point $(2, -3)$ in the direction of **E**.

23. Point A $(1, 0, 0)$, point B $(0, 1, 0)$ and point C $(0, 1, 1)$ are the vertices of a triangle. Calculate the area of the triangle.

24. $\mathbf{E} = (x + y)\hat{\mathbf{a}}_x + (y + z)\hat{\mathbf{a}}_y + (x + z)\hat{\mathbf{a}}_z$. Calculate $\int \mathbf{E} \cdot \mathbf{dl}$ from $(1, 0, 1)$ to $(-1, 1, 1)$ along the paths (*a*) $(1, 0, 1)$ to $(-1, 0, 1)$ to $(-1, 1, 1)$ and (*b*) $(1, 0, 1)$ to $(1, 0, 0)$ to $(-1, 0, 0)$ to $(-1, 1, 0)$ to $(-1, 1, 1)$.

25. Transform the vector $\mathbf{F} = xy\hat{\mathbf{a}}_x + y\hat{\mathbf{a}}_y + 3\hat{\mathbf{a}}_z$ to cylindrical coordinates.

26. Transform the vector $\mathbf{E} = r\hat{\mathbf{a}}_r$ in cylindrical coordinates into rectangular coordinates.

27. Transform the vector $\mathbf{G} = r\hat{\mathbf{a}}_r$ in spherical coordinates into rectangular coordinates.

28. Determine the gradient of (*a*) $f(x, y, z) = 3x + 4yz$, (*b*) $f(r, \phi, z) = 2r \sin \phi + 6\phi z$, (*c*) $f(r, \theta, \phi) = r^2 + \tan \phi + r \sin \theta$.

29. Evaluate both sides of Stokes' theorem for the contour shown in Figure 2.24 where $\mathbf{F} = r \sin \phi\hat{\mathbf{a}}_r + r \cos \phi\hat{\mathbf{a}}_\phi + z^2\hat{\mathbf{a}}_z$.

30. Calculate the five areas enclosing the volume of Figure 2.24.

31. Find the work done in moving a test charge, $Q_t = 10^{-8}$C, from the point $(1, 1, 1)$ to the point $(0, 2, 3)$ in the vicinity of an electric field $\mathbf{E} = (10^{-9}/2\pi\epsilon_0 r)\hat{\mathbf{a}}_r$ in cylindrical coordinates.

32. Find the potential at the point $(0, 2, 2)$ if (a) $V = 0$ at $r = 1$ (b) $V = 0$ at $r = 3$. $V = -\int \mathbf{E} \cdot d\mathbf{l}$ where $\mathbf{E} = (10^{-9}/2\pi\epsilon_0 r)\hat{\mathbf{a}}_r$ in cylindrical coordinates.

33. Find the work done in moving a test charge, $Q = -10^{-8}$C, from $(2, 1, 3)$ to the point $(0.5, 1, 0.5)$ in the vicinity of an electric field caused by a point charge $Q = 10^{-8}$C at the origin.

34. What is the potential at the point $(-1, -1, -1)$ caused by a point charge $Q = 3.2$ nC at $(2, 0, 1)$ if $V = -\int \mathbf{E} \cdot d\mathbf{l}$ where $\mathbf{E} = (Q/4\pi\epsilon_0 r^2)\hat{\mathbf{a}}_r$ in spherical coordinates?

35. Two vectors are defined in free space as $\mathbf{L} = x\hat{\mathbf{a}}_x + (x + y)\hat{\mathbf{a}}_y + 3z\hat{\mathbf{a}}_z$ and $\mathbf{N} = y\hat{\mathbf{a}}_x + (x + y)\hat{\mathbf{a}}_y + 2\hat{\mathbf{a}}_z$. Determine (a) $\mathbf{L} \cdot \mathbf{N}$ (b) the component of $\mathbf{L}$ in the direction of $\mathbf{N}$ at the point $(-2, 2, 3)$.

36. $\mathbf{F} = r \sin \phi \hat{\mathbf{a}}_r + r \cos \phi \hat{\mathbf{a}}_\theta + \sin \theta \hat{\mathbf{a}}_\phi$ in spherical coordinates represents a force field. Determine the component of $\mathbf{F}$ which is normal to a sphere of radius $r = 3$ m at the point $(3, \frac{\pi}{3}, \frac{\pi}{4})$.

Figure 2.24. See problems 29 and 30.

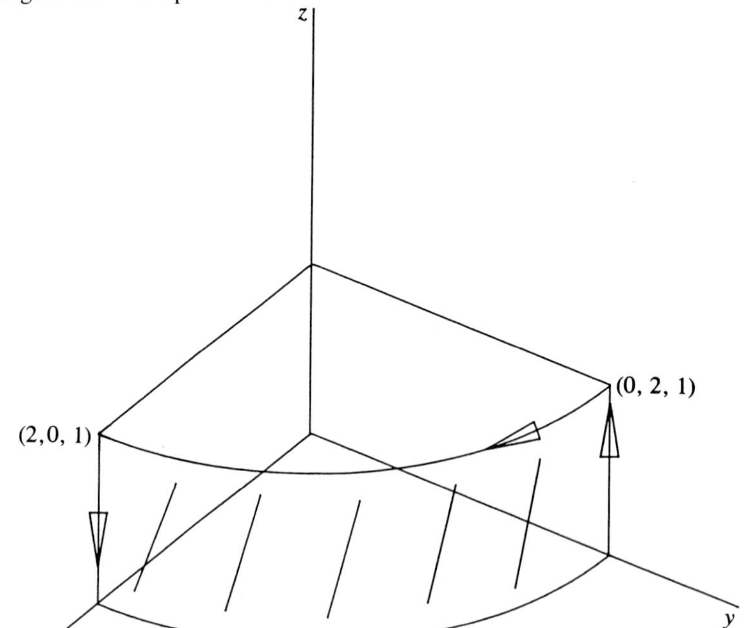

3

Electrostatics

3.1 Introduction

In this chapter we examine the electric field. The electric field is produced by stationary charges. We will examine the fields produced by various configurations such as point charges, line charges and plane charges. We define streamlines and electric potential. The first two of Maxwell's four equations are introduced. Several applications are illustrated. The chapter concludes with a discussion of electric fields in resistors and capacitors.

3.2 Coulomb's law

During the seventeenth century Colonel Charles Coulomb discovered experimentally that the force between two small charges is proportional to the product of their charges and inversely proportional to the square of the distance between them, or

$$F \propto \frac{Q_1 Q_2}{R^2} \tag{3.1}$$

where Q_1 and Q_2 are positive or negative charges and R is the separation. Under the SI units equation (3.1) now reads

$$F = \frac{Q_1 Q_2}{4\pi\epsilon_0 R^2} \tag{3.2}$$

where ϵ_0 is the permittivity of free space in F/m and has value

$$\begin{aligned} \epsilon_0 &= 8.854 \cdot 10^{-12}\ \text{F/m} \\ &\approx \frac{1}{36\pi} \cdot 10^{-9}\ \text{F/m} \end{aligned}$$

Since force is a vector, we write

$$\mathbf{F} = \frac{Q_1 Q_2}{4\pi\epsilon_0 R^2}\,\hat{\mathbf{a}}_R \tag{3.3}$$

However, this is a general expression which says very little about direction. Is it the force on Q_1 by Q_2 or vice versa? First, since we assume our universe is balanced (no excess charges), then the force on Q_1 by Q_2 must equal the force on Q_2 by Q_1 in magnitude. (We also make the assumption that there are no other charges in the universe!) Only the direction is different; in fact, they differ by a minus sign. Therefore we write the force on Q_2 by Q_1

$$\mathbf{F}_2 = \frac{Q_1 Q_2}{4\pi\epsilon_0 R_{12}{}^2} \hat{\mathbf{a}}_{R12} \tag{3.4}$$

and the force on Q_1 by Q_2

$$\mathbf{F}_1 = \frac{Q_1 Q_2}{4\pi\epsilon_0 R_{21}^2} \hat{\boldsymbol{a}}_{R21} \tag{3.5}$$

In other words, $\mathbf{F}_1 = -\mathbf{F}_2$ and $R_{12} = R_{21}$. To simplify notation, we drop the subscripts on R and write

$$\mathbf{F} = \frac{Q_t Q}{4\pi\epsilon_0 R^2} \hat{\mathbf{a}}_R \tag{3.6}$$

where Q_t is a test charge. A test charge is used to measure the force.

D3.1 Calculate the force, $\mathbf{F}$, on a 10 nC test charge placed at the origin caused by a -30 nC charge at $(1, 2, -3)$ where R points from $(1, 2, -3)$ to $(0, 0, 0)$.
Answer. $5.2 \cdot 10^{-8}[\hat{\mathbf{a}}_x + 2\hat{\mathbf{a}}_y - 3\hat{\mathbf{a}}_z]$ N.

3.3 Electric field

Now define an electric field equal to force per unit charge

$$\mathbf{E} \triangleq \frac{\mathbf{F}}{Q_t}$$

By dividing both sides of equation (3.6) by Q_t, we define an electric field. We see that the electric field caused by a charge Q (on a test charge, Q_t) is also inversely proportional to the square of the separation, or

$$\mathbf{E} = \frac{Q}{4\pi\epsilon_0 R^2} \hat{\mathbf{a}}_R \tag{3.7}$$

where

$$\hat{\mathbf{a}}_R = \frac{\mathbf{R}}{|\mathbf{R}|}$$

and

$$R \equiv |\mathbf{R}|$$

D3.2 Calculate the electric field at the origin in D3.1.
Answer. $5.15\hat{\mathbf{a}}_x + 10.3\hat{\mathbf{a}}_y - 15.45\hat{\mathbf{a}}_z$ N/C.

3.4 Point charge

Next let us try an example.

Example 3.1 What is the force in free space caused by a charge, Q, of 10^{-9} coulombs at the origin on a test charge located at the point $(2, 4, 3)$?

First we find

$$\hat{\mathbf{a}} = \frac{\mathbf{R}}{|\mathbf{R}|}$$

Figure 3.1. Effect of a point charge at the origin from example 3.1.

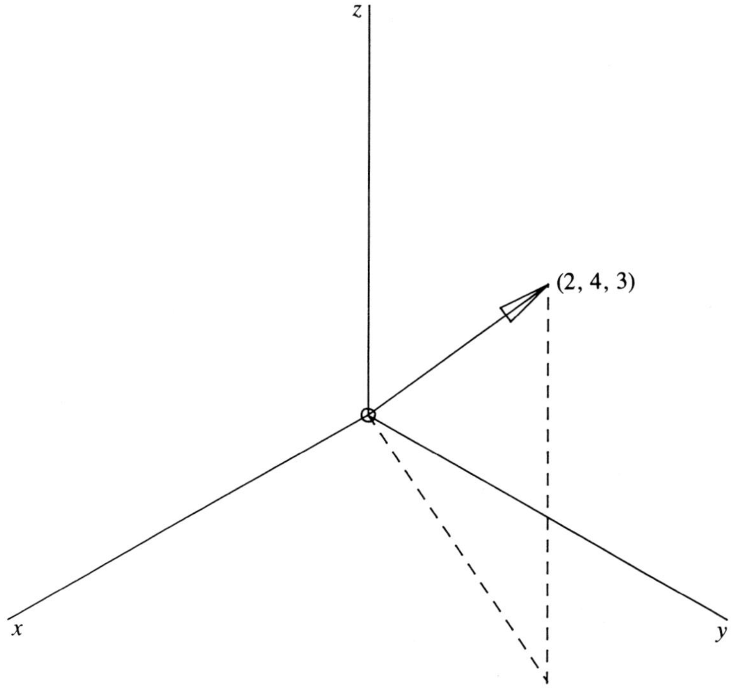

Now

$$\mathbf{R} = (2-0)\hat{\mathbf{a}}_x + (4-0)\hat{\mathbf{a}}_y + (3-0)\hat{\mathbf{a}}_z$$

and

$$|\mathbf{R}| = \sqrt{2^2 + 4^2 + 3^2}$$

Therefore

$$\mathbf{a}_R = \frac{2\hat{\mathbf{a}}_x + 4\hat{\mathbf{a}}_y + 3\hat{\mathbf{a}}_z}{\sqrt{29}}$$

and

$$\mathbf{E} = 0.058[2\hat{\mathbf{a}}_x + 4\hat{\mathbf{a}}_y + 3\hat{\mathbf{a}}_z]\ \text{V/m}$$

$$\mathbf{F} = Q_t\mathbf{E}$$

D3.3 Calculate the electric field at (2, 4, 6) caused by a point charge of 10 nC at (0, 0, 0).
Answer. $0.2148[2\hat{\mathbf{a}}_x + 4\hat{\mathbf{a}}_y + 6\hat{\mathbf{a}}_z]$ V/m

---

3.5 Superposition

If there are many charges then the total force (or electric field) is the sum of the individual forces. In other words superposition can be used. If there exist n charges, then we must sum n forces. In the limit as n increases to infinity, we naturally integrate.

Now let us consider another example which utilizes superposition.

Example 3.2 What is the electric field at (2, 2, 2) in free space caused by charges $Q_1 = -4 \cdot 10^{-9}$ C at (1, 1, 1) and $Q_2 = +2 \cdot 10^{-9}$ C at (−1, 1, 0).

Since superposition is applicable

$$\mathbf{E} = \mathbf{E}_1 + \mathbf{E}_2$$

where

$$\mathbf{E}_1 = \frac{Q_1}{4\pi\epsilon_0 R_1^{\,2}}\hat{\mathbf{a}}_{R_1}$$

and

$$\mathbf{E}_2 = \frac{Q_2}{4\pi\epsilon_0 R_2^{\,2}}\hat{\mathbf{a}}_{R_2}$$

Inserting the proper numbers, we obtain

$$\mathbf{E}_1 = \frac{-4 \cdot 10^{-9}}{4\pi \dfrac{10^{-9}}{36\pi}} \left[\frac{(2-1)\hat{\mathbf{a}}_x + (2-1)\hat{\mathbf{a}}_y + (2-1)\hat{\mathbf{a}}_z}{3\sqrt{3}} \right]$$

$$\mathbf{E}_2 = \frac{2 \cdot 10^{-9}}{4\pi \dfrac{10^{-9}}{36\pi}} \left[\frac{(2+1)\hat{\mathbf{a}}_x + (2-1)\hat{\mathbf{a}}_y + 2\hat{\mathbf{a}}_z}{14\sqrt{14}} \right]$$

and

$$\mathbf{E} = -2{\cdot}3\hat{\mathbf{a}}_x - 3\hat{\mathbf{a}}_y - 3\hat{\mathbf{a}}_z \text{ V/m}$$

Now any charge configuration can be divided into a sum of the charges, that is

$$Q = \sum^{n} \Delta Q$$

Similarly the total electric field can be written as the sum of fields, that is

$$\mathbf{E} = \sum^{n} \Delta \mathbf{E}$$

We again use superposition and write for the partial electric field from equation (3.7)

$$\Delta \mathbf{E} = \frac{\Delta Q}{4\pi\epsilon_0 R^2} \hat{\mathbf{a}}_R \tag{3.8}$$

If we are to find the total electric field, we add the 'ΔQs' or

$$\mathbf{E} = \text{sum}_n \left[\frac{\Delta Q}{4\pi\epsilon_0 R_n^{\,2}} \hat{\mathbf{a}}_{R_n} \right] \tag{3.9}$$

If the charges are discrete, we sum individually; if continuous, we integrate – which is itself a summing operation.

D3.4 (a) Calculate the electric field at the origin caused by a 10 nC charge at (1, 0, 1) and a 10 nC charge at (1, 1, 0). (b) Calculate the electric field at (1, 0, 1) caused by a 10 nC charge at the origin and a 10 nC charge at (1, 1, 0).
Answer. (a) $-63.6\hat{\mathbf{a}}_x - 31.8\hat{\mathbf{a}}_y - 31.8\hat{\mathbf{a}}_z$ (b) $31.8\hat{\mathbf{a}}_x - 31.8\hat{\mathbf{a}}_y + 63.6\hat{\mathbf{a}}_z$ V/m.

Note that to find the force on a test charge in any of the above examples, we need to multiply the electric field by the value of the test charge in coulombs. In general,

$$\mathbf{F} = Q_t \mathbf{E}$$

and once we know the electric field, multiplication by a scalar Q gives the force.

Before we proceed with additional examples, we need to underline in our minds something which will make problem solving easier. We normally need to make two choices in problem solving: 1. which coordinate system should we use, and 2. where should we put our charge configuration (source)? At times one or both of these questions are given in the problem statement. However, when they are not given in the problem, wise choices make life easier. Now let us proceed with another charge configuration.

3.6 Line charge

An infinite line charge (a filament) is the sum of n uniformly distributed, discrete charges, where $n \to \infty$. To find the electric field due to an infinite line we integrate.

Consider an infinite line charge with a uniform distribution of charges ρ_l C/m as shown in Figure 3.2. Find the electric field as a function of distance.

Recalling that answering two important questions stated above may simplify calculations, we take heart. First, if we choose the cylindrical coordinate system, we will have symmetry in the ϕ direction, and second, if we place the line charge along the z-axis, our electric field will vary with radius only – simplifying the mathematics. From equation (3.8) for a single charge

$$\Delta \mathbf{E} = \frac{\Delta Q}{4\pi\epsilon_0 R^2} \hat{\mathbf{a}}_R$$

where

$$\Delta Q = \rho_l \, \Delta l$$
$$= \rho_l \, \Delta z$$

and

$$\mathbf{R} = r\hat{\mathbf{a}}_r - z\hat{\mathbf{a}}_z$$

Now let us add all the ΔQs in our infinite line. Substituting into (3.8) and adding we obtain

$$\mathbf{E} = \int_{-\infty}^{\infty} \frac{\rho_l \, dz (r\hat{\mathbf{a}}_r - z\hat{\mathbf{a}}_z)}{4\pi\epsilon_0 (r^2 + z^2)^{3/2}} \tag{3.9a}$$

Before evaluating the integral, let us note one simplification we can make. The electric charge is infinite, which means that in Figure 3.2 we can locate a ΔQ below and a ΔQ above the xy plane which cause an electric field at point P. This electric field has components in the plus and minus $\hat{\mathbf{a}}_z$ and $\hat{\mathbf{a}}_r$ directions. Since we are adding all ΔQs the $\hat{\mathbf{a}}_z$ components from the top ΔQ and the bottom ΔQ cancel, leaving the $\hat{\mathbf{a}}_r$ components.

Therefore, integrating equation (3.9a),

$$\mathbf{E} = \frac{\rho_l}{2\pi\epsilon_0 r} \hat{\mathbf{a}}_r \tag{3.10}$$

Figure 3.2. An infinite line charge placed along the z-axis.

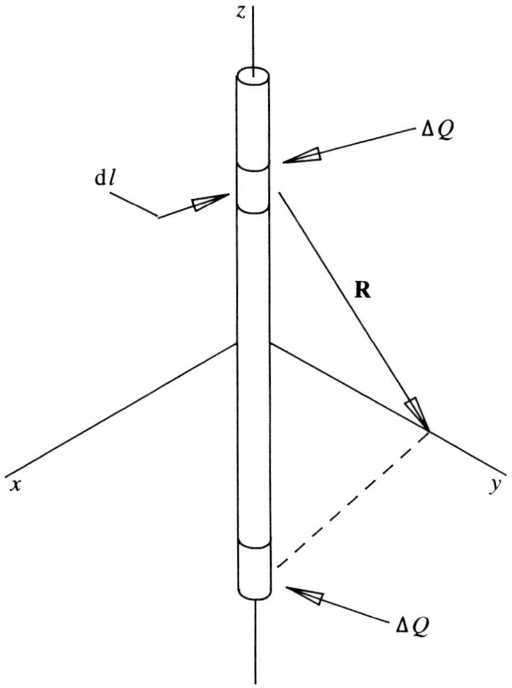

D3.5 Calculate the electric field at $(1.5, 30°, 5)$ due to an infinite line charge along the z-axis where $\rho_l = 12\,\text{nC/m}$.
Answer. $144\hat{\mathbf{a}}_r$ V/m.

3.7 Infinite sheet

Next let us find the electric field due to an infinite sheet with charges distributed uniformly. Once again we can divide the sheet into $n \times n$ charges and add the charges in both directions. But wait a minute! We have already calculated the field in one of the directions (the infinite line). Let's use those results. Thus we add infinite *line* charges for quicker results. We can use equation (3.10) to evaluate the electric field due to an infinite sheet of charge.

Example 3.3 Find the electric field at any point in space due to an infinite sheet of uniform charge ρ_s C/m^2.

Let us use Cartesian coordinates and place the sheet in the yz plane as shown in Figure 3.3.

Figure 3.3. An infinite sheet shown divided into line charges.

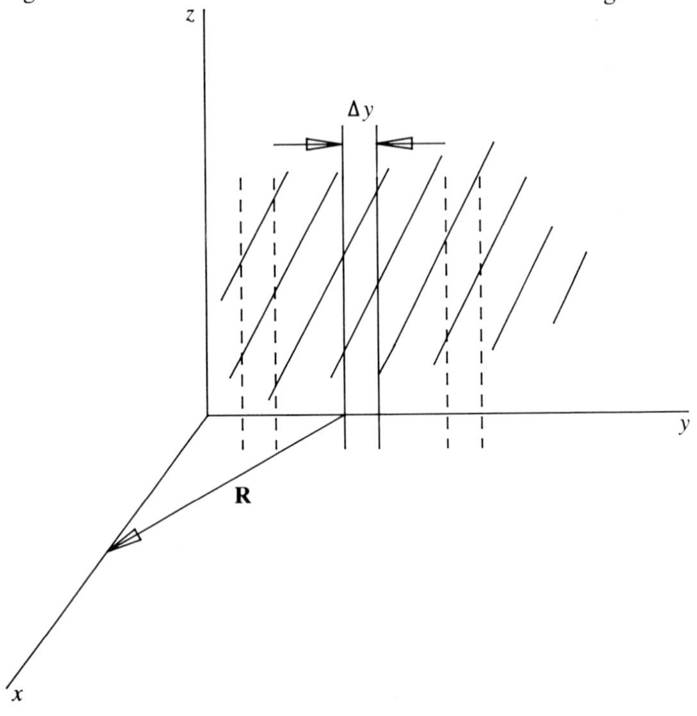

We divide the sheet into an infinite number of line charges. Since we calculated the electric field due to an infinite line charge in the above example, let us use the results in this problem. From equation (3.10), we write

$$\mathbf{E} = \frac{\rho_s \, \Delta y}{2\pi\epsilon_0 r} \hat{\mathbf{a}}_r \tag{3.11}$$

where

$$\rho_l = \rho_s \, \Delta y$$

However, we see from Figure 3.3 that the '$\mathbf{E}$' in equation (3.11) is actually part of the total electric field due to the surface charge. Therefore, let us change names. Let us change equation (3.11) to

$$\Delta\mathbf{E} = \frac{\rho_s \, \Delta y}{2\pi\epsilon_0 r} \hat{\mathbf{a}}_r \tag{3.12}$$

to signify $\Delta\mathbf{E}$ is part of the total $\mathbf{E}$. Next we note from Figure 3.3

$$|\mathbf{R}| = r = \sqrt{x^2 + y^2} \tag{3.13}$$

and we write

$$\hat{\mathbf{a}}_r = \frac{x\hat{\mathbf{a}}_x - y\hat{\mathbf{a}}_y}{\sqrt{x^2 + y^2}} \tag{3.14}$$

Using equations (3.13) and (3.14) in equation (3.12), we obtain

$$\Delta\mathbf{E} = \frac{\rho_s \, \Delta y (x\hat{\mathbf{a}}_x - y\hat{\mathbf{a}}_y)}{2\pi\epsilon_0 (x^2 + y^2)} \tag{3.15}$$

As in the preceding example, we have components which cancel and components which add, that is, the $\hat{\mathbf{a}}_y$ components on each side of the z-axis cancel, while the $\hat{\mathbf{a}}_x$ components add. Adding all contributions, we obtain

$$\mathbf{E} = \int_{-\infty}^{\infty} \frac{\rho_s x \, dy}{2\pi\epsilon_0 (x^2 + y^2)}$$

$$= \frac{\rho_s}{2\epsilon_0} \hat{\mathbf{a}}_x \tag{3.16}$$

This is a strange result. For an infinite surface the electric field is independent of distance! As an application consider the lighting of large rooms. Typically the ceiling is covered with evenly distributed fluorescent lights. The lighting is diffused or spread out 'evenly,' the lighting system resembles an infinite surface charge. The result is that every point in the room has approximately the same lighting, regardless of the distance to the lights.

What about two parallel surfaces? Before we answer this, let us return to the single surface. From a side view of our infinite surface, as in Figure 3.4, we see that half of the charges are on one side of the surface and half are on the other side. This is because of the force of repulsion. Also assume one half of the negative charges are at negative infinity and half at positive infinity. The like positive charges repel one another causing the symmetric distribution indicated in Figure 3.4. From equation (3.16)

$$\mathbf{E} = \frac{\rho_s}{2\epsilon_0}\hat{\mathbf{a}}_x \qquad x > 0$$

Figure 3.4. An infinite sheet in the *yz* plane.

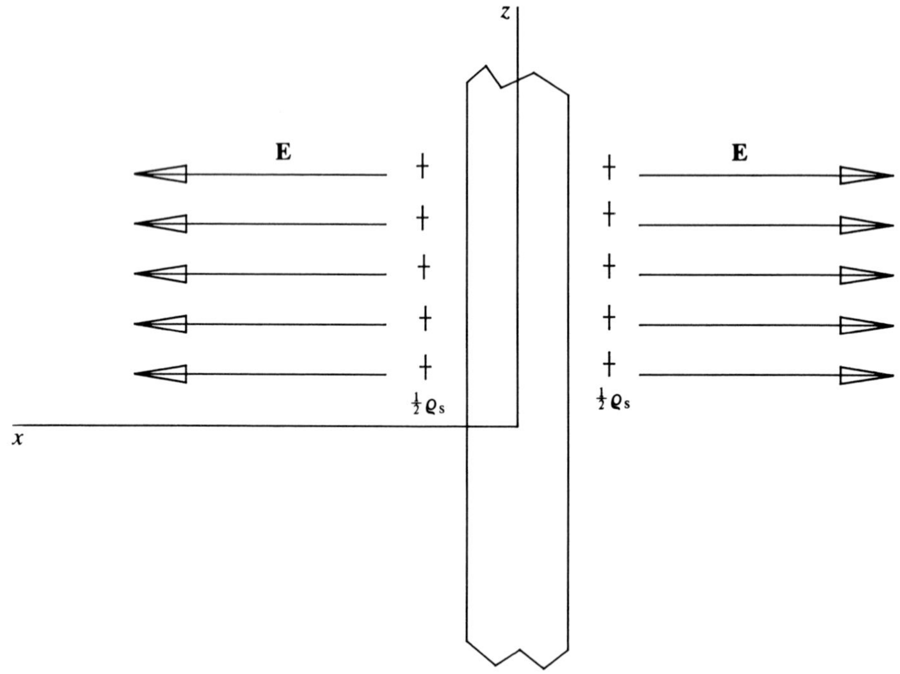

and if we calculate the electric field for $x < 0$, we obtain

$$\mathbf{E} = \frac{-\rho_s}{2\epsilon_0}\hat{\mathbf{a}}_x \qquad x < 0$$

Hence, the electric field intensity vectors are as shown in Figure 3.4.

Now if we place another infinite plane parallel to our first one above, at $x = x_0$ and give it a charge $-\rho_s$ C/m^2, we obtain

$$\mathbf{E} = \frac{\rho_s}{2\epsilon_0}\hat{\mathbf{a}}_x - \frac{-\rho_s}{2\epsilon_0}\hat{\mathbf{a}}_x$$

$$= \frac{\rho_s}{\epsilon_0}\hat{\mathbf{a}}_x \qquad 0 \le x \le x_0 \tag{3.17}$$

and

$$\mathbf{E} = 0 \qquad x < 0 \quad \text{and} \quad x > x_0$$

As shown in Figure 3.5, the only electric field is within the plates. It is the accepted convention to draw arrows from the positive charge to the

Figure 3.5. Two parallel infinite sheets.

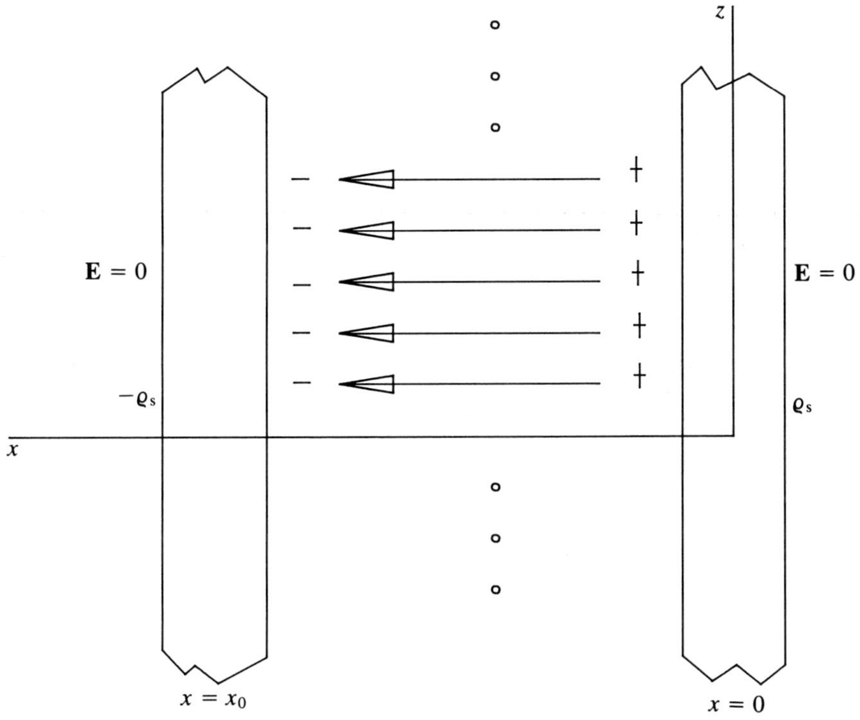

negative. The vectors are drawn at equal intervals showing a uniform charge distribution along the plates.

3.8 Streamlines

'A picture is worth a thousand words.' If we could picture lines flowing between charges or groups we could understand better electric fields. Let us imagine these lines and let us call these lines *streamlines*. The next question to be answered is the number of streamlines to utilize. Should we use one streamline per ten coulombs? Or one streamline per 100 coulombs? The choice is ours. In fact, we define flux tubes in which there are equal numbers of streamlines shown in three dimensions in Figure 3.6. No streamlines cut through the tubes. The flux tubes are imaginary also. The flux lines are also called streamlines or direction lines. They show the direction of movement of, for example, particles of dust in a heating duct. As the particles move, they have velocity components u_x and u_y. From Figure 3.7, we can write $u_x/u_y = dx/dy$ in two dimensions or $dx/u_x = \mathrm{d}y/u_y = \mathrm{d}z/u_z$ in three

Figure 3.6. Flux tubes containing streamlines or direction lines.

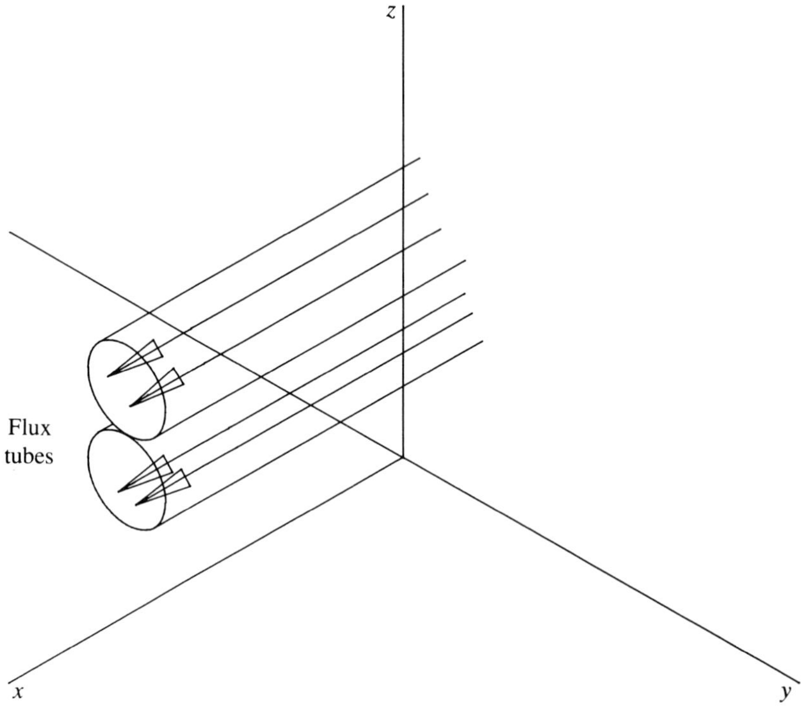

dimensions. Now consider the source of these particles. If there is no friction and if the streamlines do not cross or intersect the flux tubes, then the particle velocities must be equal. Therefore, they must have been released from the source at the same time for adjacent particles, as shown in Figure 3.7. The figure is in two dimensions; we assume no change in the z-direction. Thus the streamlines and the constant velocity lines (surfaces in three dimensions) form orthogonal lines everywhere in space.

If we consider again our two infinite planes in Figure 3.5 with charges ρ_s and $-\rho_s$, then the planes are themselves constant 'velocity' surfaces. Any imaginary plane which is parallel to the infinite planes in the region $0 < x < x_0$ is itself a constant velocity surface. The electric field vectors are streamlines. In the regions $x < 0$ and $x > x_0$ there are no streamlines and no constant velocity surfaces.

Now that we have a charge distribution, and hence an electric field, let us examine the results of moving a test charge in our electric field.

Figure 3.7. An example showing particles flowing in a duct.

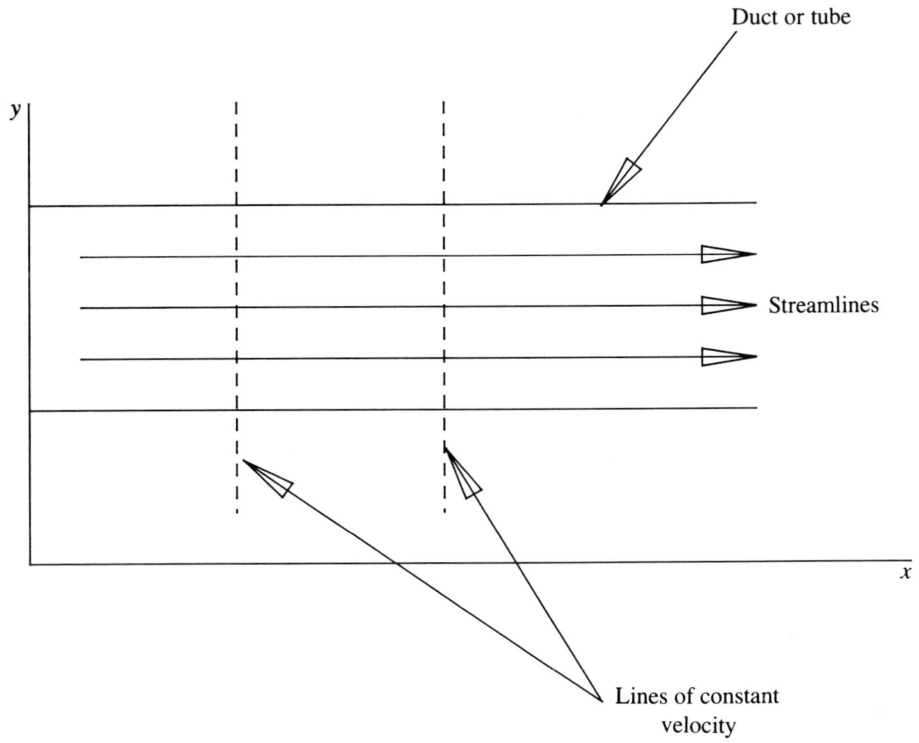

3.9 Work

We now define *work*. Work is an energy increase or decrease. If like charges are moved toward each other, energy is lost (work is done). If unlike charges are moved toward each other, energy is gained (negative work is done). We assume a charge distribution in space. This charge distribution is the source of an electric field. If we attempt to move a test charge in the electric field, we do work. The force we supply is equal and opposite to the force of the field,

$$\mathbf{F}_{\text{applied}} = -\mathbf{F}_{\text{charge dist.}}$$

Since work is force times distance, we write

$$\begin{aligned} W &= -\int_{\text{a}}^{\text{b}} \mathbf{F}_{\text{applied}} \cdot \mathbf{dl} \\ &= -\int_{\text{a}}^{\text{b}} Q\mathbf{E} \cdot \mathbf{dl} \end{aligned} \tag{3.18}$$

The integration is along some path from a to b which is in the electric field.

Figure 3.8. Showing various paths from a to b for the calculation of work done in examples 3.4 and 3.5.

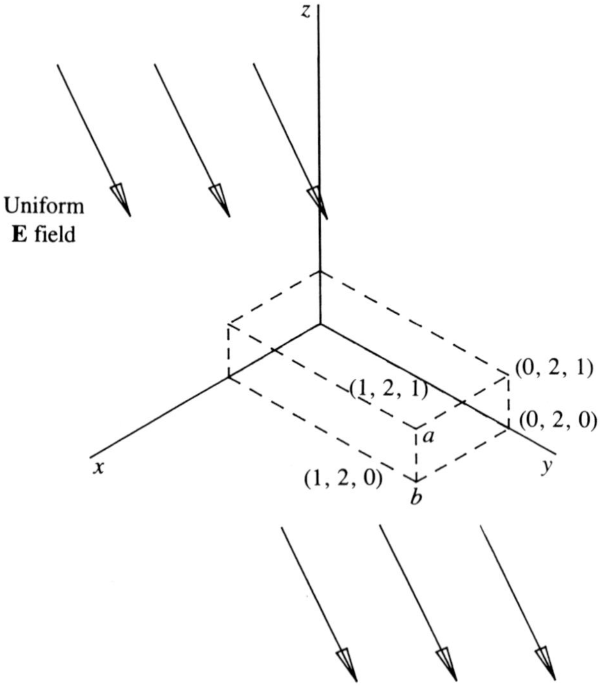

Example 3.4 In Figure 3.8 find the work done in moving a test charge of 10^{-6} C from the point $(1, 2, 1)$ to the point $(1, 2, 0)$ in the negative z direction in the electric field

$$\mathbf{E} = \hat{\mathbf{a}}_x + 4\hat{\mathbf{a}}_y - \hat{\mathbf{a}}_z$$

$$W = -\int_{(1,2,0)}^{(1,2,1)} 10^{-6}(\hat{\mathbf{a}}_x + 4\hat{\mathbf{a}}_y - \hat{\mathbf{a}}_z) \cdot (-\mathrm{d}z\, \hat{\mathbf{a}}_z)$$

$$= -10^{-6}\ \mathrm{J}$$

where $\mathbf{dl} = -\mathrm{d}z\, \hat{\mathbf{a}}_z$.

Example 3.5 In Figure 3.8 find the work done if the path a–b is changed. That is, what is the work done if we integrate from, say, $(1, 2, 1)$ to $(0, 2, 1)$ to $(0, 2, 0)$ and then to $(1, 2, 0)$?

$$W = -10^{-6}\Bigg[\int_{(0,2,1)}^{(1,2,1)} (\hat{\mathbf{a}}_x + 4\hat{\mathbf{a}}_y - \hat{\mathbf{a}}_z) \cdot (-dx\, \hat{\mathbf{a}}_x)$$

$$+ \int_{(0,2,0)}^{(0,2,1)} (\hat{\mathbf{a}}_x + 4\hat{\mathbf{a}}_y - \hat{\mathbf{a}}_z) \cdot (-\mathrm{d}z\, \hat{\mathbf{a}}_z)$$

$$+ \int_{(0,2,0)}^{(1,2,0)} (\hat{\mathbf{a}}_x + 4\hat{\mathbf{a}}_y - \hat{\mathbf{a}}_z) \cdot (\mathrm{d}x\, \hat{\mathbf{a}}_x)\Bigg]$$

$$= -10^{-6}\ \mathrm{J}$$

In the above two examples we have moved from point a to point b. In the first we have used the shortest route and in the second, we have purposely chosen a longer path. In both cases the work done has been the same. Now if we move from point b, back to point a, we do 10^{-6} J of work. Summing, we obtain

$$\oint \mathbf{E} \cdot \mathbf{dl} = \int_a^b + \int_b^a = -10^{-6} + 10^{-6} = 0$$

The work is then independent of the path a–b. The field is a conservative field.

D3.6 Find the approximate work done in moving a charge $Q = 2\ \mu\mathrm{C}$ a distance of $3 \cdot 10^{-12}$ m in the vicinity of the point $(2, 1, -2)$ in the electric field $\mathbf{E} = -2x\hat{\mathbf{a}}_x + y\hat{\mathbf{a}}_y + 3z^2\hat{\mathbf{a}}_z$ V/m.
Answer. $76 \cdot 10^{-18}$ J.

D3.7 Calculate the work done in moving a charge $Q = 15\ \mu\mathrm{C}$ along the circular path shown in Figure 3.9.
Answer. $1.2 \cdot 10^{-4}$ J.

3.10 Potential

We define the potential difference V as the work done on a test charge in an electric field,

$$V_{ba} \triangleq \frac{W}{Q_t} = -\int_a^b \mathbf{E} \cdot d\mathbf{l} \tag{3.19}$$

When discussing potential we need to clarify whether the potential is 'absolute' or relative. Absolute potential is a potential with reference to some common ground. Relative potential – sometimes called potential difference – is a potential relative to any 'nearby' voltage. Now the potential difference is

$$V = V_b - V_a$$

Example 3.6 As an example of potential difference, let us calculate the potential difference between two points $r = r_a$ and $r = r_b$ in the electric field of a point charge. We know from equation (3.7) that the

Figure 3.9. See **D3.7**.

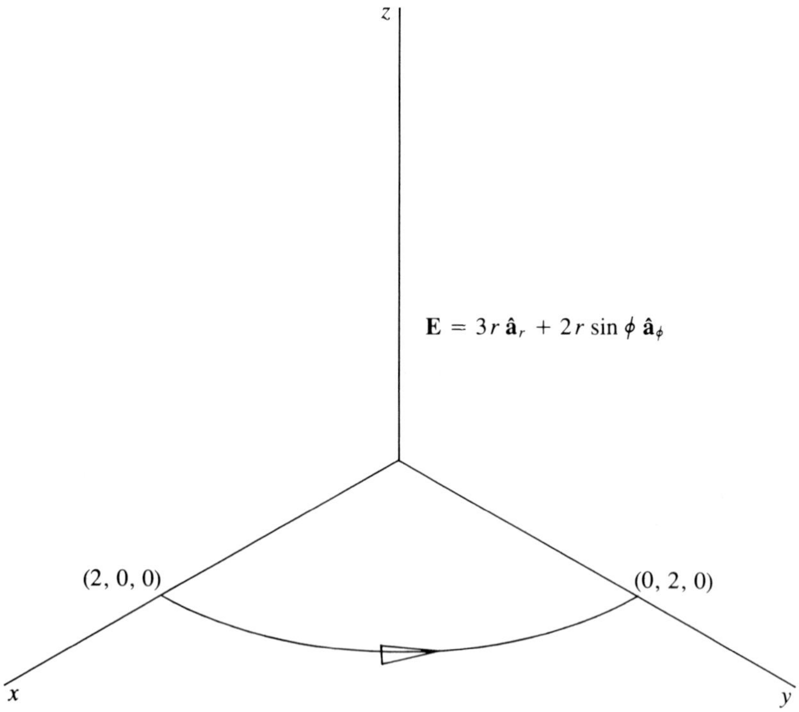

electric field of a point charge is in spherical coordinates

$$\mathbf{E} = \frac{Q}{4\pi\epsilon_0 r^2}\hat{\mathbf{a}}_r \tag{3.7}$$

Also, from Chapter 2 for spherical coordinates.

$$\mathrm{d}\mathbf{l} = \mathrm{d}r\hat{\mathbf{a}}_r + r\,\mathrm{d}\theta\hat{\mathbf{a}}_\theta + r\sin\theta\,\mathrm{d}\phi\hat{\mathbf{a}}_\phi$$

Now the potential difference is

$$\begin{aligned} V_b - V_a &= -\int_a^b \frac{Q}{4\pi\epsilon_0 r^2}\hat{\mathbf{a}}_r \cdot (\mathrm{d}r\hat{\mathbf{a}}_r + r\,\mathrm{d}\theta\hat{\mathbf{a}}_\theta + r\sin\theta\,\mathrm{d}\phi\hat{\mathbf{a}}_\phi) \\ &= -\int_a^b \frac{Q}{4\pi\epsilon_0 r^2}\,\mathrm{d}r \\ &= \frac{Q}{4\pi\epsilon_0}\left(\frac{1}{r_b} - \frac{1}{r_a}\right) \end{aligned} \tag{3.20}$$

Note that the potential difference varies inversely with the radius, as shown by equation (3.20). In addition, equation (3.20) shows that the equipotential surfaces are normal to the electric field. The potential is a scalar.

A word needs to be said about absolute potential. This is a rather philosophical area of science. Is there an absolute potential on a spacecraft? Where is ground? As the spacecraft is about to be launched, it is clear that the potentials on the craft are relative to *earth* ground. However, as soon as the craft leaves the surface of the earth, what happens to the reference (earth) ground? All voltages are now measured with respect to the craft's own ground (floating ground?). If the craft lands on the moon's surface, the reference ground changes. Is there an absolute ground somewhere? If so, can we attach one side of our voltmeter to make absolute measurements? Since our lead lengths are limited, let us take an easier approach. Let us define our reference points and make our measurements using these references.

Consider again the potential difference due to a point charge,

$$V_{ba} = V_b - V_a = \frac{Q}{4\pi\epsilon_0}\left(\frac{1}{r_b} - \frac{1}{r_a}\right) \tag{3.20}$$

A very convenient reference in this case is infinity. We choose $V = 0$ at

$r_a = \infty$ and obtain

$$V_{ba} = \frac{Q}{4\pi\epsilon_0 r_b}$$

This potential V_{ba} is the work done on a positive unit charge in carrying it from infinity to the point r_b. Thus, by choosing $V_a = 0$ at $r_a = \infty$, we have established the potential expression for moving a positive charge in the field of a point charge situated at the origin. Figure 3.10 shows a plot of the absolute potential of a point charge at the origin plotted against radius. Each radius has a unique potential. So there are an infinite number of potential surfaces; all are spheres concentric about the origin. Recall

$$\mathbf{E} = \frac{Q}{4\pi\epsilon_0 r^2}\hat{\mathbf{a}}_r$$

for a point charge. We conclude that the electric field intensity, **E**, is in the radial direction, and the equipotential surfaces are everywhere normal.

Figure 3.10. The potential due to a point charge located at the origin.

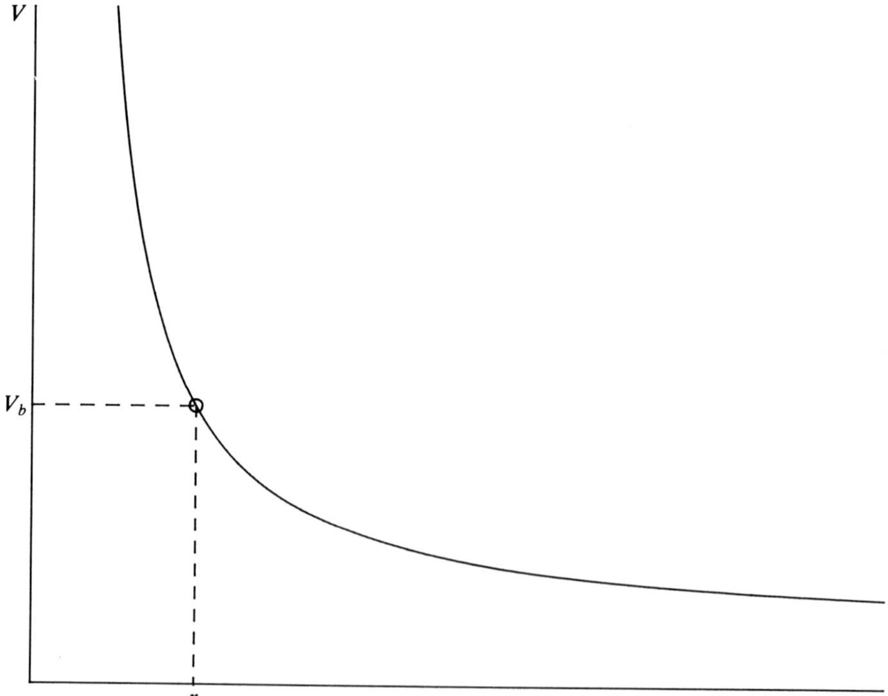

Now consider an infinite line charge with uniform charge density ρ_l C/m. From equation (3.10) for a line charge

$$\mathbf{E} = \frac{\rho_l}{2\pi\epsilon_0 r}\hat{\mathbf{a}}_r \tag{3.10}$$

and (using cylindrical coordinates)

$$\mathbf{dl} = \mathrm{d}r\hat{\mathbf{a}}_r + r\,\mathrm{d}\phi\hat{\mathbf{a}}_\phi + \mathrm{d}z\hat{\mathbf{a}}_z$$

Substituting into (3.19)

$$V = -\int_a^b \mathbf{E}\cdot\mathbf{dl}$$

$$= \frac{\rho_l}{2\pi\epsilon_0}\ln\left(\frac{a}{b}\right)$$

In this case we cannot choose the origin or infinity as a reference, so we choose some arbitrary in-between point.

D3.8 Calculate the potential at (1.5, 30°, 2.2) due to an infinite line charge where $\rho_l = 3.2\cdot 10^{-12}$ C/m
Answer. $2.34\cdot 10^{-2}$ V; ref = 1 m

3.11 Gradient

The gradient operation is useful in calculating the electric field. We can calculate the electric field from the potential. Previously we calculated the electric field knowing the charge distribution. Now we have a second method for finding **E**.

From the definition of potential difference,

$$V_{ba} = -\int_a^b \mathbf{E}\cdot\mathbf{dl}$$

we can write

$$\Delta V = -\mathbf{E}\cdot\Delta\mathbf{l}$$

$$= -|\mathbf{E}|\,|\Delta\mathbf{l}|\cos\alpha$$

or

$$|\mathbf{E}| \approx \left.\frac{\Delta V}{\Delta l}\right|_{\alpha=0}$$

In general

$$\mathbf{E} = -\nabla V \tag{3.21}$$

where

$$\nabla = \frac{\partial}{\partial x}\hat{\mathbf{a}}_x + \frac{\partial}{\partial y}\hat{\mathbf{a}}_y + \frac{\partial}{\partial z}\hat{\mathbf{a}}_z \quad \text{in rectangular coordinates}$$

$$\nabla = \frac{\partial}{\partial r}\hat{\mathbf{a}}_r + \frac{\partial}{r\,\partial \phi}\hat{\mathbf{a}}_\phi + \frac{\partial}{\partial z}\hat{\mathbf{a}}_z \quad \text{in cylindrical coordinates}$$

$$\nabla = \frac{\partial}{\partial r}\hat{\mathbf{a}}_r + \frac{\partial}{r\,\partial \theta}\hat{\mathbf{a}}_\theta + \frac{1}{r\sin\theta}\frac{\partial}{\partial \phi}\hat{\mathbf{a}}_\phi \quad \text{in spherical coordinates}$$

These were introduced in Chapter 2 and are repeated here for convenience. We say the electric field intensity is the negative of the gradient of the potential.

D3.9 Given the potential field $V = \ln(xyz)$, determine the electric field at $(3, 4, 2)$.
Answer. $-\frac{1}{3}\hat{\mathbf{a}}_x - \frac{1}{4}\hat{\mathbf{a}}_y - \frac{1}{2}\hat{\mathbf{a}}_z$ V/m.

3.12 Maxwell's equation – divergence

Before we attempt some applications let us introduce two of Maxwell's equations. They appear in both integral and differential forms. Recall Gauss' law which says

$$\oint \mathbf{D} \cdot \mathrm{d}\mathbf{A} = Q$$

For a volume charge density, ρ, we write

$$\oint \mathbf{D} \cdot \mathrm{d}\mathbf{A} = Q = \iiint \rho \, \mathrm{d}v$$

This is Maxwell's equation in integral form. Now using the divergence theorem

$$\oint \mathbf{D} \cdot \mathrm{d}\mathbf{A} = \iiint \nabla \cdot \mathbf{D} \, \mathrm{d}v$$

We obtain

$$\iiint \nabla \cdot \mathbf{D} \, \mathrm{d}v = \iiint \rho \, \mathrm{d}v$$

or

$$\nabla \cdot \mathbf{D} = \rho$$

This is Maxwell's equation in point form. For a small volume the divergence of the electric flux density equals the volume charge density, ρ C/m^3.

D3.10 Given $\mathbf{D} = (x + y)\hat{\mathbf{a}}_x + (y + z)\hat{\mathbf{a}}_y + (x + z)\hat{\mathbf{a}}_z$, evaluate $\nabla \cdot \mathbf{D}$ at the point $(-3, 2.5, 1.3)$.
Answer. 3 C/m^3

3.13 Maxwell's equation

Recall for a conservative field

$$\oint \mathbf{E} \cdot \mathrm{d}\mathbf{l} = 0$$

and from Stokes' theorem

$$\oint \mathbf{E} \cdot \mathrm{d}\mathbf{l} = \iint \nabla \times \mathbf{E} \cdot \mathrm{d}\mathbf{A}$$

In other words

$$\nabla \times \mathbf{E} = 0$$

This is another of Maxwell's equations, in point form (as applied to electrostatics). Those two Maxwell equations are shown in integral and point forms in Table 3.1.

Table 3.1. *Two of Maxwell's equations in integral and point forms*

Integral	Point
$\oint \mathbf{D} \cdot \mathrm{d}\mathbf{A} = \int_{\text{vol.}} \rho \, \mathrm{d}v$	$\nabla \cdot \mathbf{D} = \rho$
$\oint \mathbf{E} \cdot \mathrm{d}\mathbf{l} = 0$	$\nabla \times \mathbf{E} = 0$

3.14 Applications

Gauss' law

Now let us illustrate the use of Gauss' law by applying it to several problems. The law facilitates problem solving if, and only if, it is symmetrical. (For the nonsymmetrical problems other methods will be used.) Gauss' law is repeated here

$$\Phi \equiv Q = \oint \mathbf{D} \cdot \mathrm{d}\mathbf{A} \tag{3.22}$$

Equation (3.22) says that the flux Φ emanating from a source equals the flux density $\mathbf{D}$ intersecting the area which encloses the source. It almost seems to be a trivial statement, yet its importance cannot be overemphasized. A limitation of Gauss' law is that it is most useful with symmetric problems. The problem is a symmetrical problem if $\mathbf{D}$ is normal to the Gaussian surface everywhere. Let us try Gauss' law on a spherical problem.

Example 3.7 Find the electric flux density, $\mathbf{D}$, for a sphere of radius a meters with a uniform density ρ C/m^3. Remembering that we need to make two wise choices before proceeding with any problem, we choose the spherical system and place the charged sphere at the origin. Knowing the flux Φ is the result of the charge, we write

$$\Phi \equiv Q = \oint \mathbf{D} \cdot \mathrm{d}\mathbf{A}$$

$$= \int_0^{2\pi} \int_0^{\pi} D_r \hat{\mathbf{a}}_r \cdot r^2 \sin\theta \, \mathrm{d}\theta \, \mathrm{d}\phi \hat{\mathbf{a}}_r$$

$$= 4\pi r^2 D_r$$

or

$$D_r = \frac{Q}{4\pi r^2}$$

In vector form

$$\mathbf{D} = \frac{Q}{4\pi r^2} \hat{\mathbf{a}}_r \qquad r > a \tag{3.23}$$

Now for the density inside the sphere, $r < a$. Recall

$$Q = \iiint \rho \, \mathrm{d}v \tag{3.24}$$

and for a uniform density

$$Q = \rho \int_0^{2\pi} \int_0^{\pi} \int_0^{a} r^2 \sin\theta \, \mathrm{d}r \, \mathrm{d}\theta \, \mathrm{d}\phi$$

or

$$\rho = \frac{Q}{4/3\pi a^3} \qquad \mathrm{C/m^3}$$

Is there a flux density, **D**, inside the solid sphere? We now move our Gaussian surface inside the sphere and include less charge. Once again from Gauss' law

$$Q_{\text{enclosed}} = \iiint \rho \mathrm{d}v$$
$$= \int_0^{2\pi} \int_0^{\pi} \int_0^{r} \rho r^2 \sin\theta \, \mathrm{d}r \, \mathrm{d}\theta \, \mathrm{d}\phi$$

where $r < a$

$$Q_{\text{enclosed}} = \frac{Q}{4/3\pi a^3} \left(\frac{r^3}{3}\right) 4\pi$$

where

$$\rho = \frac{Q}{4/3\pi a^3}$$

Now

$$Q_{\text{enclosed}} = \oint \mathbf{D} \cdot \mathrm{d}\mathbf{A}$$

or

$$D_r = \frac{Q_{\text{enclosed}}}{4\pi r^2}$$

$$\mathbf{D} = \frac{Qr}{4\pi a^3} \hat{\mathbf{a}}_r \ \mathrm{C/m^2} \qquad r < a$$

This then is the flux density caused by the charge in Figure 3.11.

We can see from this result that the sphere of radius $r = a$ can be replaced by a point charge of Q coulombs and the flux density will not be affected.

Next we recall that the electric field for a point charge is

$$\mathbf{E} = \frac{Q}{4\pi\epsilon_0 r^2}\hat{\mathbf{a}}_r \tag{3.25}$$

Combining equations (3.23) and (3.25), we obtain

$$\mathbf{D} = \epsilon_0\mathbf{E}$$

Finally, we can find the potential outside our sphere by using

Figure 3.11. A sphere of radius a with a uniform charge density ρ for example 3.7.

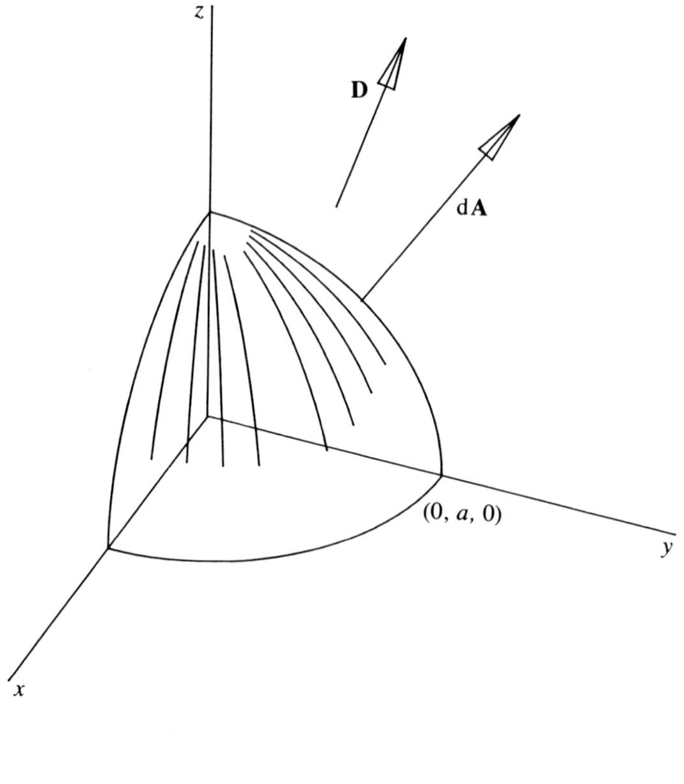

equation (3.19),

$$V = -\int \mathbf{E} \cdot \mathrm{d}\mathbf{l}$$

$$= -\int \frac{Q}{4\pi\epsilon_0 r^2} \hat{\mathbf{a}}_r \cdot \mathrm{d}r\hat{\mathbf{a}}_r$$

$$= \frac{Q}{4\pi\epsilon_0 r} \tag{3.26}$$

where we assume the potential $V = 0$ at $r \rightarrow \infty$

Example 3.8 Find the electric flux density due to an infinite line charge, shown in Figure 3.12, located along the z-axis. Using cylindrical coordinates, since the line is infinitely long, we cannot integrate to find Q and get meaningful results, that is,

$$Q = \int_{-\infty}^{\infty} \rho_l \, \mathrm{d}l \rightarrow \infty$$

Figure 3.12. An infinite line charge showing flux lines for example 3.8.

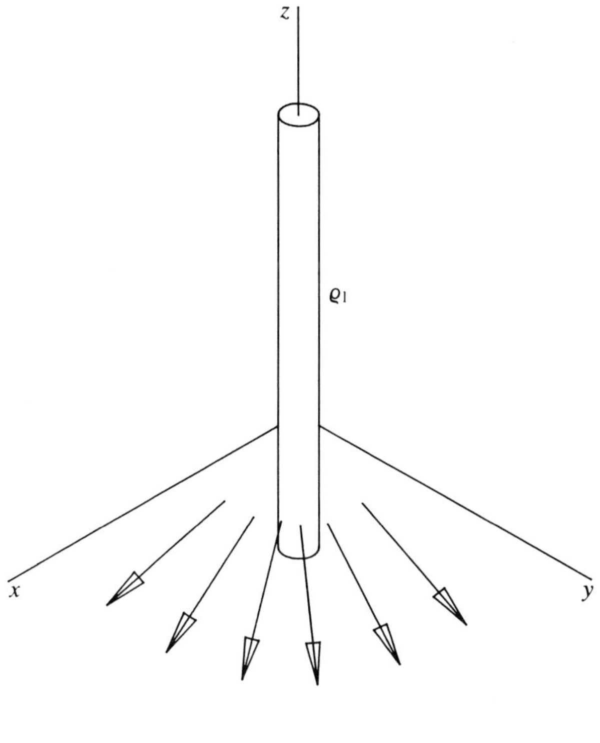

So we write

$$\int_0^L \rho_l \, dl = \oint \mathbf{D} \cdot d\mathbf{A}$$

$$= \int_0^L \int_0^{2\pi} D_r \hat{\mathbf{a}}_r \cdot r \, d\phi \, dz \hat{\mathbf{a}}_r$$

$$\rho_l L = D_r 2\pi r L$$

and

$$\mathbf{D} = \frac{\rho_l}{2\pi r} \hat{\mathbf{a}}_r \tag{3.27}$$

Recall from the previous example

$$\mathbf{D} = \epsilon_0 \mathbf{E}$$

and the electric intensity is

$$\mathbf{E} = \frac{\rho_l}{2\pi\epsilon_0 r} \hat{\mathbf{a}}_r \tag{3.28}$$

Now we calculate the potential due to the line charge

$$V = -\int_a^b \frac{\rho_l}{2\pi\epsilon_0 r} dr = \frac{\rho_l \ln(a/b)}{2\pi\epsilon_0} \tag{3.29}$$

We make two important observations at this point: 1. the potential of a line charge varies as the natural log of the radius, and 2. we cannot choose our reference at infinity (or at zero). Now let us assume the line charge has some finite radius $r = a$ so that it becomes a volume charge ρ C/m^3 as shown in Figure 3.13.

Example 3.9 Determine the flux density for an infinite line charge with radius $r = a$.

We again have the problem of infinite charge so we consider a length L in the z-direction.

$$\int_0^L \int_0^{2\pi} \int_0^a \rho r \, dr \, d\phi \, dz = \int_0^L \int_0^{2\pi} D_r r \, d\phi \, dz$$

$$\rho \left(\frac{a^2}{2}\right) 2\pi L = D_r r 2\pi L$$

$$\mathbf{D} = \frac{\rho a^2}{2r} \hat{\mathbf{a}}_r \tag{3.30}$$

However, we can write the volume charge ρ as

$$\rho = \frac{Q}{\pi a^2 L}\ \mathrm{C/m^3} \tag{3.31}$$

Substituting equation (3.31) into equation (3.30), we obtain

$$\begin{aligned}\mathbf{D} &= \left(\frac{Q}{\pi a^2 L}\right)\frac{a^2}{2r}\,\hat{\mathbf{a}}_r \\ &= \frac{Q/L}{2\pi r}\,\hat{\mathbf{a}}_r \end{aligned} \tag{3.32}$$

which gives us the same result as equation (3.27) with $\rho_l = Q/L$ C/m.

At this point we make an observation similar to that made in the previous example, that is, the filamentary charge can be replaced by a volumetric cylindrical charge without disturbing the flux lines (outside the cylinder). The opposite is also true – if we replace the cylinder with a filament, the flux density is in either case

$$\mathbf{D} = \frac{\rho_l}{2\pi r}\,\hat{\mathbf{a}}_r \qquad r > a \tag{3.33}$$

Figure 3.13. A finite line charge with uniform density.

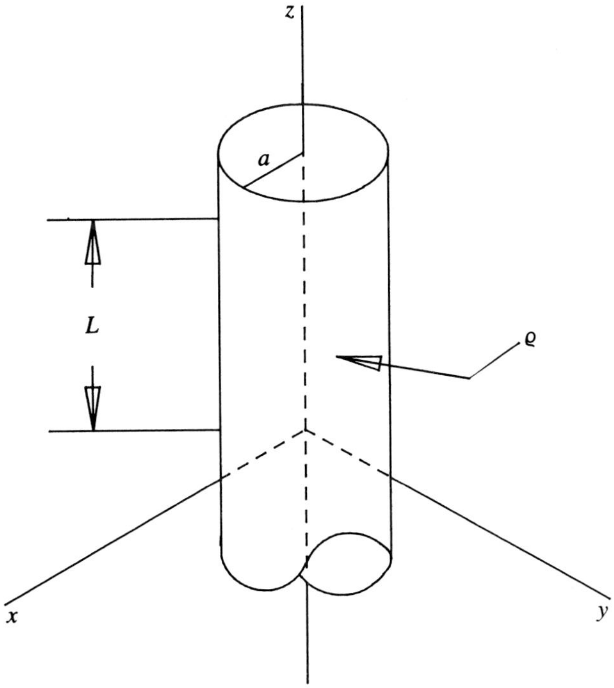

In addition,

$$\mathbf{E} = \frac{\rho_l}{2\pi\epsilon_0 r}\hat{\mathbf{a}}_r \qquad r > a \tag{3.34}$$

and

$$V = \frac{\rho_l}{2\pi\epsilon_0}\ln r \tag{3.35}$$

Now let us ask a more difficult question: what is the flux density for $r < a$, or inside the cylinder?

We again use Gauss' law since we have a symmetric problem.

$$Q = \oint \mathbf{D} \cdot \mathrm{d}\mathbf{A} \tag{3.22}$$

From our understanding of Gauss' law, we realize that the left-hand side of equation (3.22) deals with the 'real' charge, Q, while the right-hand side deals with the imaginary Gaussian surface. Figure 3.14 shows a cross-sectional view of the volume charge with the dotted line

Figure 3.14. Cross section of finite line charge with uniform volume density.

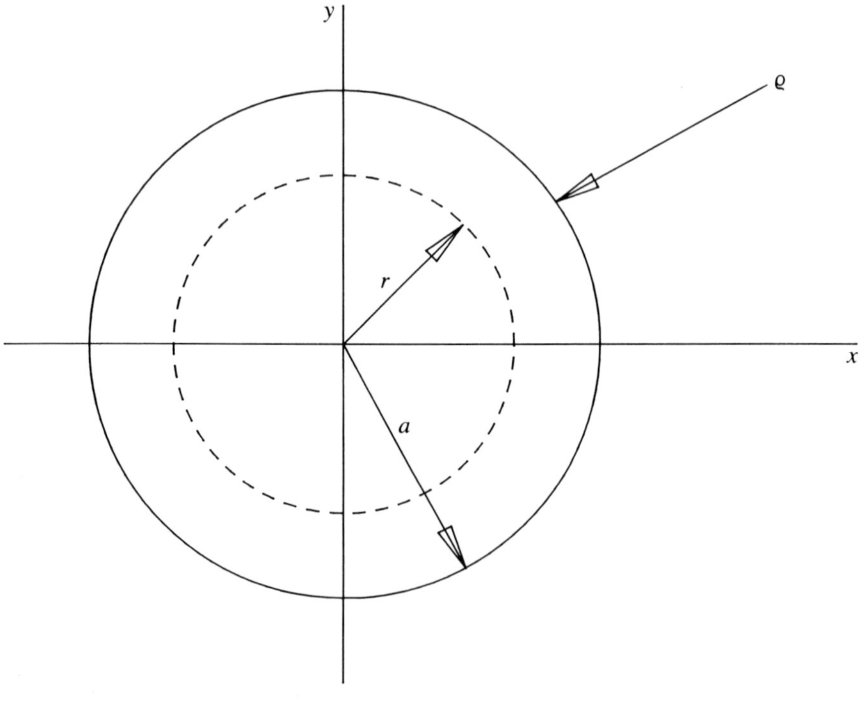

representing the Gaussian surface. Since only the charge within the Gaussian surface produces flux density at the surface, we need to find the total charge up to the radius r.

$$\int_0^L \int_0^{2\pi} \int_0^r \frac{Q}{\pi a^2 L} r \, \mathrm{d}r \, \mathrm{d}\phi \, \mathrm{d}z = \int_0^L \int_0^{2\pi} D_r \hat{\mathbf{a}}_r \cdot r \, \mathrm{d}\phi \, \mathrm{d}z \hat{\mathbf{a}}_r$$

$$\mathbf{D} = \frac{(Q/L)r}{2\pi a^2} \hat{\mathbf{a}}_r \qquad r < a$$

and

$$\mathbf{E} = \frac{\mathbf{D}}{\epsilon_0} = \frac{(Q/L)r}{2\pi\epsilon_0 a^2} \hat{\mathbf{a}}_r \qquad r < a \tag{3.36}$$

As a check on our results, let us find the flux density at the surface of the cylinder $r = a$. Substituting $r = a$ into equation (3.36), we obtain

$$\mathbf{D} = \frac{(Q/L)a}{2\pi a^2} \hat{\mathbf{a}}_r = \frac{Q/L}{2\pi a} \hat{\mathbf{a}}_r \tag{3.37}$$

and

$$\mathbf{E} = \frac{Q/L}{2\pi\epsilon_0 a} \hat{\mathbf{a}}_r$$

which agrees with equations (3.27) and (3.32).

D3.11 Given a line charge with uniform charge density $\rho = 3.4$ pC/m^3 and radius $r = 0.25$ m. Calculate the flux density at (a) $r = 2.5$ m (b) $r = 0.025$ m.
Answer. (a) and (b) $4.25 \cdot 10^{-14}$ C/m^2.

3.15 Nonsymmetric problems

For symmetric configurations such as a sphere or cylinder Gauss' law is extremely helpful in finding flux, electric field or potential. For nonsymmetric problems some other method must be found. The following application problems are considered nonsymmetric because Gauss' law cannot be used without some difficulty. In actuality, these problems have some symmetry as we will see. We will examine the electrical dipole and the electric doublet. First let us look at the dipole.

3.16 The electric dipole

Consider two charges of equal magnitude and opposite polarity separated by a distance d as shown in Figure 3.15(a). We can obtain the potential using superposition. Then using

$$\mathbf{E} = -\nabla V$$

we can calculate the electric field.

For an isolated charge, the potential is, from equation (3.26),

$$V = \frac{Q}{4\pi\epsilon_0 R}$$

and for two equal charges of opposite polarity

$$V = \frac{Q}{4\pi\epsilon_0 R_1} + \frac{-Q}{4\pi\epsilon_0 R_2}$$

$$= \frac{Q}{4\pi\epsilon_0} \frac{R_2 - R_1}{R_1 R_2} \tag{3.38}$$

where from Figure 3.15(a)

$$R_1 = \sqrt{x^2 + y^2 + (z - d/2)^2}$$

Figure 3.15. The electric dipole with P (a) in the vicinity of the dipole, (b) at a distance R, where $R \gg d$.

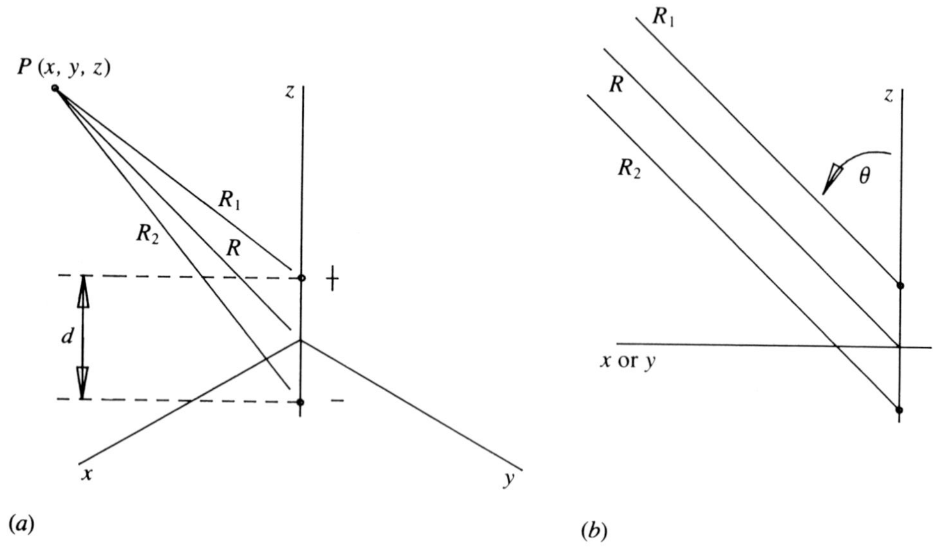

and

$$R_2 = \sqrt{x^2 + y^2 + (z + d/2)^2}$$

Now as d decreases or as the point in Figure 3.15(a) is removed to a distant point, R is approximately parallel to R_1 and R_2 in Figure 3.15(b) and

$$R_2 - R_1 \approx d \cos\theta; \qquad R_1 R_2 \approx r^2$$

$$V = \frac{Q}{4\pi\epsilon_0} \frac{d \cos\theta}{r^2} \tag{3.39}$$

where Qd = constant.

D3.12 Calculate the potential due to an electric dipole. The positive charge $Q_+ = 2.2$ nC is at (0, 0, 0.01) and the negative charge $Q_- = -2.2$ nC is at (0, 0, −0.01). Find V at (a) (0, 0.1, 0.1), (b) (0, 1, 0), (c) (0, 100, 100).
Answer. (a) 13.98 (b) 0 (c) 0 V.

Example 3.10 Determine the loci of equipotentials for an electric dipole. From equation (3.39), we obtain the equation for equipotential surfaces: Let $V = V_0$ and rewrite equation (3.39) as

$$\frac{V_0 4\pi\epsilon_0}{Qd} = \frac{\cos\theta}{r^2} \tag{3.40}$$

The left side of equation (3.40) is constant, so we write

$$K = \frac{\cos\theta}{r^2} \tag{3.41}$$

Equation (3.41) is plotted in Figure 3.16 for several values of potential. As potential increases, radius decreases and vice versa. Notice that the horizontal axis can be x or y: there is symmetry for the dipole along the z-axis. In other words the product stays constant for an electric dipole as the separation changes.

Next let's determine the electric field for an electric dipole. In spherical coordinates from equation (2.17)

$$\nabla = \frac{\partial}{\partial r}\hat{\mathbf{a}}_r + \frac{\partial}{r\,\partial\theta}\hat{\mathbf{a}}_\theta + \frac{1}{r\sin\theta}\frac{\partial}{\partial\phi}\hat{\mathbf{a}}_\phi \tag{2.17}$$

Combining equations (3.39) and (2.17), we have

$$\nabla V = \frac{Qd\cos\theta}{4\pi\epsilon_0}\left(\frac{-2}{r^3}\right)\hat{\mathbf{a}}_r + \frac{Qd(-\sin\theta)}{4\pi\epsilon_0 r^3}\hat{\mathbf{a}}_\theta \tag{3.42}$$

and

$$\mathbf{E} = -\nabla \mathrm{V}$$

or

$$\mathbf{E} = \frac{Qd}{4\pi\epsilon_0 r^3}(2\cos\theta\hat{\mathbf{a}}_r + \sin\theta\hat{\mathbf{a}}_\theta) \tag{3.43}$$

D3.13 Plot several values of the vector field from equation (3.43). Let $Qd/4\pi\epsilon_0 = 1$.

Figure 3.17 shows a sketch of typical streamlines in two dimensions ($E_z = 0$). We can write the differential equation

$$\frac{E_r}{E_\theta} = \frac{\mathrm{d}r}{r\,\mathrm{d}\theta} \tag{3.44}$$

Figure 3.16. Equipotential surfaces for a dipole with symmetry mounted along the z-axis for example 3.10.

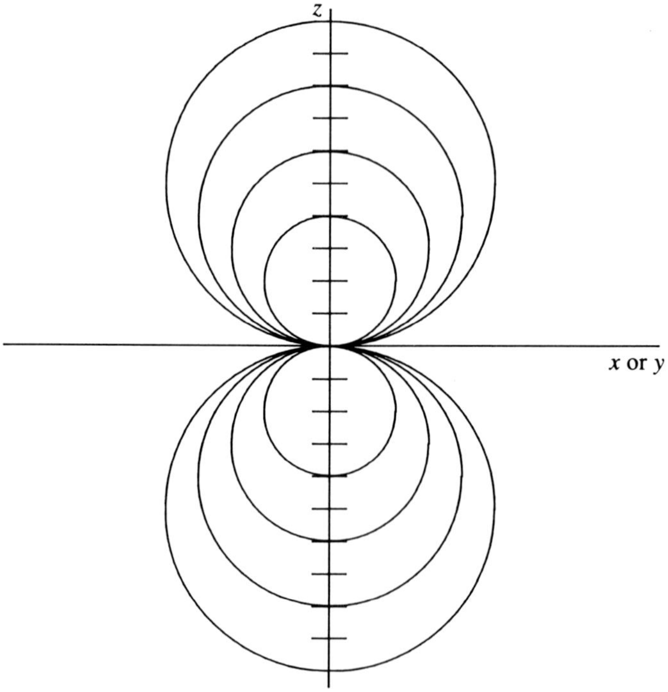

From equations (3.43) and (3.44), we write

$$\frac{\mathrm{d}r}{r\,\mathrm{d}\theta} = \frac{2\cos\theta}{\sin\theta}$$

or

$$\frac{\mathrm{d}r}{r} = 2\cot\theta\,\mathrm{d}\theta$$

from which we obtain

$$r = k\sin\theta \tag{3.45}$$

Several streamlines are plotted using equation (3.45) in Figure 3.18.

We define the dipole as

$$\mathbf{p} \triangleq Q\mathbf{d} \tag{3.46}$$

where $\mathbf{d}$ is the vector distance between $+Q$ and $-Q$ and since

$$\mathbf{d}\cdot\hat{\mathbf{a}}_r = d(1)\cos\theta$$

Figure 3.17. Streamlines in a two-dimensional field with ϕ symmetry.

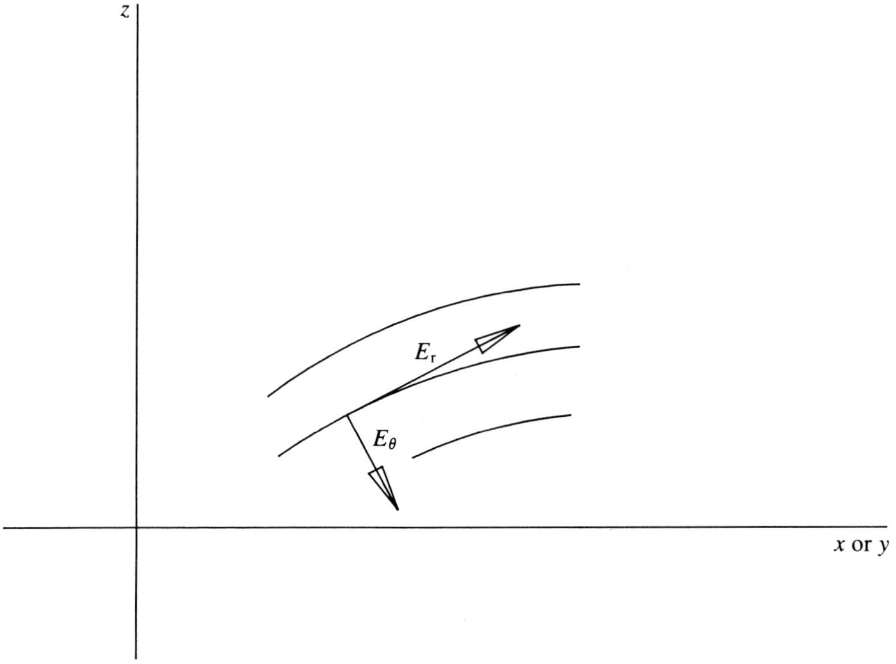

we can write

$$V = \frac{\mathbf{p} \cdot \hat{\mathbf{a}}_r}{4\pi\epsilon_0 r^2} \tag{3.47}$$

Equation (3.47) is the potential expression for an electric dipole. Now let us proceed to the electric doublet.

3.17 Electric doublet

An electric doublet is a pair of parallel, infinite line charges separated by a distance d, where d is small. Our approach will be similar to the electric dipole. (Recall the electric dipole was a pair of point charges.) We first consider two line charges, then allow the separation to become small – or view them from a distance, where $d \ll R$. Consider two infinite filamentary charges of opposite polarity, $+\rho_l$ and $-\rho_l$, separated by a distance as shown in Figure 3.19(a). To find the electric field, it is much easier to find the potential first then take derivatives (than to find $\mathbf{E}$ and integrate to find V).

Figure 3.18. Streamlines for an electric dipole with ϕ symmetry mounted along the z-axis from example 3.10.

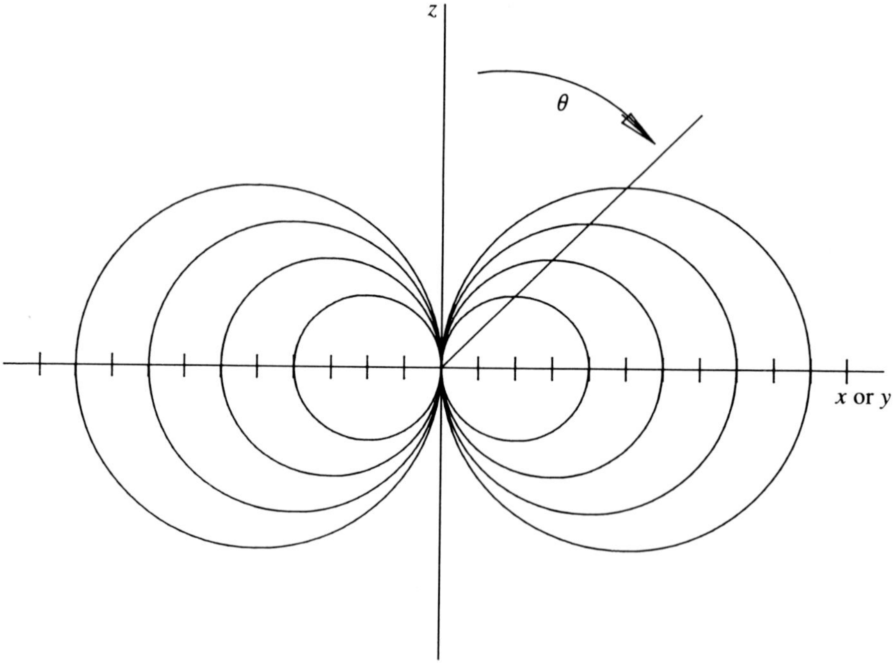

Figure 3.19. Infinite line charges mounted parallel to the z axis separated by a distance d. (a) The two lines in the $y = 0$ plane, (b) the equipotentials in the $x = 0$ plane.

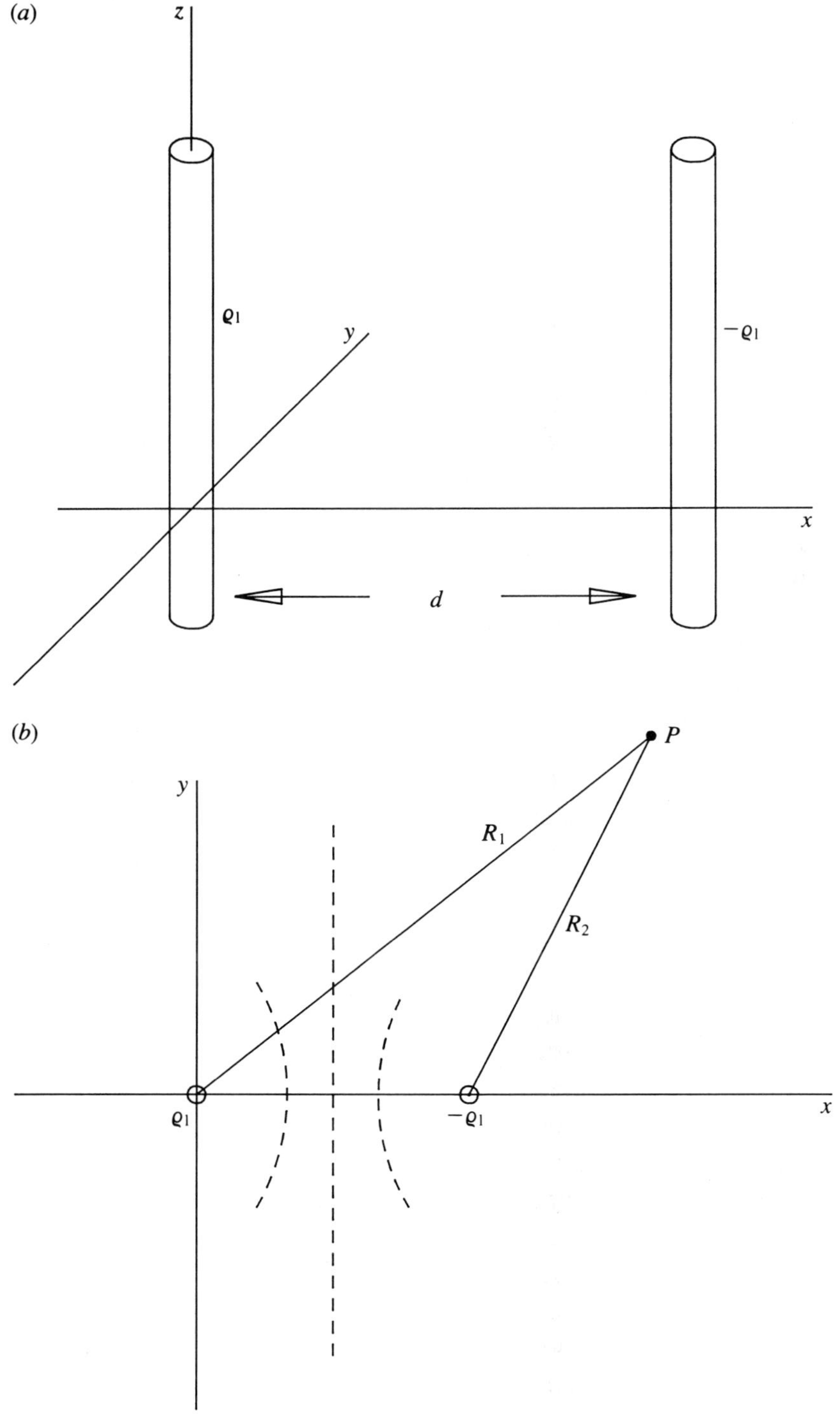

Recall from equation (3.29) the potential due to a line charge was

$$V = \frac{\pm\rho_l}{2\pi\epsilon_0} \ln \frac{a}{b} \tag{3.29}$$

Let the potential due to ρ_l be V_{source} and let the potential due to $-\rho_l$ be V_{sink}. Therefore, using superposition, we write for the potential at point P

$$\begin{aligned} V &= V_{\text{source}} + V_{\text{sink}} \\ &= \frac{\rho_l}{2\pi\epsilon_0} \ln \frac{R_{\text{ref}_1}}{R_1} + \frac{-\rho_l}{2\pi\epsilon_0} \ln \frac{R_{\text{ref}_2}}{R_2} \end{aligned} \tag{3.48}$$

where R_{ref_1} and R_{ref_2} are the points at which the potentials are referenced. We choose the references for the source/sink at the same point, that is, $R_{\text{ref}_1} = R_{\text{ref}_2}$ and write

$$V = \frac{\rho_l}{2\pi\epsilon_0} \ln(R_2/R_1) \tag{3.49}$$

From Figure 3.19(b) we substitute

$$R_1 = \sqrt{x^2 + y^2}$$

and

$$R_2 = \sqrt{(d-x)^2 + y^2}$$

We substitute R_1 and R_2 into equation (3.49) so that

$$V = \frac{\rho_l}{4\pi\epsilon_0} \ln \left[\frac{(d-x)^2 + y^2}{x^2 + y^2} \right] \tag{3.50}$$

Example 3.11 Determine the loci of equipotentials for two parallel lines. From equation (3.50) we write

$$V = V_0 = \frac{\rho_l}{4\pi\epsilon_0} \ln \left[\frac{(d-x)^2 + y^2}{x^2 + y^2} \right]$$

Solving the natural logarithm, we obtain

$$V_0 \frac{4\pi\epsilon_0}{\rho_l} = \ln \left[\frac{(d-x)^2 + y^2}{x^2 + y^2} \right] \tag{3.51}$$

Taking the exponential of both sides of equation (3.51), we get

$$e^{(4\pi\epsilon_0 V_0)/\rho_l} = \frac{(d-x)^2 + y^2}{x^2 + y^2} \tag{3.52}$$

Let

$$e^{(4\pi\epsilon_0 V_0)/\rho_l} \equiv K$$

The left-hand side of equation (3.52) is constant, so

$$(x^2 + y^2)K = (d - x)^2 + y^2$$

Since both x and y seem to have quadratic form let us try to obtain equations of circles.

$$Kx^2 + Ky^2 = d^2 - 2xd + x^2 + y^2$$

$$x^2(K - 1) + 2xd + y^2(K - 1) = d^2$$

$$x^2 + 2x\frac{d}{K-1} + y^2 = \frac{d^2}{K-1} \tag{3.53}$$

Recalling the form of equations of circles, we add a term to both sides of equation (3.53) which gives

$$\left(x + \frac{d}{K-1}\right)^2 + (y - 0)^2 = \frac{d^2K}{(K-1)^2} = \frac{d^2K}{(1-K)^2} \tag{3.54}$$

Equation (3.54) is the equation of circles with the center on the x-axis. This tells us that the equipotential surfaces for a source/sink pair are circles (cylinders really) with radius of $d\sqrt{K}/(1-K)$ and with centers at $(-d/(K-1), 0)$.

Note if $V_0 < 0$ then $K < 1$ and the circles are on the positive x, actually $x > d/2$, region. If $V_0 > 0$ then $K > 1$ and the circles are in the negative x, actually $x < d/2$, region. If $V_0 = 0$ then $K = 1$ and the circle has infinite radius with the center at infinity. The $V_0 = 0$ potential is shown in Figure 3.19(b) as a vertical line through the point $(d/2, 0)$. In addition the circles on one side of the $V_0 = 0$ surface have identical images on the opposite side. More will be said about the *method of images* later.

In summary, the equipotential surfaces are circles centered on the x-axis. Now let us calculate the electric field for the electric doublet.

$$\mathbf{E} = -\nabla V$$

$$= -\frac{\partial V}{\partial x}\hat{\mathbf{a}}_x - \frac{\partial V}{\partial y}\hat{\mathbf{a}}_y - \frac{\partial V}{\partial z}\hat{\mathbf{a}}_z \tag{3.55}$$

Recall we said above that there is no variation of potential in the z direction. Thus, the last derivative on the right side of equation (3.55) is zero. Calculating each term independently, we have

$$\frac{\partial V}{\partial x} = \frac{\rho_l}{4\pi\epsilon_0} \frac{2d(x^2 - y^2 - xd)}{[(d-x)^2 + y^2][x^2 + y^2]} \tag{3.56}$$

and similarly

$$\frac{\partial V}{\partial y} = \frac{\rho_l}{4\pi\epsilon_0} \left\{ \frac{2yd(2x - d)}{[(d-x)^2 + y^2][x^2 + y^2]} \right\} \tag{3.57}$$

Combining we obtain

$$\mathbf{E} = \frac{-\rho_l d}{2\pi\epsilon_0} \left\{ \frac{(x^2 - xd - y^2)\hat{\mathbf{a}}_x + y(2x - d)\hat{\mathbf{a}}_y}{[(d-x)^2 + y^2][x^2 + y^2]} \right\} \tag{3.58}$$

This vector, equation (3.58), describes the electric field intensity for two infinite line charges.

D3.14 Plot several electric field vectors for two infinite line charges. $\rho_l^+ = 1.6\,\text{nC/m}$ at $(0, 0, 0)$ and $\rho_l^- = -1.6\,\text{nC/m}$ at $(1, 0, 0)$.

The graphical method

From observing equation (3.58) or after doing D3.14 we may wonder if there isn't an easier solution to the problem of plotting the flux lines (or electric field lines or lines of force) for two parallel line charges. Actually there is. It is the graphical method. We illustrate with two parallel lines.

Figure 3.20. The graphical solution to the problem of plotting flux lines due to two parallel line charges from example 3.12.

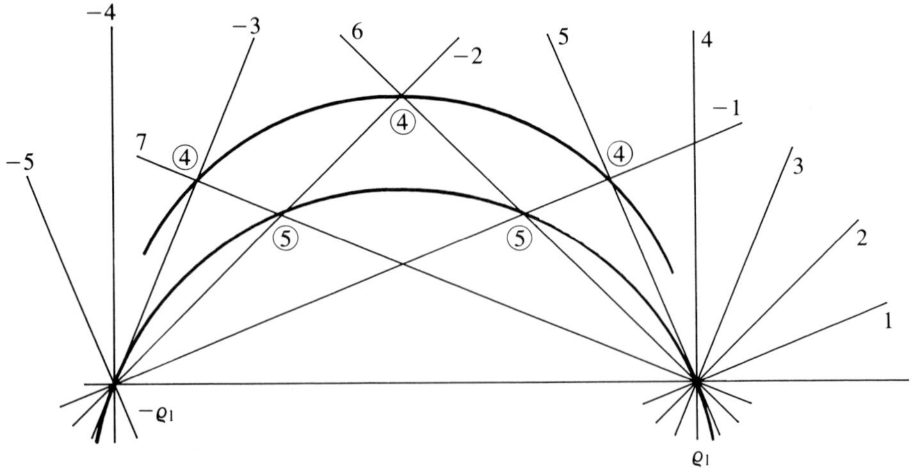

Example 3.12 Sketch the streamlines for two parallel lines. These are shown in Figure 3.19 as ρ_l and $-\rho_l$. The flow from each line is divided into 16 flux tubes. These are shown as $1, 2, 3, \ldots$ from ρ_l and -1, -2, $-3, \ldots$ from $-\rho_l$ in Figure 3.20. For simplicity not all flux tubes are shown. Using superposition we add the values of the flux tube at each intersection. These are shown as circled numbers in Figure 3.20. All that remains is to sketch streamlines through like numbers.

Now let us define the doublet in the same way we define the electric dipole. We allow the line charges ρ_l and $-\rho_l$ to come together by decreasing d or by removing the point P in Figure 3.21 to a distance such that R_1 and R_2 are approximately parallel. We need to observe that if we remove P too far, then $R_1 = R_2$ and $\ln(R_2/R_1) = 0$ or $V = 0$. In other words, if we remove ourselves far enough away from the line charges, they have a canceling effect. We remain at some intermediate distance – not too close, not too far – such that the angles between R_1, R and R_2 and the x-axis are approximately equal, as shown in Figure 3.21.

Figure 3.21. An electric doublet.

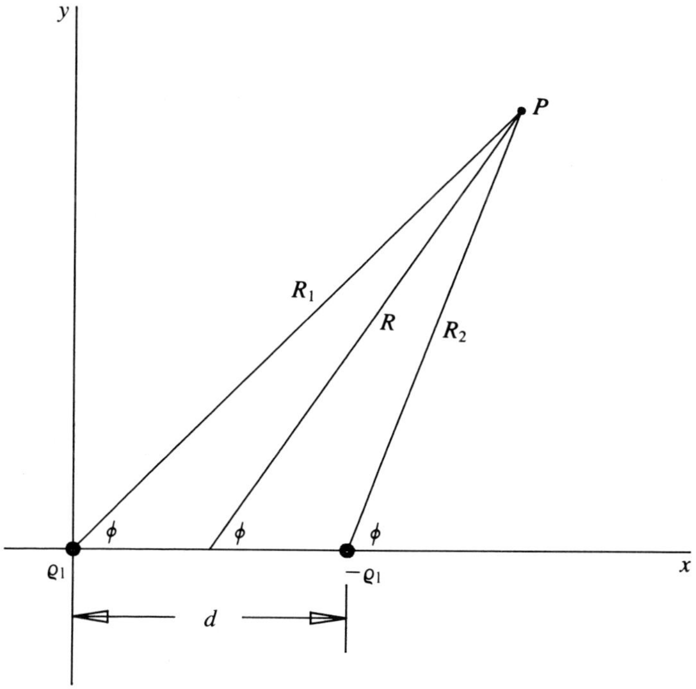

Using the law of cosines, we obtain from Figure 3.21

$$R_1^2 = R^2 + \left(\frac{d}{2}\right)^2 - 2\left(\frac{d}{2}\right)R\cos(180° - \phi)$$

$$= R^2 + \frac{d^2}{4} + \mathrm{d}R\cos\phi$$

and

$$R_2^2 = R^2 + \left(\frac{d}{2}\right)^2 - \mathrm{d}R\cos\phi$$

$$R_1^2 - R_2^2 = 2\,\mathrm{d}R\cos\phi$$

or

$$(R_2/R_1)^2 = 1 - \frac{2\,\mathrm{d}R}{R_1^2}\cos\phi$$

We assume $R_1 \approx R$, so that

$$(R_2/R_1)^2 = 1 - \frac{2d}{R}\cos\phi \tag{3.59}$$

Now we substutute equation (3.59) into equation (3.49) to obtain

$$V = \frac{\rho_l}{2\pi\epsilon_0}\ln\left(\frac{R_2}{R_1}\right)$$

$$= \frac{\rho_l}{2\pi\epsilon_0}\ln\left(1 - \frac{2d}{R}\cos\phi\right)^{1/2} \tag{3.60}$$

We use the series for the natural log, that is,

$$\ln(1+x) = x - \frac{x^2}{2} + \frac{1}{3}x^3 - \cdots \qquad -1 < x < 1$$

and note that the 'x' term which is $-(2d/R)\cos\phi$ is very small ($R \gg d$) compared to 1. In other words,

$$\ln\left(1 - \frac{2d}{R}\cos\phi\right) \approx -\frac{2d}{R}\cos\phi \tag{3.61}$$

Neglecting second order and higher terms, we substitute equation

(3.61) into equation (3.60)

$$V = \frac{\rho_l}{4\pi\epsilon_0}\left(-\frac{2d}{R}\cos\phi\right)$$

$$= \left[\frac{-\rho_l d}{2\pi\epsilon_0 R}\right]\cos\phi \tag{3.62}$$

Refer to Figure 3.22. We let $R = R_1$, and write

$$\cos\phi = \frac{x}{\sqrt{x^2+y^2}} \tag{3.63}$$

and

$$R = R_1 = \sqrt{x^2+y^2} \tag{3.64}$$

We substitute equations (3.63) and (3.64) into equation (3.62) and the

Figure 3.22. The radial distance from the doublet when $R \gg d$.

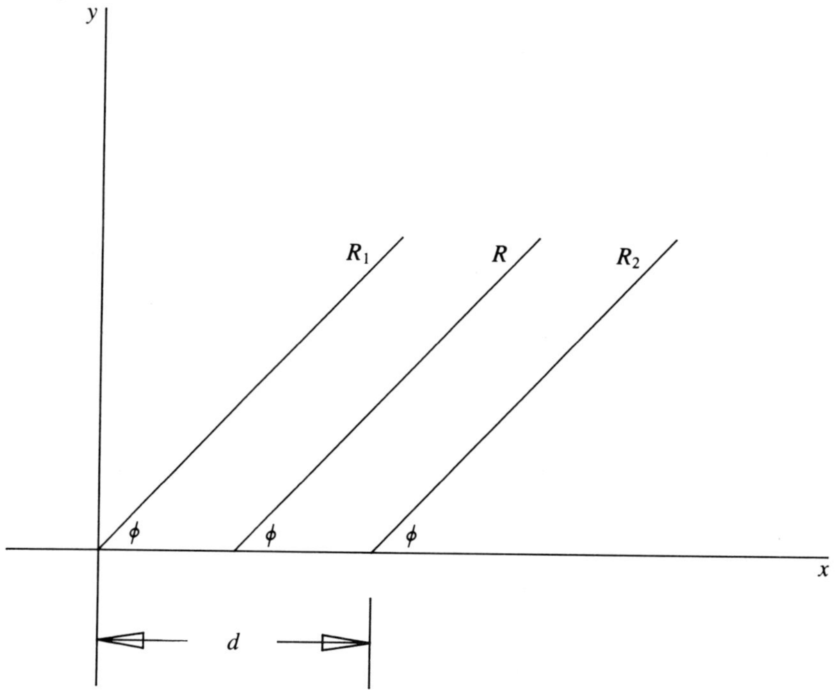

potential expression becomes

$$V = -\frac{\rho_l d \frac{x}{\sqrt{x^2 + y^2}}}{2\pi\epsilon_0 \sqrt{x^2 + y^2}} \tag{3.65}$$

$$= -\frac{\rho_l dx}{2\pi\epsilon_0 (x^2 + y^2)} \tag{3.66}$$

D3.15 Calculate the potential due to an electric doublet where $\rho_l = 4.42$ pC/m at the point (*a*) (1, 1, 1) (*b*) (10, 30°, 10) where $d = 0.01$ m. *Answer.* (*a*) $3.8 \cdot 10^{-4}$ (*b*) $6.5 \cdot 10^{-5}$ V.

Example 3.13 Determine the equipotentials for an electric doublet. Once again, we set $V = V_0$ and manipulate equation (3.66).

$$-\frac{2V_0\pi\epsilon_0}{\rho_l d} = \frac{x}{x^2 + y^2} \tag{3.67}$$

Let

$$K = -\frac{2V_0\pi\epsilon_0}{\rho_l d}$$

in equation (3.67) so that

$$Kx^2 + Ky^2 = x$$

$$x^2 - \frac{1}{K}x + y^2 = 0$$

$$\left(x - \frac{1}{2K}\right)^2 + y^2 = \left(\frac{1}{2K}\right)^2 \tag{3.68}$$

Recall equation (3.68) is based on equation (3.66), which is the potential expression for a doublet. A change in K implies a change in potential. We plot several circles in Figure 3.23. It is important to remember these circles represent equipotential surfaces normal to the plane of the page. The doublet is parallel to these surfaces; it is also normal to the plane of the page. What about the flux lines?

We know $\mathbf{E} = -\nabla V$ so a straight-forward differentiation should give us the electric field (also the flux lines). We proceed.

$$\mathbf{E} = -\left(\frac{\partial V}{\partial r}\hat{\mathbf{a}}_r + \frac{1}{r}\frac{\partial V}{\partial \phi}\hat{\mathbf{a}}_\phi + \frac{\partial V}{\partial z}\hat{\mathbf{a}}_z\right)$$

$$= -\frac{\rho_1 d \cos\phi}{2\pi\epsilon_0 r^2}\hat{\mathbf{a}}_r - \frac{\rho_1 d \sin\phi}{2\pi\epsilon_0 r^2}\hat{\mathbf{a}}_\phi \tag{3.69}$$

We recall the procedure we followed previously to obtain the stream-lines of the dipole. From Figure 3.17 we write, in spherical coordinates,

$$\frac{E_r}{E_\theta} = \frac{\mathrm{d}r}{r\,\mathrm{d}\theta} \tag{3.44}$$

For the doublet we write, in cylindrical coordinates,

$$\frac{\mathrm{d}r}{r\,\mathrm{d}\phi} = \left[\frac{\rho_1 d \cos\phi}{2\pi\epsilon_0 r^2}\right]\left[\frac{\rho_1 d \sin\phi}{2\pi\epsilon_0 r^2}\right]^{-1} = \frac{\cos\phi}{\sin\phi}$$

$$r = K \sin\phi \tag{3.70}$$

Equation (3.70) is plotted in Figure 3.24 for several values. It is a plot of the E or flux lines for a doublet. Thus the electric field lines (or flux lines) are orthogonal to the equipotentials everywhere. Figure 3.25

Figure 3.23. Equipotential surfaces due to an electric doublet.

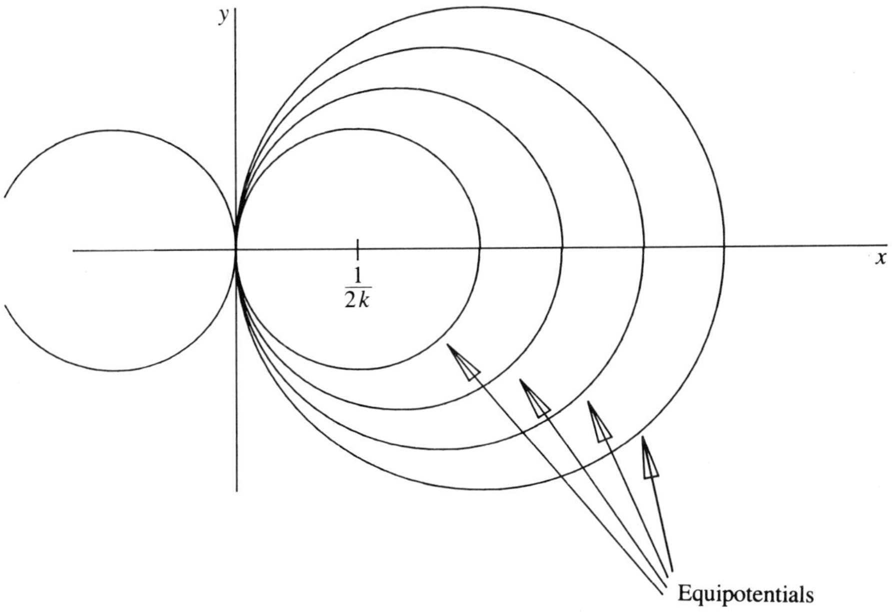

Figure 3.24. Flux surfaces due to an electric doublet.

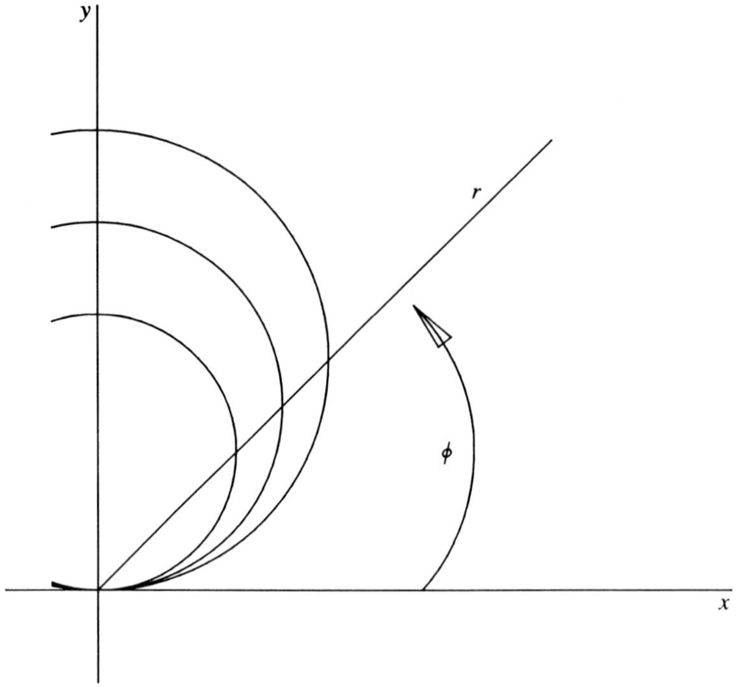

Figure 3.25. Combined potential and flux lines for an electric doublet

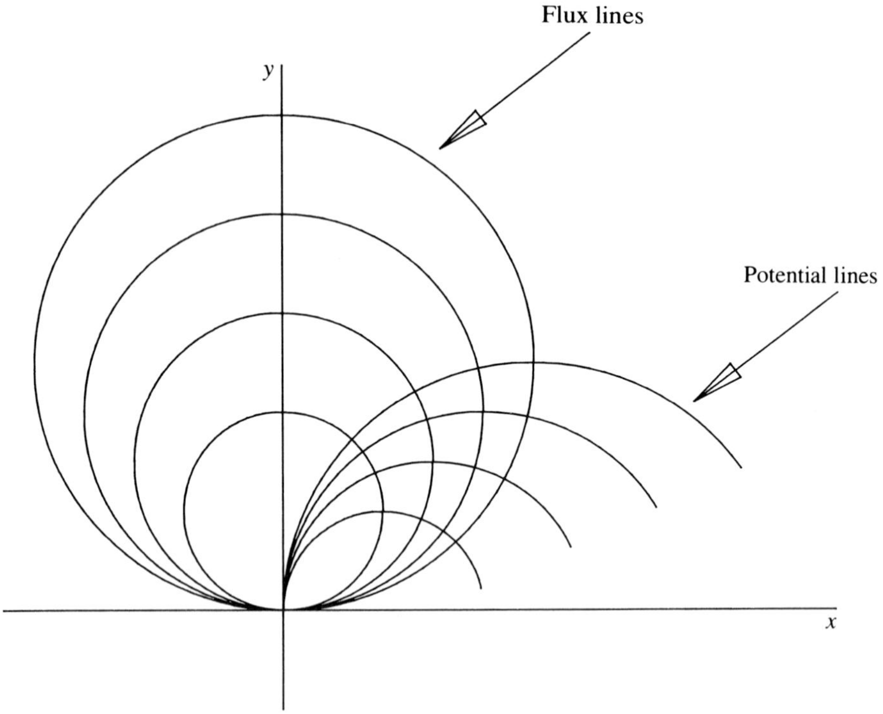

shows the potential and flux lines combined. Now we consider materials other than free space.

3.18 Dielectrics

A dielectric is a material which contains free and bound charges. A perfect dielectric is a material which contains all bound charges with no free charges to conduct electric current. If a dielectric is subjected to an (external) electric field, the bound charges will distort somewhat. If the field is strong enough, bonds will be broken and conduction will result.

Consider a parallel plate capacitor. Figure 3.26 shows the capacitor with two different dielectrics within the plates. We will discuss the varying flux density due to the dielectrics. In Figure 3.26(*a*) the dielectric is free space and $\mathbf{D} = \epsilon_0\mathbf{E}$. The density **D** and intensity **E** are shown. In Figure 3.26(*b*), with the dielectric in place, the source must supply more flux. Inside the dielectric there is a shifting of bound charges (electric dipoles). The sum of these dipoles is called the polarization vector **P**.

$$\mathbf{P} = \epsilon_0\chi_e\mathbf{E}$$

where χ_e is electric susceptibility. The total flux density is the sum of the flux density in free space and the flux density in the dielectric or

$$\mathbf{D} = \epsilon_0\mathbf{E} + \epsilon_0\chi_e\mathbf{E}$$
$$= \epsilon\mathbf{E}$$

Figure 3.26. A parallel plate capacitor with two different dielectrics, (*a*) dielectric free space, (*b*) dielectric insulator.

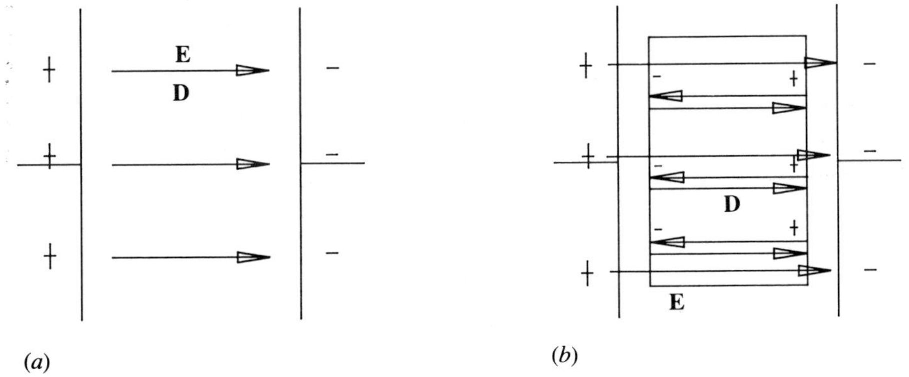

where $\epsilon = \epsilon_0(1+\chi_e) = \epsilon_0\epsilon_r$ and ϵ_r is the relative permittivity of the dielectric.

To summarize, recall Maxwell's first equation

$$\nabla \cdot \mathbf{D} = \rho$$

where ρ is the volume charge density in C/m^3. Up to this point free space was assumed or

$$\nabla \cdot \epsilon_0 \mathbf{E} = \rho$$

When dielectrics are present ρ is a total charge density or

$$\rho = \rho_{\text{free}} + \rho_{\text{bound}}$$

and

$$\nabla \cdot \epsilon_0 \mathbf{E} = \rho_{\text{free}} + \rho_{\text{bound}}$$

Again where dielectrics are present we write

$$\rho_{\text{bound}} = -\nabla \cdot \mathbf{P}$$

or

$$\nabla \cdot \epsilon_0 \mathbf{E} = \rho_{\text{free}} - \nabla \cdot \mathbf{P}$$

thus

$$\nabla \cdot \epsilon_0 \mathbf{E} + \nabla \cdot \mathbf{P} = \rho_{\text{free}}$$

and

$$\nabla \cdot \mathbf{D} = \rho_{\text{free}} \equiv \rho$$

where

$$\mathbf{D} = \epsilon_0 \mathbf{E} + \mathbf{P}$$

For convenience we omit the subscript 'free'.

D3.16 If $\mathbf{E} = r\hat{\mathbf{a}}_r + r\sin\phi\hat{\mathbf{a}}_\phi + z\hat{\mathbf{a}}_z$, calculate $\mathbf{D}$ at $(2.1, 15°, 1.3)$ where the dielectric is styrofoam for which $\epsilon_r = 1.5$.
Answer. $[27.8\hat{\mathbf{a}}_r + 7.21\hat{\mathbf{a}}_\phi + 17.2\hat{\mathbf{a}}_z]10^{-11}\ C/m^2$.

3.19 Boundary conditions

How do we explain the fact that light seems to 'bend' when it enters water? Why does a straight stick appear crooked from above a pond when it is stuck into the water? It is necessary to study boundary conditions between various media. For example, if two dielectrics are used in a capacitor, how does the electric field behave at the boundary? We use two familiar equations to find the boundary conditions between two media:

$$\oint \mathbf{E} \cdot \mathrm{d}\mathbf{l} = 0$$

and

$$\oint \mathbf{D} \cdot \mathrm{d}\mathbf{A} = Q$$

From Figure 3.27(a) we have the boundary between a medium with permittivity ϵ_{r_1} and a medium with permittivity ϵ_{r_2}. We divide the electric flux density into the tangential and the normal components. For the normal components of electric flux density, we use Gauss' law and write

$$\oint \mathbf{D} \cdot \mathrm{d}\mathbf{A} = Q$$

or

$$\oint \mathbf{D} \cdot \mathrm{d}\mathbf{A} = \int_{\text{top}} + \int_{\text{side}} + \int_{\text{bottom}} \tag{3.71}$$

From Figure 3.27(a) we force $\Delta h \to 0$ which causes $\int_{\text{side}}$ in equation (3.71) to be zero, and

$$\int \mathbf{D} \cdot \mathrm{d}\mathbf{A} \approx -D_{n_1}\,\Delta A + D_{n_2}\,\Delta A = Q$$

If $Q = 0$ at the boundary

$$D_{n_1} = D_{n_2} \tag{3.72}$$

or

$$\epsilon_{r_1} E_{n_1} = \epsilon_{r_2} E_{n_2} \tag{3.73}$$

in the limit as our closed surface shrinks.

If we assume charge ρ_s exists at the boundary

$$-D_{n_1}\,\Delta A + D_{n_2}\,\Delta A = \rho_s\,\Delta A$$

$$-D_{n_1} + D_{n_2} = \rho_s \tag{3.74}$$

or

$$-\epsilon_{r_1}\epsilon_0 E_{n_1} + \epsilon_{r_2}\epsilon_0 E_{n_2} = \rho_s$$

For the tangential component of electric field, we write, from Figure 3.27(*b*),

$$\oint \mathbf{E} \cdot \mathbf{dl} \approx \int_a^b + \int_b^c + \int_c^d + \int_d^a = 0$$

Figure 3.27. (*a*) $\oint \mathbf{D} \cdot \mathrm{d}\mathbf{A} = Q$ used for finding the normal components, (*b*) $\oint \mathbf{E} \cdot \mathbf{dl} = 0$ used for finding the tangential components.

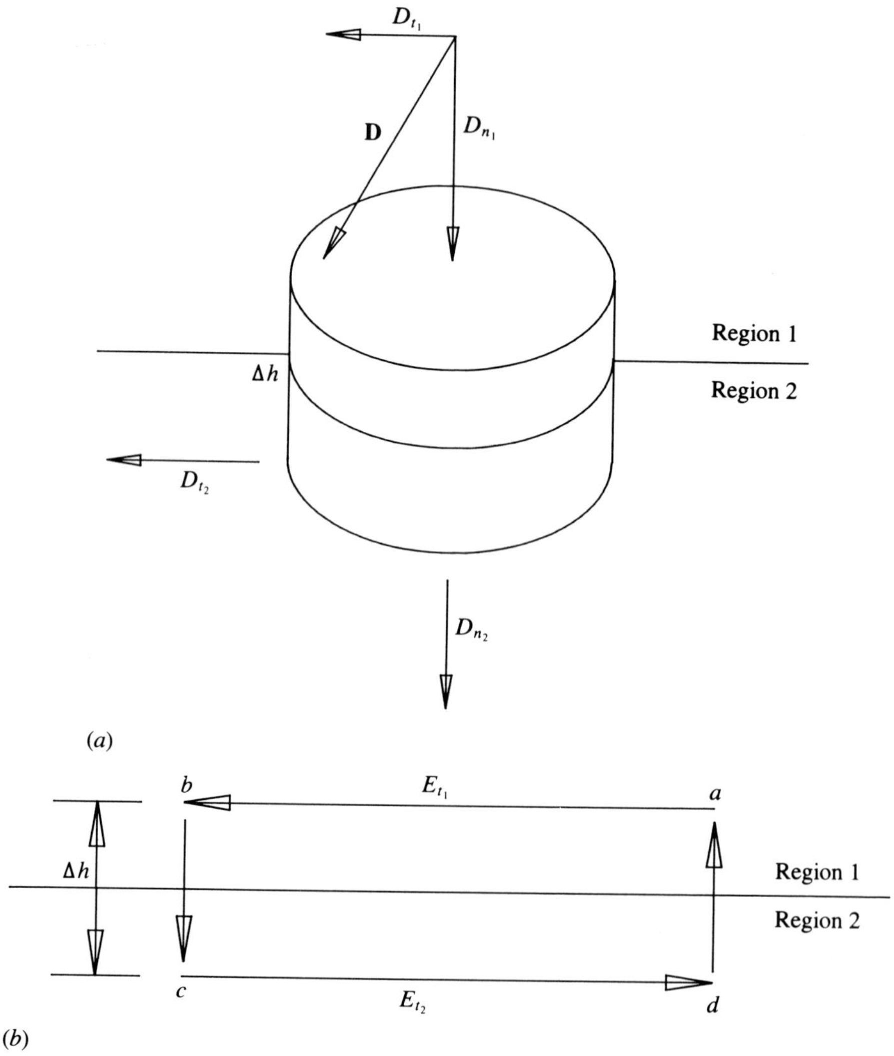

for a conservative field. In the limit as $\Delta h \to 0$

$$\oint \mathbf{E} \cdot d\mathbf{l} \approx E_{t_1} \Delta l + 0 - E_{t_2} \Delta l + 0 = 0 \tag{3.75}$$

The second and fourth terms on the right side of equation (3.75) are zero because the dot product is zero and

$$E_{t_1} = E_{t_2} \tag{3.76}$$

or

$$\frac{D_{t_1}}{\epsilon_{r_1}} = \frac{D_{t_2}}{\epsilon_{r_2}} \tag{3.77}$$

D3.17 The electric field in air above an air–salt water interface is $\mathbf{E}_1 = 3\hat{\mathbf{a}}_x + 2.5\hat{\mathbf{a}}_y - \hat{\mathbf{a}}_z$. Find $\mathbf{D}_1$, $\mathbf{E}_2$ and $\mathbf{D}_2$, where 1 refers to air and 2 refers to sea water. The relative permittivity of salt water is 81. $z > 0$, $\epsilon_{r_1} = 1$; $z < 0$, $\epsilon_{r_2} = 81$.

Answer.

$\mathbf{D}_1 = 26.6\hat{\mathbf{a}}_x + 22.1\hat{\mathbf{a}}_y - 8.85\hat{\mathbf{a}}_z$ pC/m^2

$\mathbf{E}_2 = 37\hat{\mathbf{a}}_x + 30.9\hat{\mathbf{a}}_y - 10^3\hat{\mathbf{a}}_z$ mV/m

$\mathbf{D}_2 = 2155\hat{\mathbf{a}}_x + 1790\hat{\mathbf{a}}_y - 8.85\hat{\mathbf{a}}_z$ pC/m^2

3.20 Resistance/capacitance

There is an analogy between the current density in a conductor and the electric flux density in a dielectric. From Ohm's law, we write

$$R = \frac{V}{I} = \frac{\text{potential}}{\text{flux}} \tag{3.78}$$

and from the definition of capacitance, we write

$$C \triangleq \frac{Q}{V} = \frac{\text{flux}}{\text{potential}} \tag{3.79}$$

From equation (3.78), we write

$$R = \frac{V}{J(\text{Area})}$$

for a linear resistor

$$= \frac{V}{\sigma E(\text{Area})} = \frac{d}{\sigma A} \tag{3.80}$$

where σ is the conductivity in Siemens, E is the electric field in volts/meter, d is the plate separation in meters, J is current density in amperes/meter^2 and A is the cross-sectional area in meters^2. From equation (3.79) we write

$$\begin{aligned} C &= \frac{\oint \mathbf{D} \cdot \mathrm{d}\mathbf{A}}{V} \\ &= \frac{\epsilon E(\text{Area})}{V} \\ &= \frac{\epsilon A}{d} \end{aligned} \tag{3.81}$$

where ϵ is the permittivity of the dielectric, E is the electric field, d is the plate separation, and A is the cross-sectional area.

Taking the product of equations (3.80) and (3.81), we write

$$\begin{aligned} RC &= \left(\frac{d}{\sigma A}\right)\left(\frac{\epsilon A}{d}\right) \\ &= \epsilon/\sigma \end{aligned} \tag{3.82}$$

We see from Figure 3.28 that the resistor and the capacitor have the

Figure 3.28. The analogy (*a*) current density in a conductor and (*b*) flux density in a dielectric.

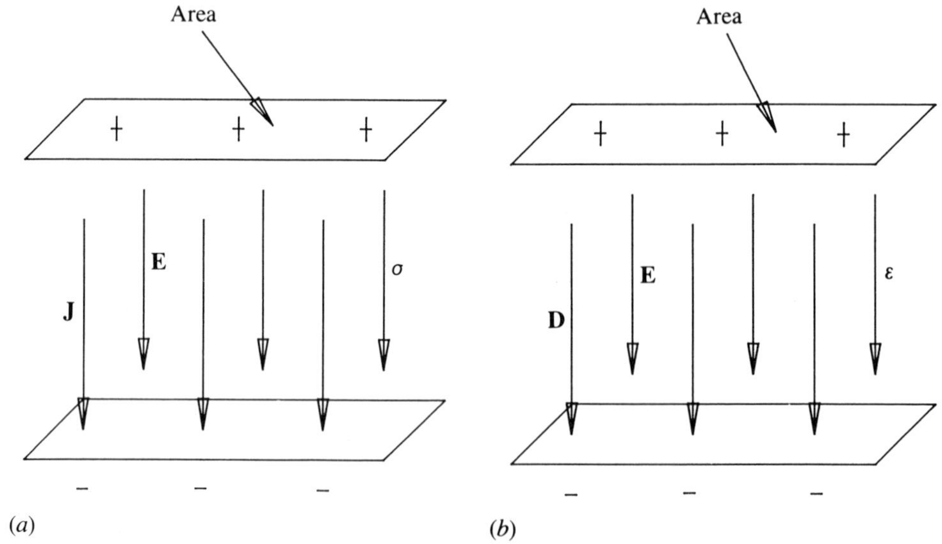

same configuration. The product of ohms (resistance) and farads (capacitance) has the units of seconds; we recall from linear circuits the time constant of an *RC* circuit. In fields and flow we speak of ϵ/σ as the time constant – with units of seconds.

D3.18 Calculate the capacitance between the ends of a can of Pepsi (the metal sides have been replaced by plastic whose $\epsilon_r = 1$) filled with sea water. Neglect fringing.
Answer. 20.9 pF for $r \approx 3.35$ cm; $d \approx 12.1$ cm.

D3.19 Calculate the resistance of the configuration in **D3.18** if the sea water is replaced by carbon.
Answer. 1.20 mΩ.

3.21 Energy density in an electrostatic field

What is the energy in an electric field? Let us consider the parallel plates (Figure 3.28) of the previous section. How is the capacitor charged? What is the energy density? (Please do not ask *where* the energy is stored). Recall equation (3.19) in which the potential was defined as

$$V \triangleq \frac{W}{Q} \tag{3.19}$$

For a small element of charge we write

$$\Delta W = V\,\Delta Q$$

since

$$V = Q/C$$

$$\Delta W = Q/C\,\Delta Q$$

Integrating we obtain

$$W = \frac{1}{C}\int Q\,\mathrm{d}Q$$

$$= \tfrac{1}{2}CV^2$$

Now using equation (3.81) we obtain

$$\begin{aligned} W &= \frac{1}{2}\frac{\epsilon A}{d}(Ed)^2 \\ &= \tfrac{1}{2}\epsilon E^2(Ad) \\ &= \tfrac{1}{2}DE(Ad) \end{aligned}$$

where (Ad) is the volume between the plates. In integral form this is written

$$W = \frac{1}{2}\iiint D(E)\,\mathrm{d}v \tag{3.83}$$

Finally, if the medium (in this case dielectric) is not isotropic, we must write

$$W = \frac{1}{2}\iiint \mathbf{D}\cdot\mathbf{E}\,\mathrm{d}v \tag{3.84}$$

D3.20 The half space $z > 0$ is filled with Teflon. $\epsilon_r = 2.25$. Calculate the energy stored in the cylindrical region $0 < r < 0.25$, $1.3 < z < 1.6$. For $z > 0$ $\mathbf{E} = r\hat{\mathbf{a}}_r$ in cylindrical coordinates.
Answer. $1.83 \cdot 10^{-14}$ J

This concludes the chapter on electrostatics. The next chapter will involve our study of magnetostatics – the magnetic field under static conditions.

Problems

1. A 50 μC point charge is located at $(1, 2, 0)$ and a -75 μC point charge is located at $(2, 2, 0)$. Find $\mathbf{E}$ at $(0, 0, 0)$.

2. Three point charges $+10$ μC each are located at $(-3, 0, 0)$, $(3, 0, 0)$ and $(0, 3, 0)$. Find $\mathbf{E}$ at $(3, 3, 0)$.

3. A 25 μC point charge is located at $(0, 2, 2)$ and a -30 μC point charge is located at $(0, -2, 1)$. Find at least one location where $E_z = 0$.

4. Two uniform lines ρ_l are located on the z-axis from $z = 1$ to infinity and $z = -1$ to minus infinity. Find (a) $\mathbf{E}$ at the point $(2, 3, 0)$ and (b) $\mathbf{E}(x, y, z)$ on the $z = 0$ plane.

5. An infinite line filament has uniform charge density of $\rho_l = 10^{-3}$ C/m. Determine the electric field intensity at a radial distance of 10 m from the line.

6. A cylinder with uniform surface density $\rho_s = -10^{-3}$ C/m^2 with radius $r = 10^{-2}$ m is placed concentric to the line filament of problem 5. Determine the electric field intensity at a radial distance of 10 m from the line.

7. Determine the surface charge on the cylinder of problem 6 such that the electric field is zero at $r = 10$ m.

8. A uniform surface charge is distributed on the surface in the $z = 0$ plane from $x = -\infty$ to $x = \infty$ and $y = 1$ to $y = 2$. Find $\mathbf{E}(x, y, z)$.

9. Find the **E** field on the z-axis due to uniform surface charge $\rho_s = 10$ nC/m^2 (see Figure 3.29).

Figure 3.29. See problem 9.

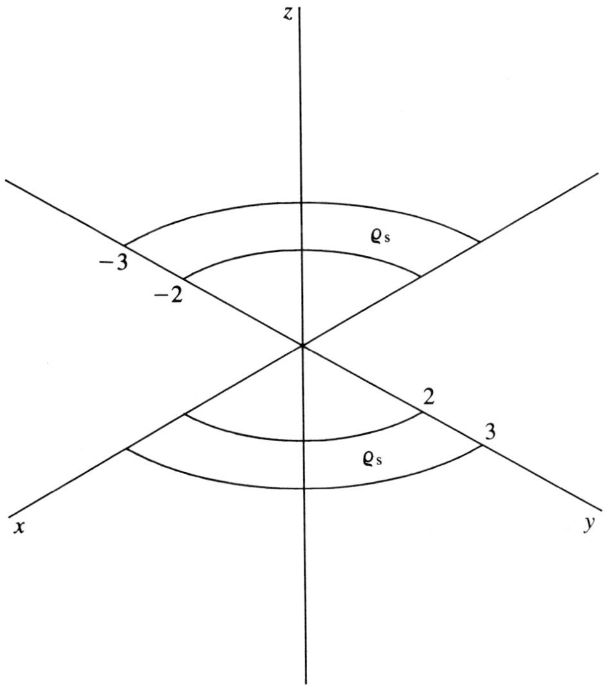

10. $\mathbf{E} = 3x\hat{\mathbf{a}}_x + 3y\hat{\mathbf{a}}_y$. Find the equation of streamlines and sketch several.

11. $\mathbf{E} = 2x^2y\hat{\mathbf{a}}_x + 2xy^2\hat{\mathbf{a}}_y + xyz\hat{\mathbf{a}}_z$. Is the field conservative in the region shown in Figure 3.30?

12. A point charge of 10^{-9} C is situated at the origin. Another point charge of -10^{-8} C is moved from $\theta = \pi/2$, $\phi = 0$, $r = 2$ m to $r = 4$ m. How much work is done? Now it is moved $+\pi/4\ \hat{\mathbf{a}}_\theta$. How much work is done?

13. Find flux density, $\mathbf{D}$, due to a uniformly charged sphere of radius 1.5 m. The total charge is $12 \cdot 10^{-9}$ C. Plot $|\mathbf{D}|$ versus r everywhere.

14. Find the electric field intensity at a distance of 3 m from the center of a sphere with a charge of 3 μC uniformly distributed on its surface. The sphere has a radius of (a) 1 m (b) 2 m.

15. What is the electric field intensity for the spheres in problem 14 at a distance of 0.5 m?

Figure 3.30. See problem 11.

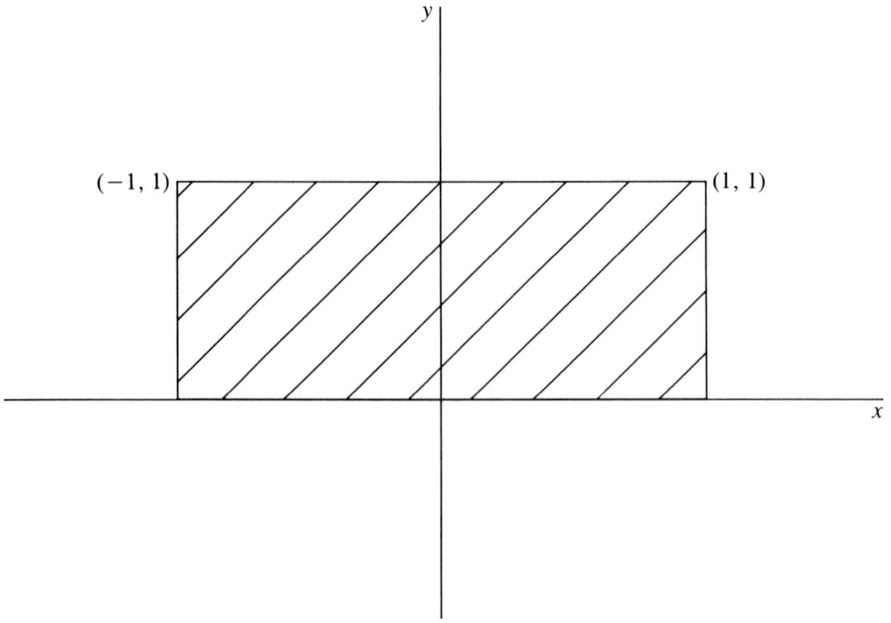

16. A sphere of 0.5 m radius has been eaten away from inside the sphere of problem 13 by coulomb-loving termites. Find **E** everywhere.

17. An infinite plane has a uniform surface charge density of 10^{-10} C/m^2. Calculate the flux density on the face of the equilateral triangle shown in Figure 3.31 at a distance d from the infinite plane.

18. There is a potential field of $V = 6x^2 + 3y + 4z$ volts. The point (1, 1, 1) is on a conductor surface. Find the equation of the conductor surface. Find the electric field, **E**.

19. $V = r^2 + 3\phi$ volts in cylindrical coordinates. Find the electric field intensity and electric flux density.

20. Given that $V = r^2 + \frac{1}{2}\theta + 4\phi$ volts in spherical coordinates. Determine at the point (3, 4, 5): (*a*) V, (*b*) **E**, (*c*) **D**, (*d*) ρ.

21. (*a*) Plot several potential lines for an electric dipole. (*b*) Plot several electric field lines for an electric dipole.

22. There exist two dielectric regions, $\epsilon_{r_1} = 2$ for $z > 0$ and $\epsilon_{r_2} = 3$ for $z < 0$. In the region $z > 0$, $\mathbf{E}_1 = 3\hat{\mathbf{a}}_x + 4\hat{\mathbf{a}}_y - \hat{\mathbf{a}}_z$ V/m. Find $\mathbf{E}_2$, $\mathbf{D}_1$, and $\mathbf{D}_2$.

23. Repeat problem 22 if $\rho_s = 10^{-9}$ C/m^2 exists on the boundary $z = 0$.

Figure 3.31. See problem 17.

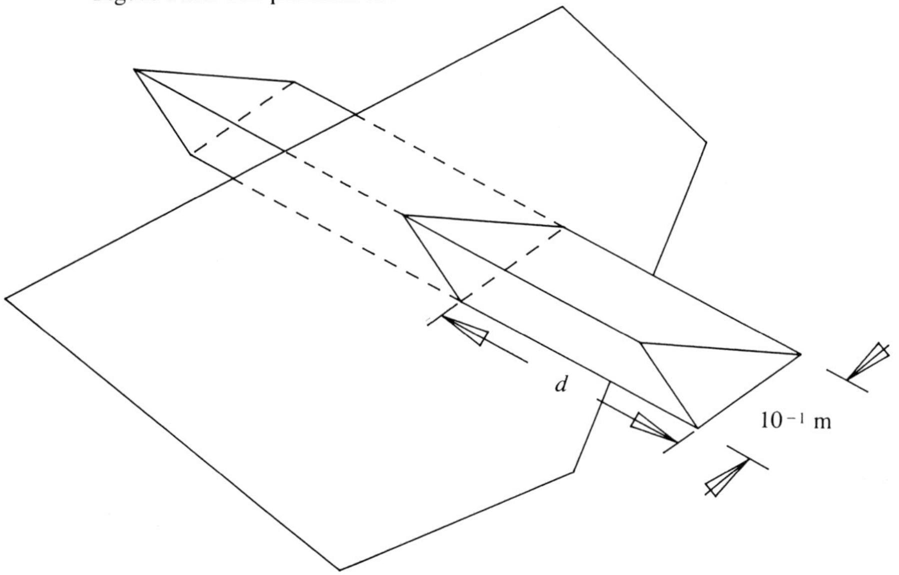

4

Magnetostatics

4.1 Introduction

Magnetic fields can be caused by currents, magnets and time varying fields. Most of what we do in this chapter will deal with fields caused by currents and magnets. We will look at two important laws – Biot–Savart and Ampere. First let us look at some similarities between Chapters 3 and 4. Table 4.1 compares the most important ones.

4.2 Duality

In Chapter 3 on electrostatics, we discovered that the electric field intensity was the result of a single charge or many charges. In addition, we imagined a flux and flux tubes extending from positive to negative charges. This naturally led to a flux density **D**. Analogous quantities exist in magnetostatics. Analogous to the electric field intensity, **E**, is the magnetic field intensity, **H**, and analogous to the electric flux density, **D**, is the magnetic flux density, **B**. In electrostatics, we write

$$\mathbf{D} = \epsilon_r \epsilon_0 \mathbf{E}$$

where the permittivity

$$\epsilon_0 = 8.854 \cdot 10^{-12}\ \mathrm{F/m}$$

while in magnetostatics, we write

$$\mathbf{B} = \mu_r \mu_0 \mathbf{H}$$

where the permeability

$$\mu_0 = 4\pi \cdot 10^{-7}\ \mathrm{H/m}$$

B is the magnetic flux density in Tesla and **H** is the magnetic field intensity in A/m (see Table 4.1).

Now that we have discussed a few analogies, let us discuss differences. One major difference is that magnetic charges have not been isolated. From Coulomb's experiments we can write

$$F = \frac{Q_1 Q_2}{4\pi\epsilon_0 R^2}$$

for two isolated point electric charges. No similar experiment has been performed in magnetics yet. Consider a magnet as shown in Figure 4.1(*a*). If we cut the magnet in half as shown in Figure 4.1(*b*), there will be two magnets with north and south poles. If we continue to halve each magnet, we continue to get smaller magnets with N–S poles – but never isolate a pole.

Recall Gauss' law

$$Q = \oint \mathbf{D} \cdot \mathrm{d}\mathbf{A}$$

and remember that the integration was about a *closed* surface. If we

Table 4.1. *A comparison of electric and magnetic quantities*

	Electrostatics	Magnetostatics
Field intensity	$\mathbf{E}$ V/m	$\mathbf{H}$ A/m
Flux density	$\mathbf{D}$ C/m^2	$\mathbf{B}$ T
Constitutive constant	ϵ_0 F/m	μ_0 H/m
Energy storage component	C F	L H

Figure 4.1. A magnet showing the lines of flux (*a*) whole and (*b*) halved.

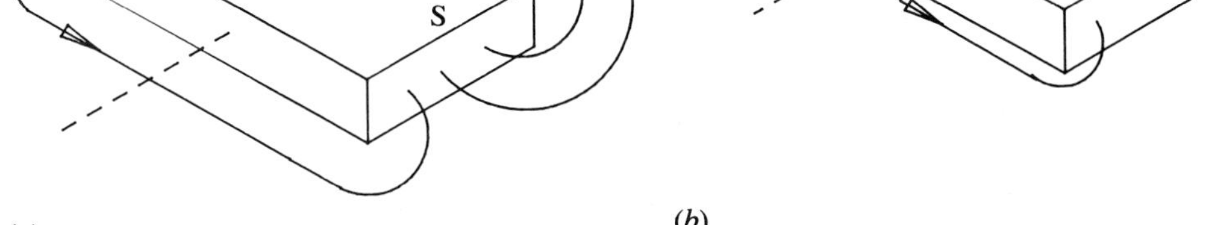

(*a*) (*b*)

invent a Gauss' law for magnetism, we must write

$$Q_{\text{mag}} = \oint \mathbf{B} \cdot \mathrm{d}\mathbf{A} = 0 \tag{4.1}$$

4.3 Maxwell's equation – divergence

In Chapter 3 we encountered two of Maxwell's equations. In this chapter we meet the remaining two. The equation

$$\oint \mathbf{B} \cdot \mathrm{d}\mathbf{A} = 0$$

is the integral form of Maxwell's equation. Using the divergence theorem, we write

$$\oint \mathbf{B} \cdot \mathrm{d}\mathbf{A} = \int_{\text{Vol.}} \nabla \cdot \mathbf{B}\, \mathrm{d}v = 0 \tag{4.2}$$

or

$$\nabla \cdot \mathbf{B} = 0$$

which is the derivative or point form of the third Maxwell equation. This states mathematically what was stated above: that magnetic charges cannot be isolated. Stating it yet another way, we say that magnetic lines do not diverge. They close upon themselves. Now let us introduce the last equation of Maxwell.

4.4 Maxwell's equation (Ampere's law)

There is another notable difference between electrostatics and magnetostatics. Recall, for a conservative field,

$$\oint \mathbf{E} \cdot \mathrm{d}\mathbf{l} = 0$$

In magnetics

$$\oint \mathbf{H} \cdot \mathrm{d}\mathbf{l} = I \tag{4.3}$$

where I is the current enclosed, in amperes, by the path $\mathrm{d}\mathbf{l}$. This is a statement of Ampere's law.

Equation (4.3) can be written

$$\oint \mathbf{H} \cdot \mathrm{d}\mathbf{l} = \iint \mathbf{J} \cdot \mathrm{d}\mathbf{A}$$

where **J** is current density in A/m^2. Using Stokes' theorem we can write

$$\oint \mathbf{H} \cdot d\mathbf{l} = \iint \nabla \times \mathbf{H} \cdot d\mathbf{A}$$

which gives

$$\iint \nabla \times \mathbf{H} \cdot d\mathbf{A} = \iint \mathbf{J} \cdot d\mathbf{A} \quad \text{and} \quad \nabla \times \mathbf{H} = \mathbf{J}.$$

Table 4.2 shows Maxwell's equations for statics.

From equation (4.3) we note that d**l** is integrated around *any* path enclosing the current I. The path could be a circle, square or any unorthodox path which encloses the current. As d**l** varies **H** varies. Just as Gauss' law is useful for symmetric electric problems so Ampere's law is useful for symmetric magnetic problems. As the Gaussian surface encloses the charge, the Amperian path encloses the current. Using the 'right hand rule', we say if the thumb points in the direction of current the fingers point in the direction of **H**. Let us try an example; let us find the magnetic field intensity due to an infinite line filament carrying current.

Example 4.1 Find the magnetic field intensity, **H**, due to an infinite line filament. We choose cylindrical coordinates and place the line along the z-axis as shown in Figure 4.2.

Using the right-hand rule, we see that d**l** is in the $\hat{\mathbf{a}}_\varphi$ direction for I flowing in the $+z$-direction. Since **H** must be in the same direction, we

Table 4.2. *Maxwell's equations for statics*

Integral	Point
$\oint \mathbf{E} \cdot d\mathbf{l} = 0$	$\nabla \times \mathbf{E} = 0$
$\oint \mathbf{D} \cdot d\mathbf{A} = \int_{\text{Vol.}} \rho \, dv$	$\nabla \cdot \mathbf{D} = \rho$
$\oint \mathbf{B} \cdot d\mathbf{A} = \int_{\text{Vol.}} \nabla \cdot \mathbf{B} \, dv$	$\nabla \cdot \mathbf{B} = 0$
$\oint \mathbf{H} \cdot d\mathbf{l} = I$	$\nabla \times \mathbf{H} = \mathbf{J}$

write

$$\oint \mathbf{H} \cdot \mathrm{d}\mathbf{l} = \int_0^{2\pi} H_\phi \hat{\mathbf{a}}_\phi \cdot r \, \mathrm{d}\phi \hat{\mathbf{a}}_\phi = I_{\mathrm{enc}}$$

or

$$H_\phi = \frac{I_{\mathrm{enc}}}{2\pi r} \tag{4.4}$$

and

$$\mathbf{H} = \frac{I_{enc}}{2\pi r} \hat{\mathbf{a}}_\phi$$

D4.1 Calculate the magnetic field intensity, **H**, at the point (2, 3, 3) due to an infinite line carrying current $I = 2.5$ A.
Answer. $0.11\hat{\mathbf{a}}_\phi$ A/m.

Example 4.1 leads to another question: If the line of current has some dimension will our results be the same? Let us try another example.

Figure 4.2. Infinite line with current I.

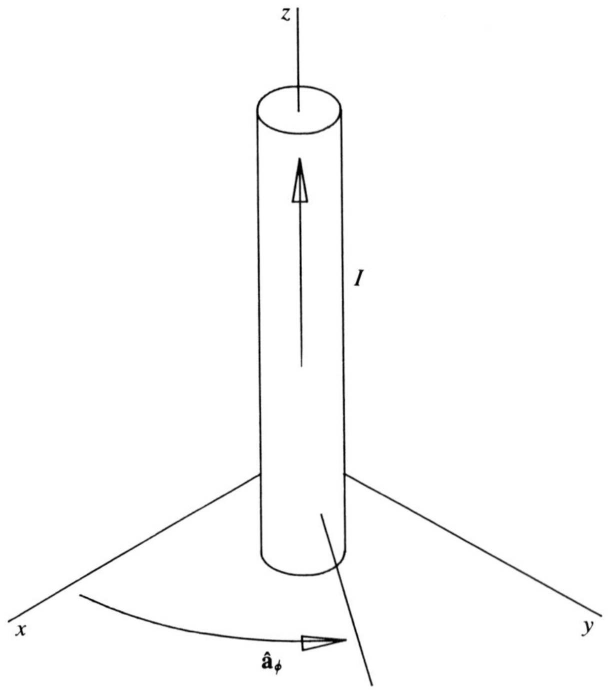

Example 4.2 Find the magnetic field intensity, H, due to an infinite cylindrical shell conductor of radius $r = a$ with the current, I, flowing on its surface as shown in Figure 4.3. If the integration is done outside the cylinder, $r > a$, the same current is enclosed as in the previous example and

$$\mathbf{H} = \frac{I_{\text{enc}}}{2\pi r}\hat{\mathbf{a}}_\phi \qquad r > a$$

If the integration is done inside the cylinder, no current is enclosed and

$$\oint \mathbf{H} \cdot \mathbf{dl} = 0$$

and $H = 0$, $r < a$.

D4.2 An infinite length of copper tubing with outer radius $r = 0.003$ m is carrying a total current $I = 2.5$ A. Calculate the magnetic field, $\mathbf{H}$, and the magnetic flux density, $\mathbf{B}$, at the point (2, 3, 3).
Answer. $0.11\hat{\mathbf{a}}_\phi$ A/m, $1.39 \cdot 10^{-7}\hat{\mathbf{a}}_\phi$ T.

Figure 4.3. An infinite cylindrical conductor with surface current K.

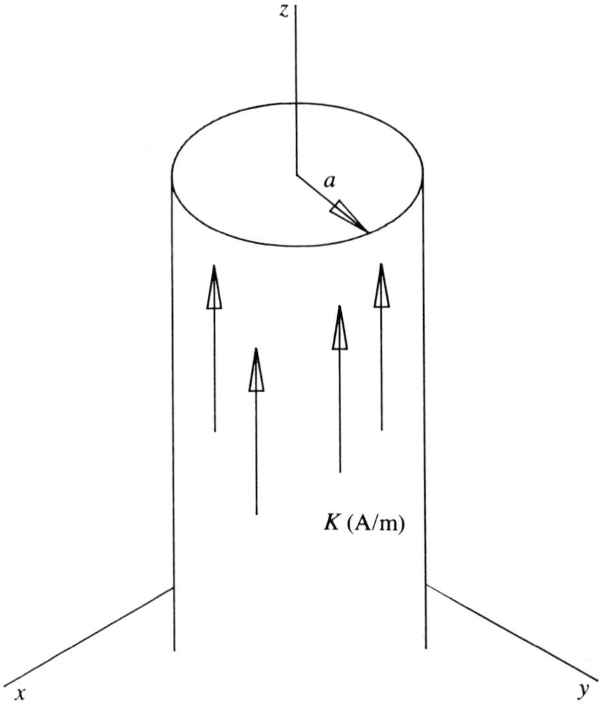

So the surface current, K A/m, causes the same magnetic field intensity as the filament. Similar results were obtained in our study of electrostatics when we observed that the electric field intensity external to a charged sphere was identical to the field intensity from a point charge with the same charge Q.

Example 4.3 Find the magnetic field intensity inside and outside a cylinder of radius $r = a$ where the current I is uniformly distributed throughout the conductor as shown in Figure 4.4.

From the previous examples for $r > a$, $\mathbf{H}$ is the same regardless of the cross-section of the conductor. For $r > a$

$$\mathbf{H} = \frac{I_{\text{enc}}}{2\pi r}\hat{\mathbf{a}}_{\phi} \tag{4.5}$$

Figure 4.4. A solid cylindrical conductor carrying uniformly distributed current I for example 4.3.

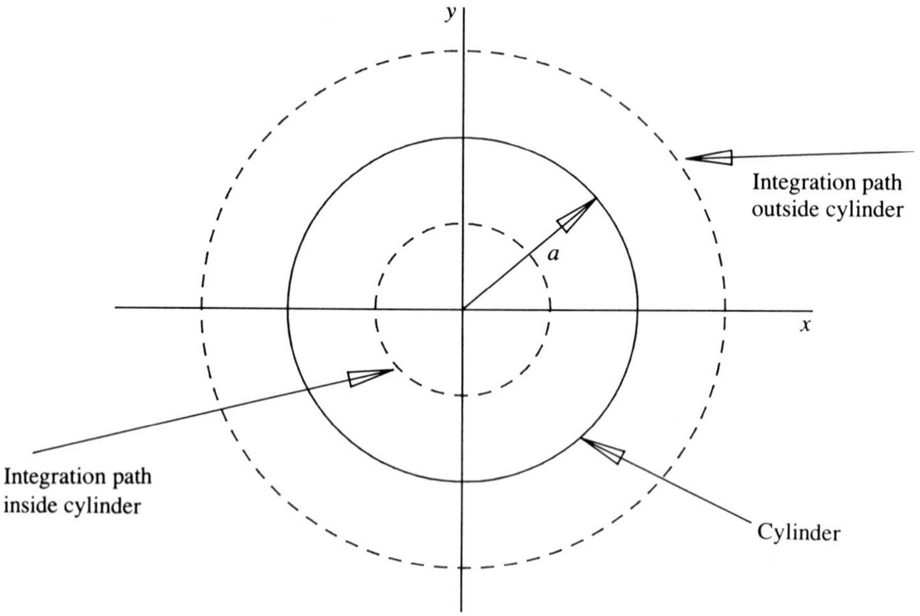

For $r < a$, we write

$$\oint \mathbf{H} \cdot \mathrm{d}\mathbf{l} = I_{\mathrm{enc}} = \iint \mathbf{J} \cdot \mathrm{d}\mathbf{A}$$

$$\int_0^{2\pi} H_\phi r \,\mathrm{d}\phi = \int_0^r \int_0^{2\pi} \frac{I}{\pi a^2} \hat{\mathbf{a}}_z \cdot r \,\mathrm{d}r \,\mathrm{d}\phi \hat{\mathbf{a}}_z$$

$$\mathbf{H} = \frac{Ir}{2\pi a^2} \hat{\mathbf{a}}_\phi \tag{4.6}$$

Equations (4.5) and (4.6) have a form which may seem familiar. Recall in Chapter 3 we found the electric field inside and outside a solid cylindrical conductor with uniform density. Table 4.3 is a comparison of analogous quantities. For both **E** and **H** for $r > a$ (outside the source), the intensities vary as r^{-1}, while for $r < a$ (inside the source) the intensities vary as r.

There is a duality which is powerful in helping us understand. Figure 4.5 is a plot of the magnitude of the magnetic field intensity as a function of radius. Notice that the intensity is a linear function of radius inside, $r < a$, while it is inversely proportional outside, $r > a$. Comparing Figure 4.5 with Table 4.3, we can see that plotting either E or H versus radius should give us identical plots. The only difference will be the ordinate designations.

D4.3 Calculate the magnetic field intensity at the point (0.003, 15°, 1.25) if a cylindrical rod along the z-axis is carrying 0.25 amps uniformly in the z-direction.
Answer. $13.3\hat{\mathbf{a}}_\phi$ A/m where $r > a$.

Table 4.3. *Comparison of E and H due to a cylindrical source*

	Electricity E	Magnetics H
$r > a$	$\frac{\rho_1}{2\pi\epsilon_0 r}$	$\frac{I}{2\pi r}$
$r < a$	$\frac{\rho_1 r}{2\pi\epsilon_0 a^2}$	$\frac{Ir}{2\pi a^2}$

4.5 Biot–Savart Law

One disadvantage of Ampere's law is it is limited to symmetrical problems for the most part just as Gauss' law was limited to symmetrical problems in Chapter 3. In the nineteenth century Professors Biot and Savart proposed the Biot–Savart law which says that the magnetic field intensity varies inversely to the square of the distance from a current. With this law we can solve problems lacking symmetry. In equation form Biot–Savart states

$$\mathbf{H} = \int \frac{I\,\mathrm{d}\mathbf{l} \times \hat{\mathbf{a}}_R}{4\pi R^2} \tag{4.7}$$

where $\mathbf{R}$ is the vector from the current to the point of interest and $\mathrm{d}\mathbf{l}$ is the vector along the source or disturbance. (Note that this differs from Ampere's law where $\mathrm{d}\mathbf{l}$ was the length in the direction of $\mathbf{H}$.) The magnetic field intensity is normal to $\mathrm{d}\mathbf{l}$ and $\hat{\mathbf{a}}_R$ in Biot–Savart's equation. When the current is along a surface, the Biot–Savart

Figure 4.5. Plot of magnetic field intensity, $|\mathbf{H}|$, with radius for a solid cylindrical conductor carrying uniform current I.

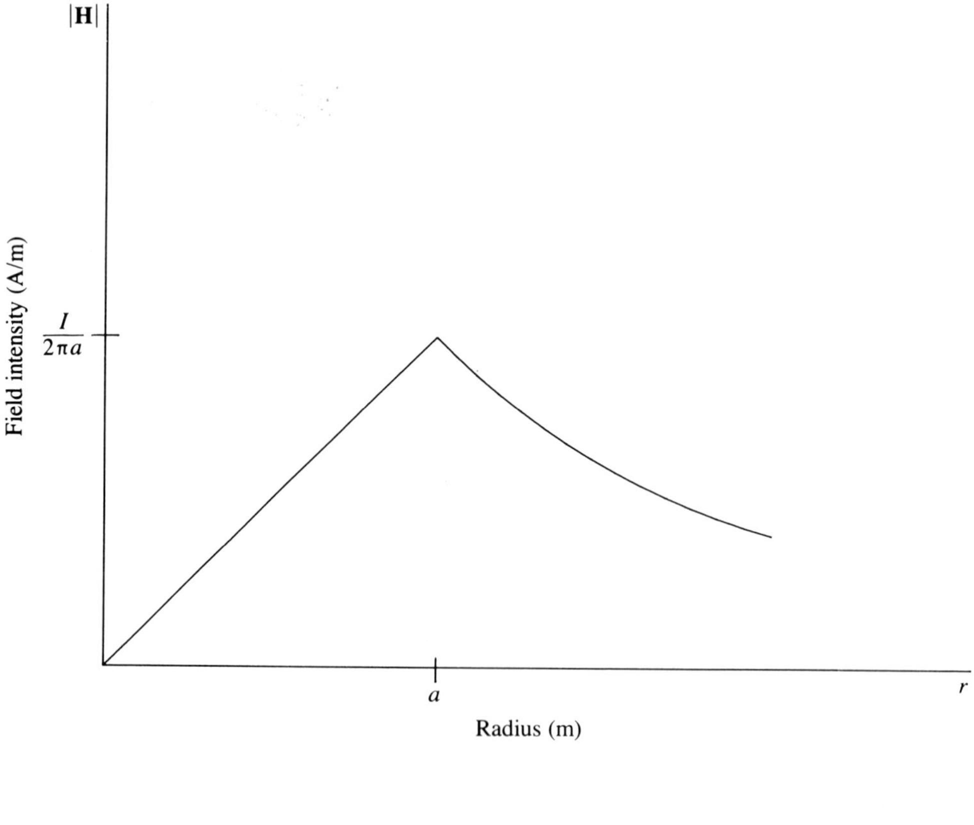

equation is

$$\mathbf{H} = \iint \frac{K\,\mathrm{d}\mathbf{A} \times \hat{\mathbf{a}}_R}{4\pi R^2}$$

where K is a surface current in A/m, and when the current is through a volume it is written

$$\mathbf{H} = \iiint \frac{\mathbf{J}\,\mathrm{d}v \times \hat{\mathbf{a}}_R}{4\pi R^2}$$

We illustrate Biot–Savart by repeating the first example: we find **H** for an infinite filament.

Example 4.4 Find the magnetic field intensity, **H** due to an infinite, filamentary line carrying current I A.

Let us take advantage of symmetry once more. We place the line along the z-axis and note that **H** will be in the $\hat{\mathbf{a}}_\phi$ direction and independent of ϕ. From Figure 4.6 we see that

$$I\,\mathrm{d}\mathbf{l} = I\,\mathrm{d}z\hat{\mathbf{a}}_z$$

Figure 4.6. An infinite filament along the z-axis carrying current I.

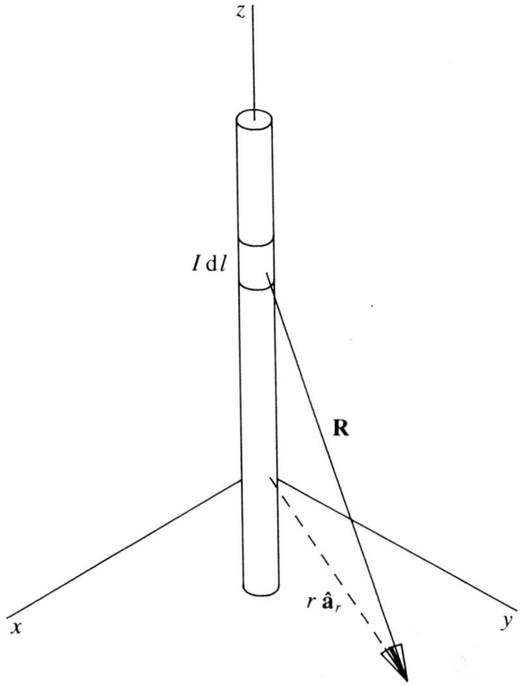

and $\mathbf{R} = r\hat{\mathbf{a}}_r - z\hat{\mathbf{a}}_z$

$$\mathbf{H} = \int_{-\infty}^{\infty} \frac{I\,\mathrm{d}z\hat{\mathbf{a}}_z \times (r\hat{\mathbf{a}}_r - z\hat{\mathbf{a}}_z)}{4\pi(r^2 + z^2)^{3/2}}$$

$$= \frac{I}{4\pi} \int \frac{r\,\mathrm{d}z}{(r^2 + z^2)^{3/2}} \hat{\mathbf{a}}_\phi$$

$$= \frac{I}{2\pi r} \hat{\mathbf{a}}_\phi$$

We note that the results are identical to those obtained using Ampere's law. This illustrates that Ampere's law and Biot–Savart are equivalent as can be proved by an analysis which is somewhat beyond the scope of this text.

D4.4 Calculate $\mathbf{H}$ at $(0.5, 0.5, 0)$ for an infinite filament carrying current $I = 0.25$ A in the z direction. Express $\mathbf{H}$ in Cartesian coordinates.
Answer. $3.98 \cdot 10^{-2}(-\hat{\mathbf{a}}_x + \hat{\mathbf{a}}_y)$ A/m.

4.6 Magnetic potential

In electricity a voltage, V, is considered a 'pressure' which causes current, I, to flow – not unlike fluid. Let's try this pressure in magnetics. Let V_{m} be a magnetic potential which causes magnetic flux to flow. Recall from Chapter 3 the electrostatic potential was written

$$V = -\int \mathbf{E} \cdot \mathrm{d}\mathbf{l}$$

By duality we replace $\mathbf{E}$, with units of volts per meter, with $\mathbf{H}$, with units of amperes per meter. Thus $\mathbf{H} \cdot \mathrm{d}\mathbf{l}$ has units of amperes and, let's call it V_{m}, $V_{\mathrm{m}} = -\int \mathbf{H} \cdot \mathrm{d}\mathbf{l}$ is a magnetic potential in amperes! Also, recall that the potential lines, V, and electric field lines, $\mathbf{E}$, are orthogonal everywhere. We can say the magnetic potential lines, V_{m}, and the magnetic field lines, $\mathbf{H}$ are orthogonal everywhere. Consider Figure 4.7. Figure 4.7 shows the cross-section of a coaxial cable with flux between inner and outer conductors. Note that the current is entering the page in the center conductor, thus the flux lines are circumferential. Orthogonal to these lines everywhere are the magnetic equipotential lines shown in Figure 4.7 as dotted lines. It should be clear that potential is a function of ϕ or $V_{\mathrm{m}} = f(\phi)$.

What can we say about conservation? Recall that electrostatic fields were conservative, that is, $\oint \mathbf{E} \cdot d\mathbf{l} = 0$. It is noteworthy that $\oint \mathbf{H} \cdot d\mathbf{l} \neq 0$. In other words the magnetostatic fields are not conservative; $\oint \mathbf{H} \cdot d\mathbf{l} = I$. This is an important difference between the two fields. More will be said about magnetic potential later.

Before we discuss magnetic materials consider the problem of **H** fields in the vicinity of a coaxial conductor. (Knowing **H** we are able to determine the magnetic potential.)

Example 4.5 Calculate the magnetic field intensity, **H**, in the four regions of the coaxial cable in Figure 4.7. Since the problem is symmetric, we can use Ampere's law.

$$\oint \mathbf{H} \cdot d\mathbf{l} = I_{\mathrm{enc}}$$

Figure 4.7. The cross section of a coaxial conductor showing flux and potential lines.

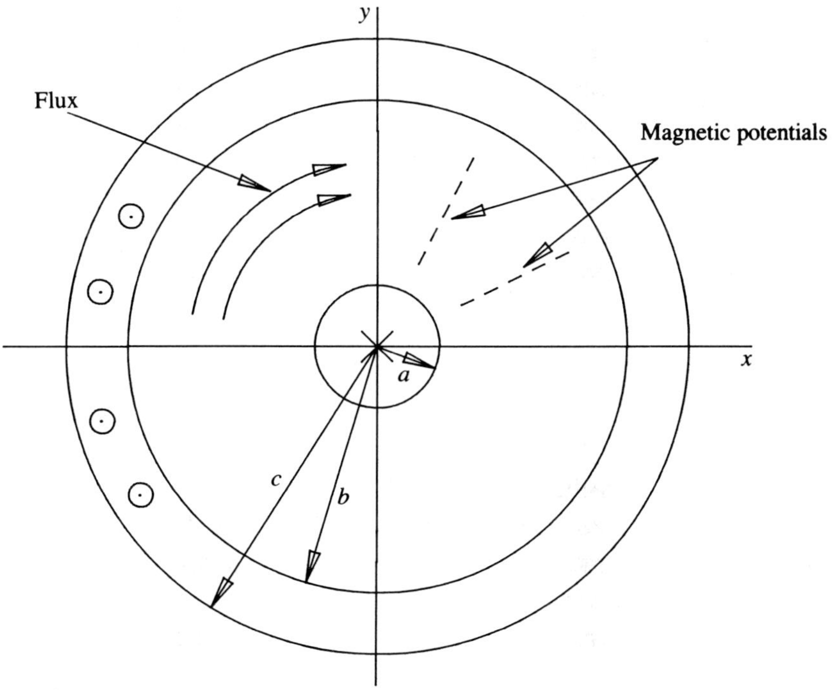

region 1 $r<a$

$$\oint \mathbf{H}\cdot \mathrm{d}\mathbf{l} = I_{\mathrm{enc}} = \iint \mathbf{J}\cdot \mathrm{d}\mathbf{A}$$

$$= \int_0^{2\pi}\int_0^r \left(-\frac{I}{\pi a^2}\right) r'\, \mathrm{d}r'\, \mathrm{d}\phi$$

$$\mathbf{H} = -\frac{Ir}{2\pi a^2}\hat{\mathbf{a}}_\phi$$

which is the same magnitude obtained in example 4.3 on page 113 for equation (4.6). The minus sign difference is due to the fact that I is in the opposite direction.

region 2 $a<r<b$

$$\oint \mathbf{H}\cdot \mathrm{d}\mathbf{l} = I_{\mathrm{enc}}$$

$$H_\phi 2\pi r = I_{\mathrm{enc}}$$

$$H_\phi = \frac{I_{\mathrm{enc}}}{2\pi r}$$

$$\mathbf{H} = -\frac{I_{\mathrm{enc}}}{2\pi r}\hat{\mathbf{a}}_\phi$$

or

$$\mathbf{H} = -\frac{I}{2\pi r}\hat{\mathbf{a}}_\phi \tag{4.8}$$

Again the result is the same as the previous example cited but for the minus sign.

region 3 $b<r<c$

Now the current enclosed is that of the center conductor less that amount enclosed by the path **dl**, or

$$\oint \mathbf{H}\cdot \mathrm{d}\mathbf{l} = I_{\mathrm{enc}} = I_{\mathrm{inner}} - I_{\mathrm{outer}}$$

$$= I - \iint \mathbf{J}\cdot \mathrm{d}\mathbf{A}$$

$$= I - \int_0^{2\pi}\int_b^r \frac{I}{\pi c^2 - \pi b^2}\hat{\mathbf{a}}_z \cdot r\, \mathrm{d}r\, \mathrm{d}\phi \hat{\mathbf{a}}_z$$

$$\mathbf{H} = \frac{-I}{2\pi r}\left[\frac{c^2 - r^2}{c^2 - b^2}\right]\hat{\mathbf{a}}_\phi \qquad b<r<c$$

region 4 $r > c$
In this region zero current is enclosed since the **H** due to the center conductor cancels the **H** due to the outer conductor. That is, the magnetic field intensities are in opposite directions, and to the observer outside, no magnetic field implies no current enclosed.

$$\oint \mathbf{H} \cdot d\mathbf{l} = I - I = 0$$

$$H_\phi = 0$$

Figure 4.8 is a plot of field intensity versus radius for a coaxial conductor.

D4.5 A coaxial conductor carrying current $I = 1.5$ A as shown in Figure 4.7 has $a = 0.001$ m, $b = 0.003$ m, and $c = 0.004$ m. Calculate **H** at (*a*) $r = 0.005$ m (*b*) $r = 0.002$ m (*c*) $r = 0.00035$ m (*d*) $r = 0.0045$ m.
Answer. (*a*) 0 (*b*) $-119.4\hat{\mathbf{a}}_\phi$ (*c*) $-36.5\hat{\mathbf{a}}_\phi$ (*d*) 0 A/m.

Figure 4.8. A magnitude plot of |**H**| versus radius for a coaxial conductor.

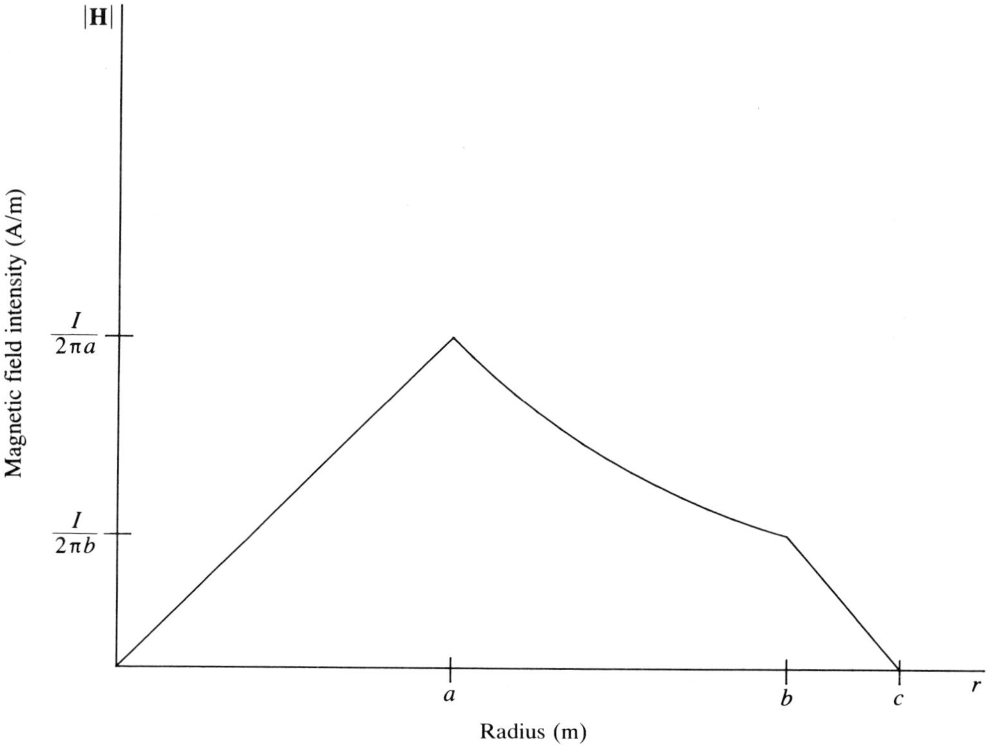

4.7 Magnetic materials

We know from linear networks that transformer coupling increases by winding the coil around a magnetic core. Let us investigate this in more detail. Figure 4.9(*a*) shows a turn of wire in free space. Using Ampere's law we write

$$\oint \mathbf{H} \cdot d\mathbf{l} = I_{air}$$

In free space

$$\mathbf{B} = \mu_0 \mathbf{H}$$

so that

$$\oint \mathbf{B} \cdot d\mathbf{l} = \mu_0 I_{air}$$

Figure 4.9(*b*) shows a turn of wire around a magnetic material. The magnetic material causes the flux to increase. This is written

$$\mathbf{B}_t = \mathbf{B}_{air} + \mathbf{B}_{mag} \tag{4.9}$$

$$\mathbf{B}_t = \mu_0(\mathbf{H} + \mathbf{M}) \tag{4.10}$$

where **M** is the magnetic polarization, defined as

$$\mathbf{M} \triangleq \chi_m \mathbf{H} \tag{4.11}$$

where χ_m is the magnetic susceptibility.

Substituting equation (4.11) into equation (4.10), we obtain

$$\mathbf{B}_t \equiv \mathbf{B} = \mu \mathbf{H} = \mu_0(1 + \chi_m)\mathbf{H}$$

where

$$\mu = \mu_0(1 + \chi_m) = \mu_0 \mu_r$$

Figure 4.9. A single turn of wire in (*a*) free space and (*b*) magnetic material.

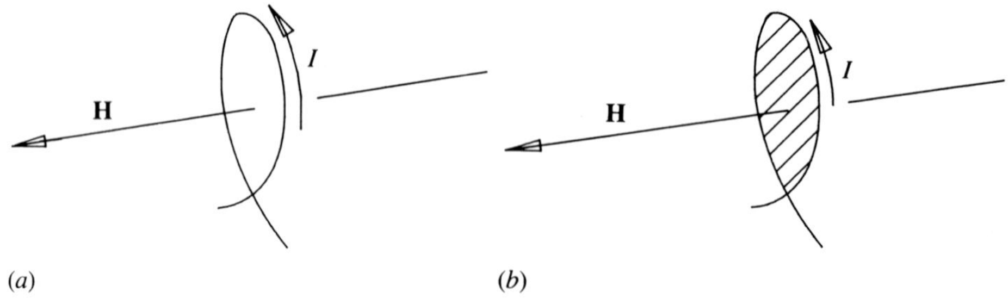

and μ_r is the relative permeability of the material. This is analogous to

$$\mathbf{D} = \epsilon_0 \epsilon_r \mathbf{E}$$

in electrostatics.

A second approach to explaining magnetic materials follows: For magnetic materials consider an expanded picture of Figure 4.9(*b*). Figure 4.10(*a*) shows the magnetic dipoles in the core material at random with no preferred direction. Each of these dipoles results in a bound or Amperian current. The induced current is zero; there is no magnetization. When the current *I* flows in the coil of wire, there are induced currents: the magnetic dipoles are in the same direction as shown in Figure 4.10(*b*). There is cancellation of currents between adjacent loops. This is the induced current that gives rise to the additional flux density.

From Stokes' theorem, we write equation (4.12)

$$\oint \mathbf{B} \cdot d\mathbf{l} = \iint \nabla \times \mathbf{B} \cdot d\mathbf{A} = \mu_0 I_{\text{air}} + \mu_0 I_{\text{ind}} \tag{4.12}$$

or

$$|\nabla \times \mathbf{B}| = \lim_{\text{Area} \to 0} \frac{\mu_0 (I_{\text{air}} + I_{\text{ind}})}{\text{Area}}$$

Figure 4.10. A single turn of wire with (*a*) no current and (*b*) current *I* flowing in the wire.

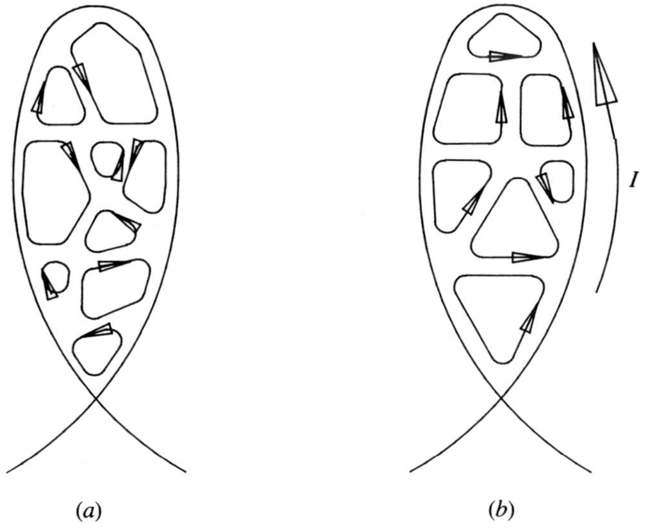

In vector form

$$\nabla \times \mathbf{B} = \mu_0(\mathbf{J}_{\text{air}} + \mathbf{J}_{\text{ind}})$$

We define

$$\nabla \times \mathbf{M} \triangleq \mathbf{J}_{\text{ind}} \tag{4.13}$$

so that

$$\nabla \times \left(\frac{\mathbf{B}}{\mu_0} - \mathbf{M}\right) = \mathbf{J}_{\text{air}} \tag{4.14}$$

Define

$$\mathbf{H} \triangleq \frac{\mathbf{B}}{\mu_0} - \mathbf{M}$$

so that

$$\nabla \times \mathbf{H} = \mathbf{J}_{\text{air}} \tag{4.15}$$

and

$$\oint \mathbf{H} \cdot \mathrm{d}\mathbf{l} = \mathbf{I}_{\text{air}} \tag{4.16}$$

Of most importance is $\mathbf{B} = \mu\mathbf{H}$ and $\mu = \mu_0\mu_r$ where $1 \leq \mu_r \leq \infty$.

D4.6 If the magnetic field intensity, $\mathbf{H} = x\hat{\mathbf{a}}_x - 3y\hat{\mathbf{a}}_y$ find the current density, $\mathbf{J}$.
Answer. $\nabla \times \mathbf{H} = 0$.

4.8 Boundary conditions

Next let us examine the boundary conditions between two media. Recall **B** and **H** are related by

$$\mathbf{B} = \mu\mathbf{H} = \mu_0\mu_r\mathbf{H}$$

where μ is the permeability, μ_r is the relative permeability and $\mu_0 = 4\pi \cdot 10^{-7}$ is the permeability of free space, in H/m.

Consider two different media and let us find the magnetic flux density, **B**, and the magnetic field intensity, **H** in both regions. First we break the vectors into two components, the tangential and the normal, or

$$\mathbf{B} = B_N\hat{\mathbf{a}}_N + B_T\hat{\mathbf{a}}_T$$

and

$$\mathbf{H} = H_N\hat{\mathbf{a}}_N + H_T\hat{\mathbf{a}}_T$$

Figure 4.11 shows a small cylinder with the flux normal to the top and bottom surfaces. Now we integrate about the cylinder. Since magnetic charges cannot be isolated and since the integration is over a closed surface – sort of a magnetic Gaussian theorem – we can write

$$\oint \mathbf{B} \cdot \mathrm{d}\mathbf{A} = 0$$

or

$$\oint \mathbf{B} \cdot \mathrm{d}\mathbf{A} = \int_{\text{top}} + \int_{\text{bottom}} + \int_{\text{side}} = 0$$

First, we see that the third integral must be zero because we force Δh to zero and the flux through the sides becomes negligible. Hence

$$\iint \mathbf{B} \cdot \mathrm{d}\mathbf{A} = 0$$

Figure 4.11. Cylinder used to evaluate **B** and **H** at the boundary between two different media.

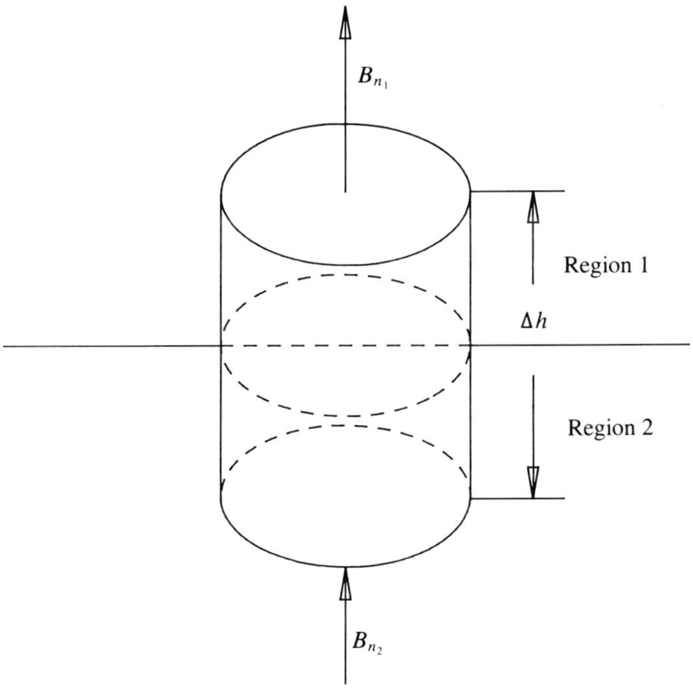

Next, as the cylinder shrinks in the limit

$$\int_{\text{top}} \approx B_{N_1}\Delta A$$

and likewise

$$\int_{\text{bottom}} \approx B_{N_2}\Delta A$$

where ΔA in both integrals points away from the enclosed area. We now can write

$$\iint \mathbf{B}\cdot \mathrm{d}\mathbf{A} \approx B_{N_1}\Delta A - B_{N_2}\Delta A = 0$$

$$B_{N_1} = B_{N_2} \tag{4.17}$$

and

$$\mu_1 H_{N_1} = \mu_2 H_{N_2} \tag{4.18}$$

That is, the normal components of flux density are continuous, but the field intensity components are not.

Example 4.6 Given a magnetic flux density normal to the x axis as shown in Figure 4.12, determine **B** and **H** in both regions. Given

$$\mathbf{B}_1 = -2.5\hat{\mathbf{a}}_y \text{ T}$$

$$\mathbf{H}_1 = \frac{-2.5}{2}\hat{\mathbf{a}}_y \text{ A/m}$$

Figure 4.12. See example 4.6.

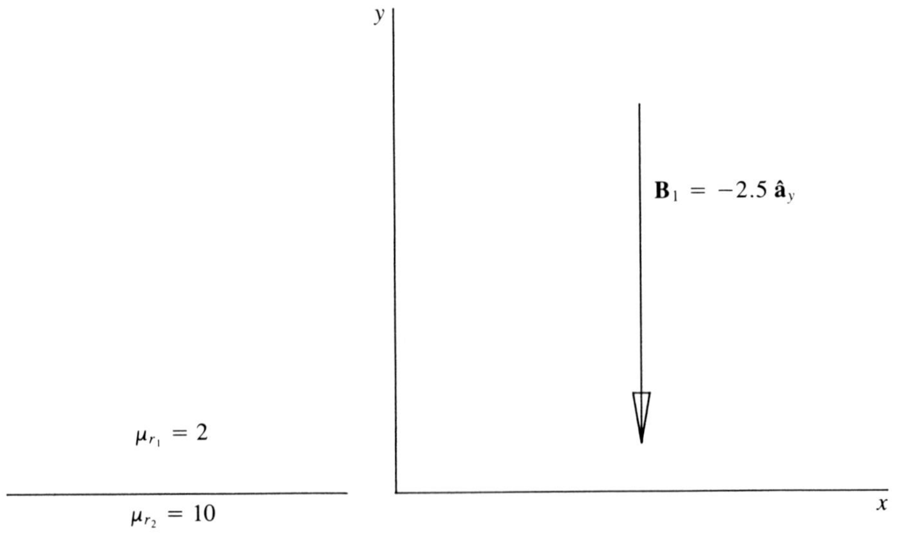

From equation (4.17)

$$\mathbf{B}_2 = -2.5\hat{\mathbf{a}}_y \text{ T}$$

and from equation (4.18)

$$\mathbf{H}_2 = \frac{-2.5}{10}\hat{\mathbf{a}}_y \text{ A/m}$$

Next we choose a closed path to evaluate the tangential components as shown in Figure 4.13. From Ampere's law

$$\oint \mathbf{H} \cdot \mathrm{d}\mathbf{l} = I$$

$$\oint \mathbf{H} \cdot \mathrm{d}\mathbf{l} = \int_a^b + \int_b^c + \int_c^d + \int_d^a \tag{4.19}$$

First, the contributions from the paths ad and bc vanish in the limit, $ad \to 0$ and $bc \to 0$,

$$\oint \mathbf{H} \cdot \mathrm{d}\mathbf{l} \approx -H_{\mathrm{T}_1}\Delta l + H_{\mathrm{T}_2}\Delta l = K\Delta l$$

where K is a surface current in A/m, and

$$-H_{\mathrm{T}_1} + H_{\mathrm{T}_2} = K \tag{4.20}$$

If there is no surface current between the two media, $K = 0$

$$H_{\mathrm{T}_1} = H_{\mathrm{T}_2} \tag{4.21}$$

and

$$\frac{B_{\mathrm{T}_1}}{\mu_1} = \frac{B_{\mathrm{T}_2}}{\mu_2} \tag{4.22}$$

Figure 4.13. Path of integration at boundary to evaluate tangential components of **B** and **H**.

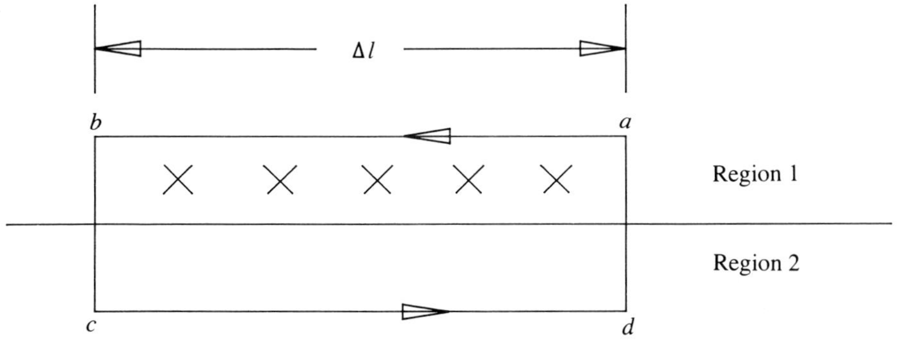

That is, the tangential components of field intensity are continuous, but the flux density components are not.

Example 4.7 Given a magnetic field intensity tangential to the x-axis as shown ($K = 0$) in Figure 4.14, determine **B** and **H** in both regions. Given

$$\mathbf{H}_1 = 4\hat{\mathbf{a}}_x \text{ A/m}$$

$$\mathbf{B}_1 = \mu \mathbf{H}_1 = 2(4\pi \cdot 10^{-7})4\hat{\mathbf{a}}_x \text{ T}$$

from equation (4.21)

$$\mathbf{H}_2 = 4\hat{\mathbf{a}}_x \text{ A/m}$$

$$\mathbf{B}_2 = 10(4\pi \cdot 10^{-7})4\hat{\mathbf{a}}_x \text{ T}$$

D4.7 Given $\mathbf{H}_1 = -2.5\hat{\mathbf{a}}_x - 2.75\hat{\mathbf{a}}_y + 1.5\hat{\mathbf{a}}_z$ in region 1 where $\mu_{r_1} = 10$, find $\mathbf{H}_2$. Region 2 has a relative permeability $\mu_{r_2} = 20$, find $\mathbf{B}_2$. The boundary is at the $z = 0$ plane.
Answer. $-2.5\hat{\mathbf{a}}_x - 2.75\hat{\mathbf{a}}_y + 0.75\hat{\mathbf{a}}_z$ A/m, $[-6.28\hat{\mathbf{a}}_x - 6.9\hat{\mathbf{a}}_y + 1.88\hat{\mathbf{a}}_z]10^{-5}$ T.

Consider the case where a surface current is flowing between two media. Equation (4.20) above gives the boundary condition which must be satisfied.

Figure 4.14. See example 4.7.

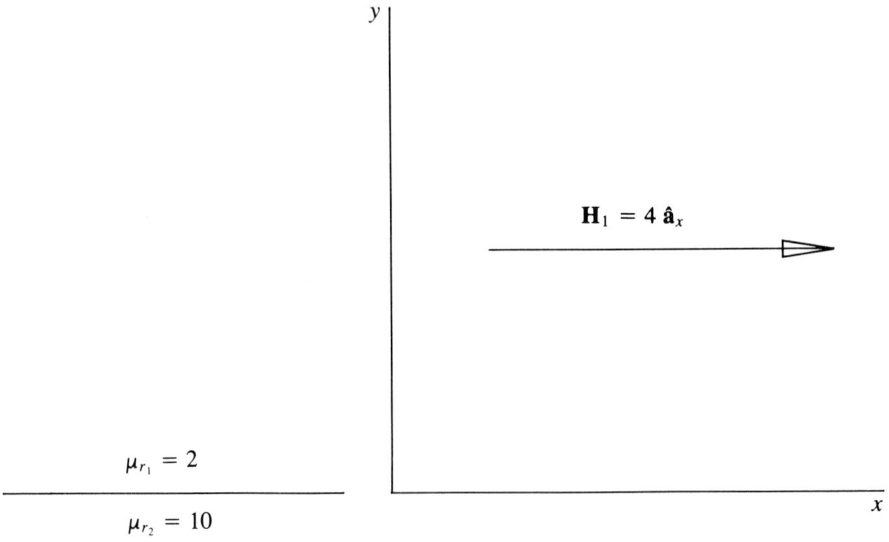

Example 4.8 Find the magnetic field intensity, **H**, due to an infinite sheet with current **K** A/m flowing in the z-direction as shown in Figure 4.15. Since we have symmetry we use Ampere's law

$$\oint \mathbf{H} \cdot \mathrm{d}\mathbf{l} = I = \int K \, \mathrm{d}l$$

where the contour integration is shown by the dashed line in Figure 4.15. Thus

$$\begin{aligned}\oint \mathbf{H} \cdot \mathrm{d}\mathbf{l} &= \int_a + \int_b + \int_c + \int_d \\ &= 0 + (-H_x)(-\mathrm{d}l) + 0 + H_x \, \mathrm{d}l \\ &= 2H_x \, \mathrm{d}l \\ &= K \, \mathrm{d}l \\ H_x &= \tfrac{1}{2}K\end{aligned}$$

In vector form

$$\mathbf{H} = \tfrac{1}{2}\mathbf{K} \times \hat{\mathbf{a}}_{\mathrm{N}}$$

Figure 4.15. An infinite sheet of current **K** flowing in the $y = 0$ plane.

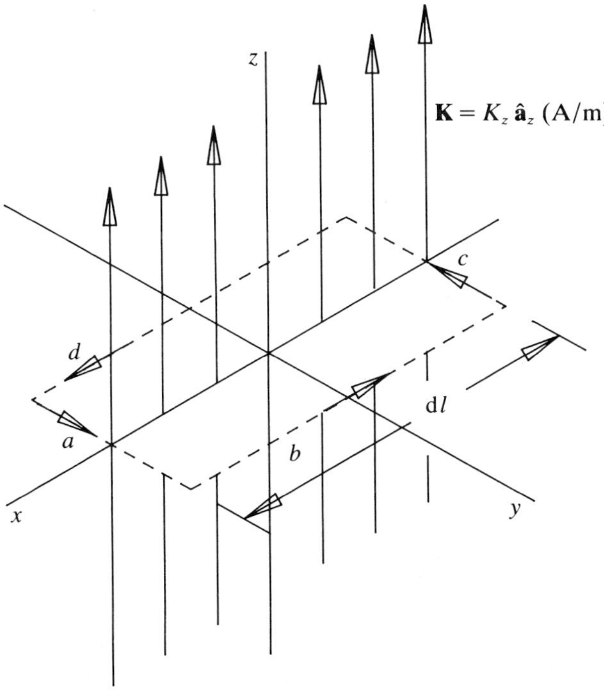

where $\hat{\mathbf{a}}_N$ is the unit vector normal to the infinite plane sheet. This is also written

$$\mathbf{H} = \tfrac{1}{2}K_z\hat{\mathbf{a}}_z \times \hat{\mathbf{a}}_y \qquad y > 0$$

$$\mathbf{H} = \tfrac{1}{2}K_z\hat{\mathbf{a}}_z \times (-\hat{\mathbf{a}}_y) \qquad y < 0$$

Compare this with the electric field due to an infinite plane surface in Chapter 3 in the, say, xz-plane: $\mathbf{E} = \tfrac{1}{2}(\rho_s/\epsilon)\hat{\mathbf{a}}_y \quad y > 0 \ \mathbf{E} = -\tfrac{1}{2}(\rho_s/\epsilon)\hat{\mathbf{a}}_y$ $y < 0$.

4.9 Lorentz force equation

One of the most important equations in power engineering is the Lorentz force equation.

$$\mathbf{F} = Q[\mathbf{E} + \mathbf{u} \times \mathbf{B}] \tag{4.23}$$

where Q is the electric charge in C, $\mathbf{E}$ is the external electric field vector in V/m, $\mathbf{u}$ is the velocity vector of the moving charge in m/s, and $\mathbf{B}$ is an external magnetic field vector in T. We can write the same equation as

$$\mathbf{F} = Q\left[\mathbf{E} + \frac{\mathbf{L}}{T} \times \mathbf{B}\right] \tag{4.24}$$

where $\mathbf{u} = \mathbf{L}/T$ is velocity in m/s. Rewriting equation (4.23), we have

$$\mathbf{F} = Q\mathbf{E} + \frac{Q}{T}\mathbf{L} \times \mathbf{B}$$

But current

$$I = \frac{Q}{T}$$

so that

$$\mathbf{F} = Q\mathbf{E} + I\mathbf{L} \times \mathbf{B} \tag{4.25}$$

Now let $E = 0$ in equation (4.25)

$$\mathbf{F} = I\mathbf{L} \times \mathbf{B} \tag{4.26}$$

This equation explains the action of a motor or a generator (see Table 4.4). In other words, if a current and a magnetic flux are provided, a force will result (a motor) or if a force and a magnetic field are provided a current will result (a generator).

Let us try equation (4.26) on two parallel conductors with currents flowing in each as shown in Figure 4.16. From Figure 4.16 we see that the currents I_1 and I_2 cause fields $\mathbf{H}_1$ and $\mathbf{H}_2$, respectively. Let us find the force on conductor number 2 caused by the current, I_1, in conductor number 1. We write formally

$$\mathbf{F}_2 = I_2\mathbf{L}_2 \times \mathbf{B}_1$$

Using the right-hand rule, we cross $\mathbf{L}_2 = L_2\hat{\mathbf{a}}_z$ into $\mathbf{B}_1 = B_\phi\hat{\mathbf{a}}_\phi$ to obtain an attractive force between conductors. More carefully, $\mathbf{B}$ is in the $-\hat{\mathbf{a}}_x$ direction at conductor number 2, and $\mathbf{L}_2$ is in the $\hat{\mathbf{a}}_z$ direction; the cross product is

$$\hat{\mathbf{a}}_z \times (-\hat{\mathbf{a}}_x) = -\hat{\mathbf{a}}_y$$

From previous work, equation (4.5), on page 112, and the fact that

Table 4.4.

Given	Result
$I\mathbf{L}$, $\mathbf{B}$	$\mathbf{F}$ motor
$\mathbf{F}$, $\mathbf{B}$	$I\mathbf{L}$ generator

Figure 4.16. Two parallel conductors carrying respective currents.

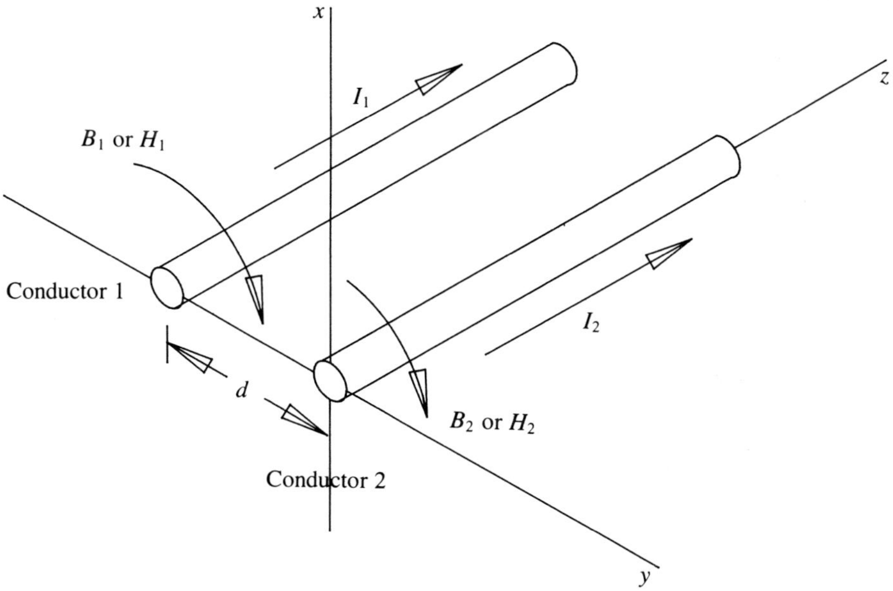

$B = \mu H$, we write

$$B_1 = \frac{\mu I_1}{2\pi d} \tag{4.27}$$

Substituting this equation into equation (4.26), we obtain

$$\mathbf{F}_2 = \mu \frac{-I_2 L_2 I_1}{2\pi d} \hat{\mathbf{a}}_y \tag{4.28}$$

If the conductors are infinite, we write this as force per unit length or

$$\frac{\mathbf{F}_2}{L_2} = -\frac{\mu I_1 I_2}{2\pi d} \hat{\mathbf{a}}_y \tag{4.29}$$

Now if we reverse the direction of either current, the direction of flux changes and the force becomes repulsive. If we reverse both currents, both fluxes would change direction and the force would be attractive.

D4.8 Two infinite, parallel conductors are carrying current $I = \frac{1}{2}$ A (each) in the $-\hat{\mathbf{a}}_z$ direction. One conductor is at $x = \frac{1}{2}$; the second is at $x = -\frac{1}{2}$. Find $\mathbf{F}$ at $(\frac{1}{2}, 0, 0)$.
Answer. $-5 \cdot 10^{-8}\hat{\mathbf{a}}_x$ N.

4.10 Magnetic potential revisited

Recall in electrostatics we write

$$\mathbf{E} = -\nabla V$$

We define a similar quantity

$$\mathbf{H} \triangleq -\nabla V_{\mathrm{m}} \tag{4.30}$$

From Maxwell's equations, Table 4.2,

$$\nabla \times \mathbf{H} = \mathbf{J}$$

or

$$\nabla \times (-\nabla V_{\mathrm{m}}) = \mathbf{J}$$

but

$$\nabla \times \nabla V_{\mathrm{m}} = 0$$

therefore

$$\mathbf{J} = 0$$

and magnetic potential is utilized *outside* a conductor. Also from Table 4.2

$$\oint \mathbf{B} \cdot \mathrm{d}\mathbf{A} = \int_{\mathrm{Vol.}} \nabla \cdot \mathbf{B}\, \mathrm{d}v = 0$$

or

$$\nabla \cdot \mathbf{B} = 0 = \mu\, \nabla \cdot \mathbf{H}$$

where μ is independent of position, and

$$\nabla \cdot \nabla V_{\mathrm{m}} = \nabla^2 V_{\mathrm{m}} = 0$$

which is Laplace's equation.

Example 4.9 Find the magnetic potential inside a coaxial conductor with current I as shown in Figure 4.17. From page 118, equation (4.8) the field intensity for the dielectric region within the coax is

$$H_\phi = \frac{I}{2\pi r}$$

Figure 4.17. A coaxial conductor showing the field and potential lines.

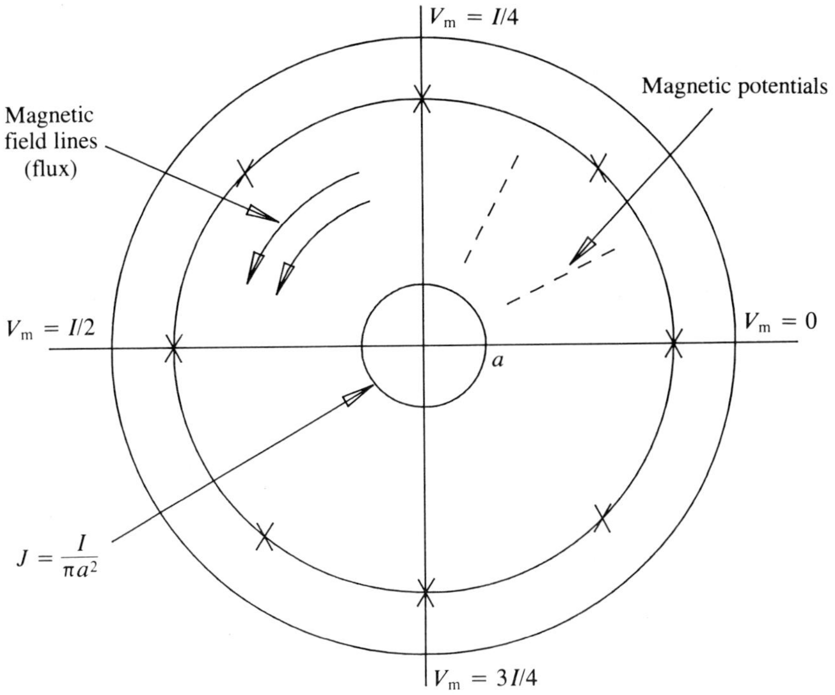

In cylindrical coordinates

$$\nabla V_{\mathrm{m}} = \frac{\partial V_{\mathrm{m}}}{\partial r}\hat{\mathbf{a}}_r + \frac{1}{r}\frac{\partial V_{\mathrm{m}}}{\partial \phi}\hat{\mathbf{a}}_\phi + \frac{\partial V_{\mathrm{m}}}{\partial z}\hat{\mathbf{a}}_z \tag{4.31}$$

Since $\mathbf{H}$ is in the $\hat{\mathbf{a}}_\phi$ direction, we keep the middle term of equation (4.31) which gives

$$\frac{I}{2\pi r} = -\frac{1}{r}\frac{\partial V_{\mathrm{m}}}{\partial \phi}$$

or

$$\mathrm{d}V_{\mathrm{m}} = -\frac{I}{2\pi}\,\mathrm{d}\phi$$

Integrating, we obtain

$$V_{\mathrm{m}} = \frac{-I}{2\pi}\phi + \text{constant}$$

Choose $V_{\mathrm{m}} = 0$ for $\phi = 0$ which implies that constant $= 0$. The potential lines are a function of ϕ. Thus, we see from Figure 4.17 that the potential and flux lines are orthogonal everywhere. Note that the usefulness of the potential will be illustrated with an example.

Example 4.10 Find the magnetic field intensity at some point P of a pair of parallel filaments carrying current I as shown in Figure 4.18. It has been shown that the field intensity due to a current carrying filament is

$$\mathbf{H} = \frac{I}{2\pi r}\hat{\mathbf{a}}_\phi$$

For a pair of filaments we use superposition and write

$$\mathbf{H} = \mathbf{H}_1 + \mathbf{H}_2$$

From Figure 4.18 we see that

$$\mathbf{H}_1 = \frac{I}{2\pi r_1}\hat{\mathbf{a}}_{\phi 1}$$

and

$$\mathbf{H}_2 = \frac{I}{2\pi r_2}\hat{\mathbf{a}}_{\phi 2}$$

where both currents are in the z direction. Thus

$$\mathbf{H} = \frac{I}{2\pi}\left[\frac{\hat{\mathbf{a}}_{\phi_1}}{r_1} + \frac{\hat{\mathbf{a}}_{\phi_2}}{r_2}\right]$$

It can be shown that in Cartesian coordinates this becomes

$$\mathbf{H} = \frac{I}{2\pi}\left\{\left[\frac{y}{\left(x+\frac{d}{2}\right)^2 + y^2} - \frac{y}{\left(x-\frac{d}{2}\right)^2 + y^2}\right]\hat{\mathbf{a}}_x + \left[\frac{x+\frac{d}{2}}{\left(x+\frac{d}{2}\right)^2 + y^2} + \frac{x-\frac{d}{2}}{\left(x-\frac{d}{2}\right)^2 + y^2}\right]\hat{\mathbf{a}}_y\right\}$$

This equation can be plotted of course but there may be an easier way – the graphical method. We can find the lines of constant intensity for each filament and add. This is illustrated by Figure 4.19. Lines of constant H are plotted. The (relative) value of each locus is added at each intersection. The final results are obtained when like values are

Figure 4.18. Two parallel filaments carrying current I for example 4.10.

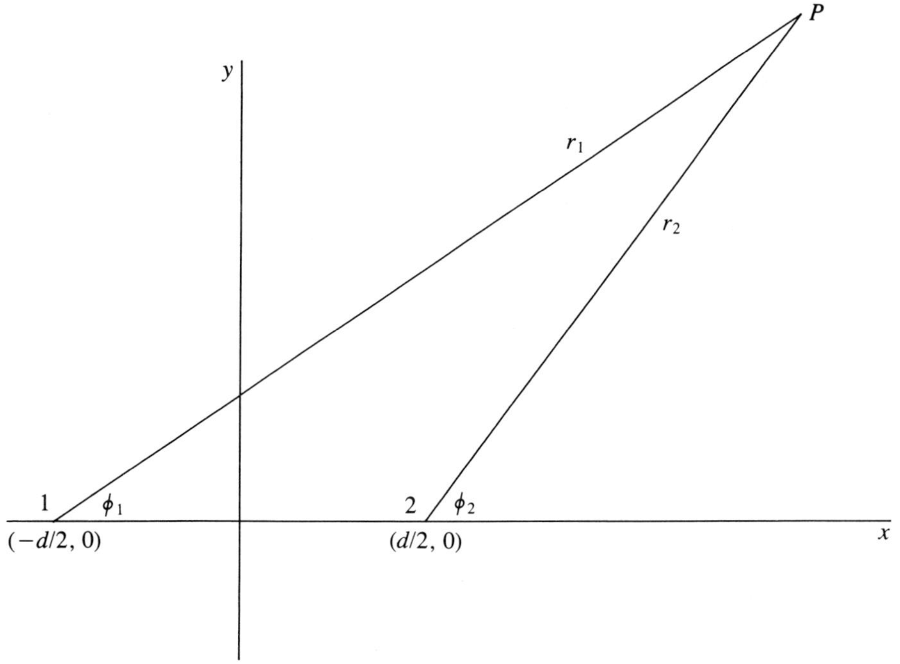

Figure 4.19. Loci for lines of equipotential for two parallel filaments, (*a*) current in same direction, (*b*) current in opposite directions.

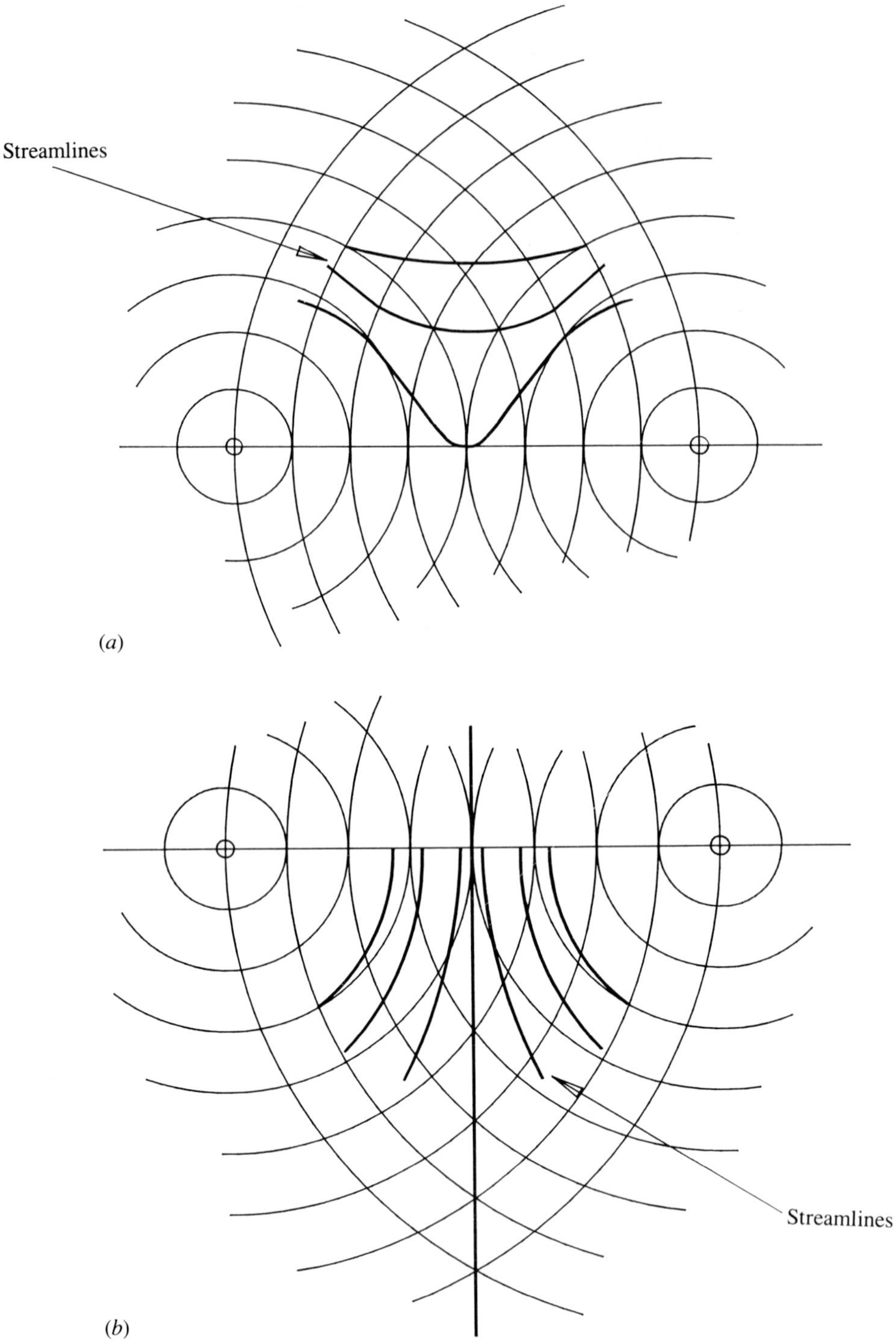

joined with the resulting loci. Figure 4.19(*a*) is the result of two filaments with currents in the *same* direction. Figure 4.19(*b*) is the result of two filaments with currents in opposite directions.

In the same way we can plot the loci for lines of magnetic potential by plotting the potential lines for each filament and adding the results. Construction is shown in Figure 4.20. Figure 4.20(*a*) is for current in the same direction in both filaments and Figure 4.20(*b*) is for current in the opposite directions. Also compare Figures 4.19 and 4.20. In particular compare Figure 4.19(*a*) with Figure 4.20(*a*) and compare Figure 4.19(*b*) with Figure 4.20(*b*). Note that the field intensity lines and the potential lines are orthogonal everywhere. Finally compare a pair of line charges with a pair of line filaments. Compare Figure 3.25 with Figures 4.19 and 4.20. Note that the potential lines in magnetics are the flux lines in electricity and the flux lines in magnetics are potential lines in electricity.

Continuing with our imaginary magnetic potential, we write

$$V_{\mathrm{m}} = +\int_b^a \mathbf{H} \cdot \mathrm{d}\mathbf{l}$$

compared with

$$V = -\int_a^b \mathbf{E} \cdot \mathrm{d}\mathbf{l}$$

or

$$V_{\mathrm{m}} = -\int_a^b \mathbf{H} \cdot \mathrm{d}\mathbf{l}$$

In addition, from $V = IR$ we write $V_{\mathrm{m}} = \Phi\mathscr{R}$ where Φ is the flux in Webers ($=\mathrm{Tm}^2$) corresponding to current I and $\mathscr{R}$ is the reluctance in A/Weber corresponding to resistance R. This is the magnetic equivalent of Ohm's law. The magnetic potential is also magnetomotive force. There is yet another analogy from electricity. Recall the resistance of a conductor is

$$\mathrm{R} = \frac{l}{\sigma A}$$

where l is the length in meters, A is the cross-sectional area in square meters and σ is the conductivity in Siemens.

The reluctance of a magnetic material is $\mathscr{R} = l/\mu A$ where μ is the permeability of the material.

Figure 4.20. Loci for lines of equipotential for two parallel filaments, (*a*) current in same direction, (*b*) current in opposite directions.

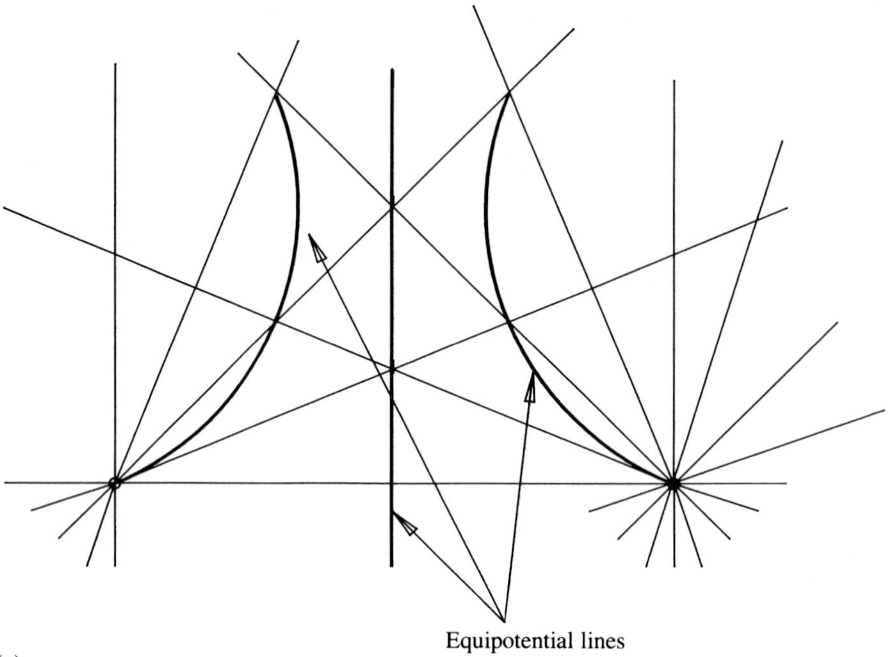

(*a*)

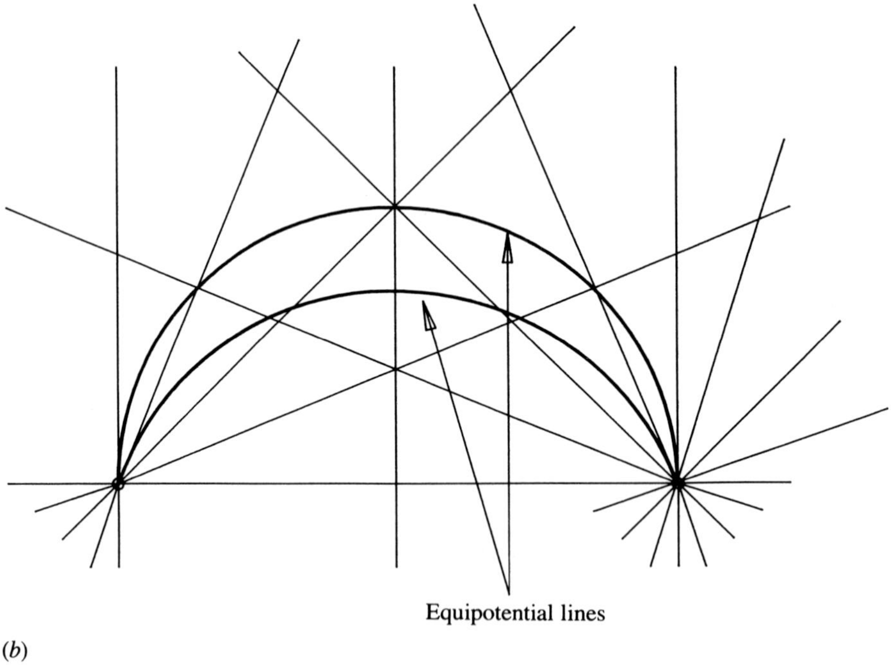

(*b*)

Example 4.11 Calculate the reluctance of a magnetic ring as shown in Figure 4.21 where $\mu_r = 350$.

$$\mathscr{R} = \frac{l}{\mu A} = \frac{2\pi \cdot 4 \cdot 10^{-2}}{350 \cdot 4\pi \cdot 10^{-7} \cdot 10^{-4}} = 5714 \cdot 10^3 \text{ A/Weber}$$

D4.9 Calculate the reluctance of the magnetic ring in Example 4.11 if half the ring (in circumference) is constructed of a magnetic material which has a permeability $\mu_r = 500$.
Answer. $4.86 \cdot 10^{+6}$ A/Weber.

4.11 Inductance

Recall

$$C = \frac{Q}{V} = \frac{\Phi}{V} = \frac{\text{flux}}{\text{potential}}$$

We realize that the inductor is the dual of the capacitor. Therefore we

Figure 4.21. See example 4.11 and **D4.9**.

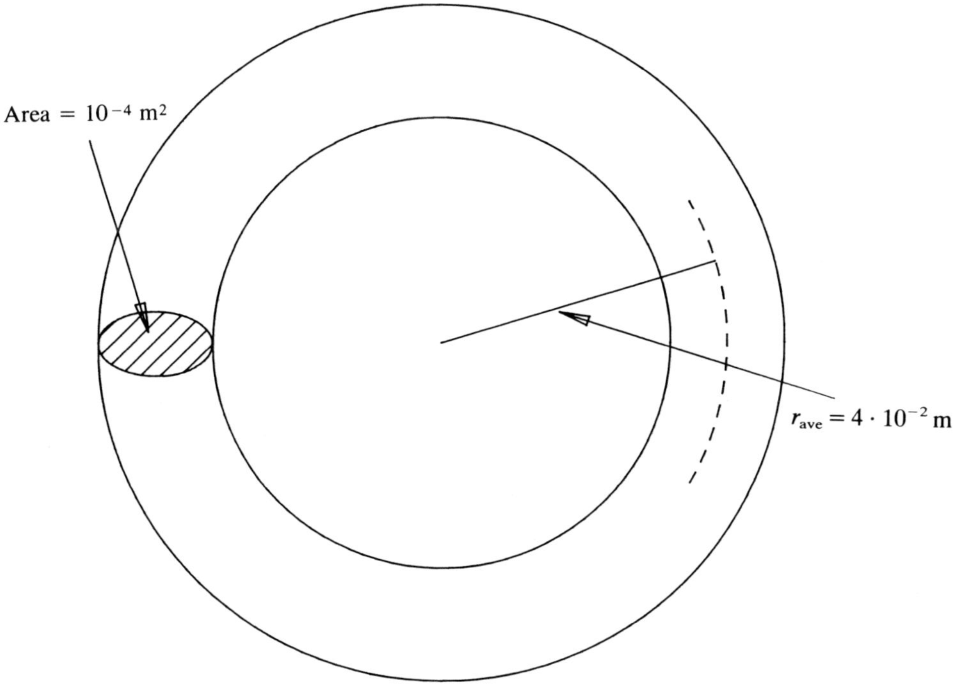

can write

$$L = \frac{\text{potential}}{\text{flux}}$$

$$= \frac{\Lambda}{I}$$

where L is inductance in Henries and Λ is the total magnetic flux linkage.

Like a capacitor an inductor stores energy. Ohm's law for magnetics is written

$$V_{\text{m}} = \Phi \mathcal{R}$$

Recall Ampere's law

$$\oint \mathbf{H} \cdot \text{d}\mathbf{l} = I$$

If there are N turns of wire enclosed

$$\oint \mathbf{H} \cdot \text{d}\mathbf{l} = NI$$

Also

$$\oint \mathbf{H} \cdot \text{d}\mathbf{l} = V_{\text{m}}$$

Also for N turns of wire

$$\Lambda = N\Phi$$

so that

$$L = \frac{N\Phi}{I}$$

$$= \frac{N\left(\frac{V_{\text{m}}}{\mathcal{R}}\right)}{I} = \frac{N\frac{(NI)}{\mathcal{R}}}{I} = \frac{N^2}{\mathcal{R}}$$

$$= \frac{\mu N^2 A}{l}$$

This formula assumes that the N turns of wire are wound such that the magnetic flux density is constant inside the turns and equal to its value at the center.

One type of inductor which is very useful is the solenoid. The solenoid is shown in Figure 4.22. A solenoid is used to obtain mechanical motion from a coil of wire. For example, a solenoid is used to start an automobile. Turning the starter switch causes the solenoid to activate. The solenoid activates a heavy current switch which connects the starter motor directly to the battery. Another type of useful inductor is the toroid; it is shown in Figure 4.23. The toroid is used in recording/playback audio and visual electronics. The gap facilitates

Figure 4.22. A magnetic solenoid.

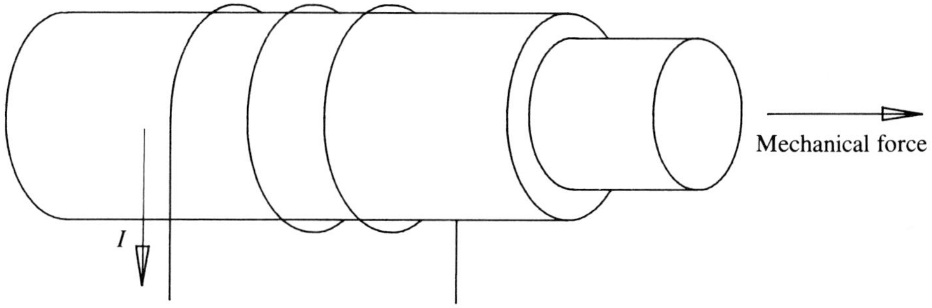

Figure 4.23. A toroidal magnetic circuit.

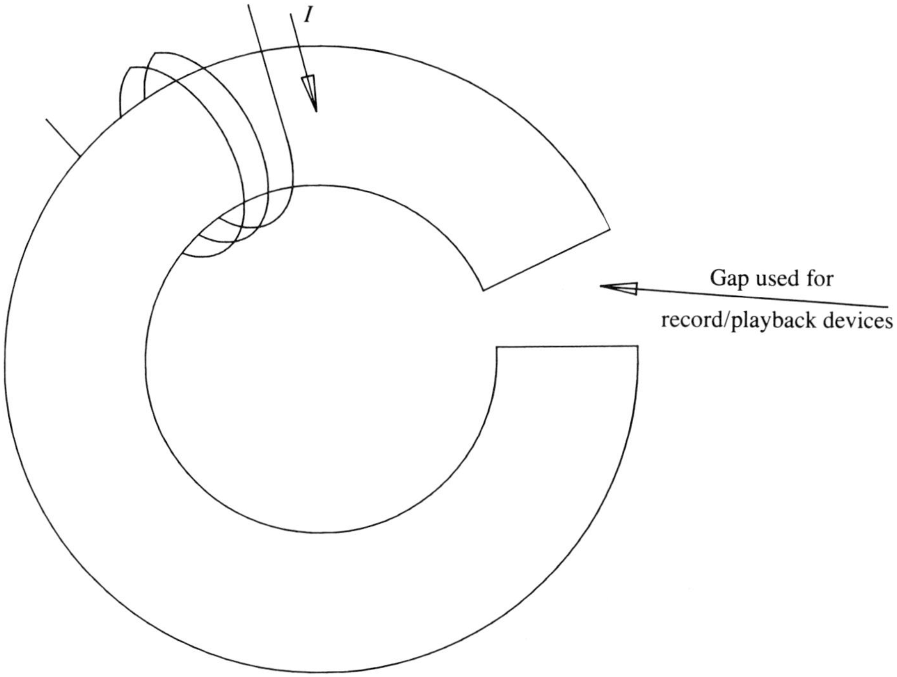

utilization of the strong B field generated in the toroidal magnetic circuit.

We can approximate the magnetic field intensity for the solenoid or the toroid with

$$H \approx \frac{I}{2\pi r}$$

This assumes no flux leakage between turns. In addition, for the solenoid, the length is large compared to the radius and, for the toroid, the average radius is large compared to the single turn radius.

A third type of useful inductor is the transformer. A typical transformer model is shown in Figure 4.24. The double bar between the primary and secondary windings indicates that the transformer is wound on an iron core. The number of turns of wire is noted by N_1 for the primary and N_2 for the secondary. Before saying more about the transformer we continue our discussion of the toroid with the following example.

Example 4.12 Find the flux density within and the inductance of a toroid of 1000 turns wound in air with current of 0.5 A as shown in Figure 4.25. From Ampere's law

$$\oint \mathbf{H} \cdot d\mathbf{l} = I_{\text{enc}} = V_{\text{m}}$$

Figure 4.24. A typical transformer.

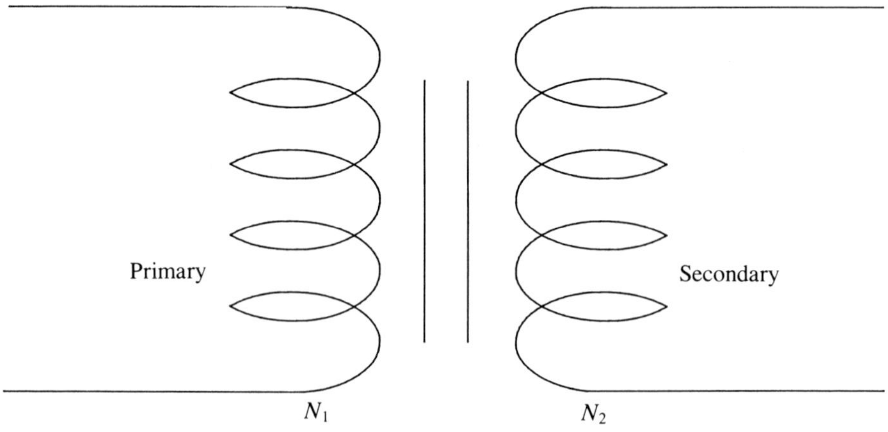

We assume there is no flux leakage between turns, or in other words, all of the flux stays within the 1000 turns of the toroid. Thus,

$$I_{\text{enc}} = NI = 500 \text{ A-Turns} = V_{\text{m}}$$

$$\mathcal{R} = \frac{l}{\mu A} = \frac{2\pi r_{\text{ave}}}{4\pi \cdot 10^{-7} \cdot 10^{-5}}$$

$$V_{\text{m}} = \Phi \mathcal{R}$$

$$\Phi = \frac{V_{\text{m}}}{\mathcal{R}} = \frac{500}{\frac{1}{2} \cdot 10^{10}} = 10^{-7} \text{ Webers}$$

If we assume uniform flux distribution

$$B = \frac{\Phi}{A} = 10^{-2} \text{ T}$$

$$H = \frac{B}{\mu} = \frac{10^5}{4\pi} \text{ A/m}$$

Since $r_{\text{ave}} = 10^{-2}$ m and $r_{\text{turn}} = 1.8 \cdot 10^{-3}$ m we use the approximation for a toroid, that is, $r_{\text{ave}} \gg r_{\text{turn}}$.

$$L = \frac{\mu N^2 A}{l} = 2 \cdot 10^{-4} \text{ H}$$

Figure 4.25. Toroid of 1000 turns with $I = 0.5$ A from example 4.12.

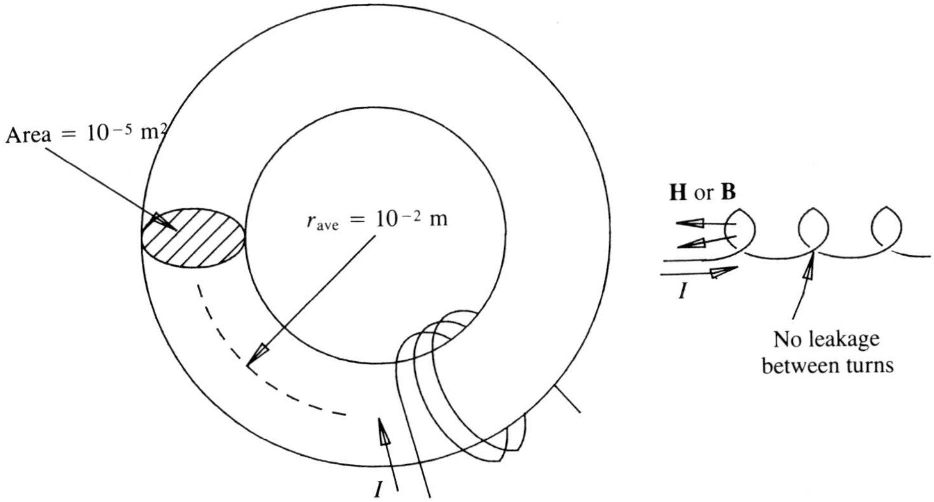

This is the case of a current producing a flux: now for the problem of flux producing a current. (Actually it is *time changing* flux which produces current.)

D4.10 Calculate the flux density within a toroid wound in air with 500 turns of wire and a current of 0.25 A. The toroid has a cross sectional area of 10^{-4} m^2 and the average circumference of the toroid is 10^{-2} m.
Answer. $1.57 \cdot 10^{-2}$ T.

D4.11 Calculate the current required to provide a flux of 10 Webers within an air-filled toroid. The average radius is 10^{-1} m and the diameter of the cross section is 10^{-3} m. There are 750 turns of wire.
Answer. $8.5 \cdot 10^{9}$ A.

Example 4.13 What is the current which will flow in a wire if uniformly distributed flux of value $\Phi = 10^{-7}$ Webers is flowing inside a toroid of the same dimensions as Example 4.12? The toroid is wound on a silicon iron core with relative permeability, $\mu_r = 7000$, $N = 500$ turns. First we calculate the magnetomotive force

$$
\begin{aligned}
V_m &= \Phi\mathcal{R} \\
&= 10^{-7}\left[\frac{2\pi r_{ave}}{\mu_r\mu_0 A}\right] \\
&= 0.07\ A\text{-turns}
\end{aligned}
$$

But $V_m = NI$

$$I = \frac{V_m}{N} = 0.14\ \text{mA}$$

Now let us introduce an air gap in the toroid as shown in Figure 4.26(a). An air gap in a magnetic circuit presents a high resistance – better reluctance – to the flow of magnetic flux. This is shown in Figure 4.26(b) by the streamlines 'escaping' at the air gap. These gaps are used in electronic recording and playback devices to transmit information – video or audio. The gap increases the total reluctance of the magnetic circuit. Reluctances may be treated like resistances in electric circuits.

If we allow an air gap in the previous example we may visualize the circuit as an electrical series circuit with a source and two resistances. We write this

$$V = I(R_1 + R_2)$$

Similarly in magnetics, for our example, we write

$$V_m = \Phi(\mathcal{R}_1 + \mathcal{R}_2)$$

Figure 4.27 shows the magnetic circuit. Notice that the coil of wire does not 'break into' the circuit as the source does in electricity. Yet it represents the source. Also it does not matter where the coil is – so long as it is not in the air gap. From Ampere's law

$$V_m = \oint \mathbf{H} \cdot d\mathbf{l} = H_{gap} l_{gap} + H_{mag} l_{mag} = NI$$

where $l = l_{gap} + l_{mag}$. Another method for solving a problem of this type is to use the magnetic Ohm's law

$$V_m = NI = \Phi(\mathcal{R}_{mag} + \mathcal{R}_{gap})$$

$$\approx BA\left[\frac{l_{gap}}{\mu_{gap}A} + \frac{l_{mag}}{\mu_{mag}A}\right]$$

This is an approximation because of the second term on the right. In particular, the permeability of magnetic material depends on the current density. This is explained in detail below.

The example on page 142 is exceptional since normally the reluctance of a magnetic material depends on the permeability, $\mu = \mu_r\mu_0$, and the permeability depends on the 'current' through the material. In the

Figure 4.26. Toroid with air gap, (*a*) toroid, (b) expanded view.

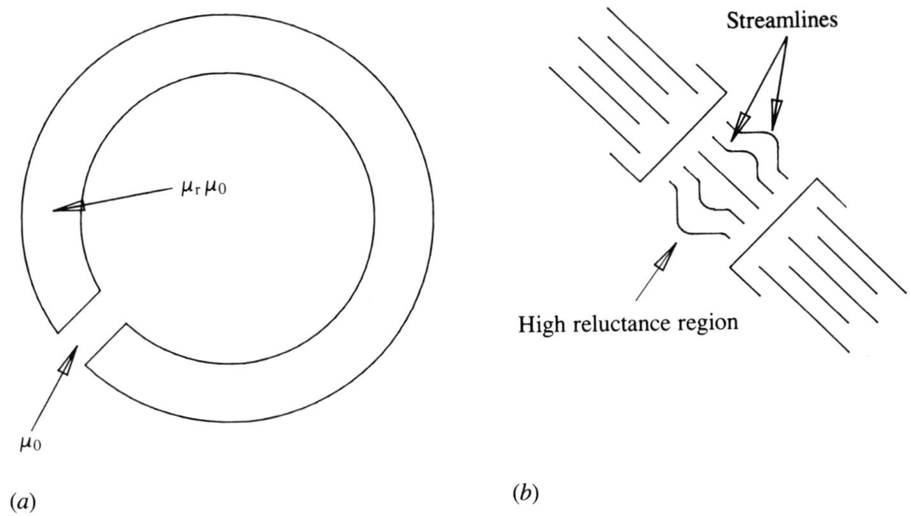

example,

$$\mathcal{R}_{\text{gap}} = \frac{l_{\text{gap}}}{\mu_{\text{gap}} A}$$

and

$$\mathcal{R}_{\text{mag}} = \frac{l_{\text{mag}}}{\mu_{\text{mag}} A}$$

Figure 4.28 shows the relationship between B and H for a linear material. In actuality magnetic materials are nonlinear and respond as Figure 4.29 illustrates.

Let us assume that the material is demagnetized. From Figure 4.29 we start at point 1. As the 'current', H, increases the material becomes magnetized and then saturates at point 2. As the current decreases to zero at point 3 a flux density remains. This is designated B_r. As we continue to decrease H the flux density decreases to zero at point 4. Thus we have 'coerced' the flux density to zero by applying a negative current, $-H_c$.

Figure 4.27. The magnetic circuit.

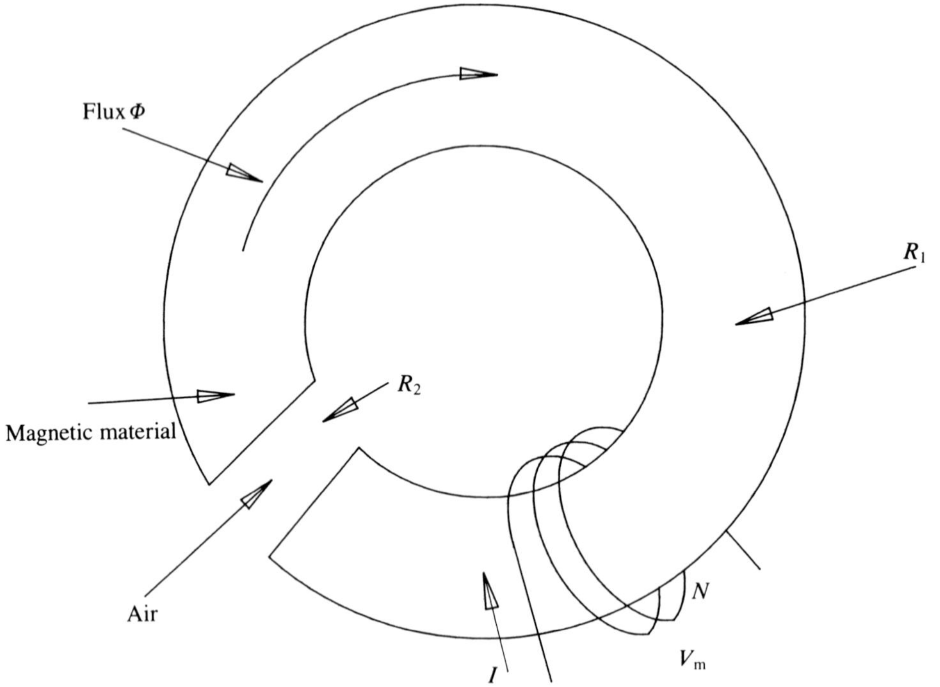

Figure 4.28. *B* versus *H* curve for linear magnetic material.

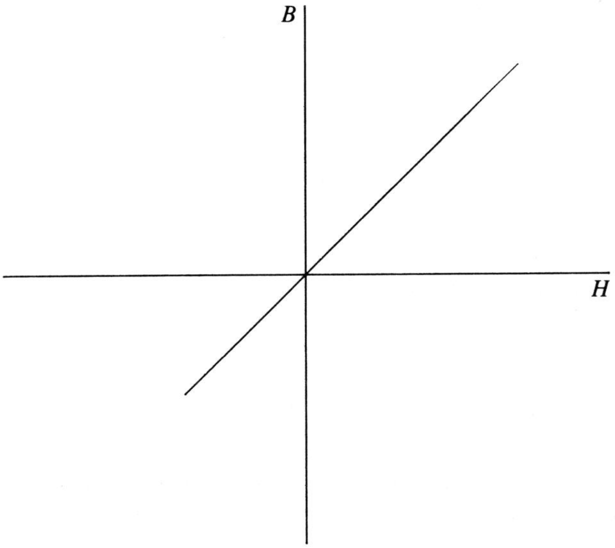

Figure 4.29. *B* versus *H* curve for nonlinear magnetic material.

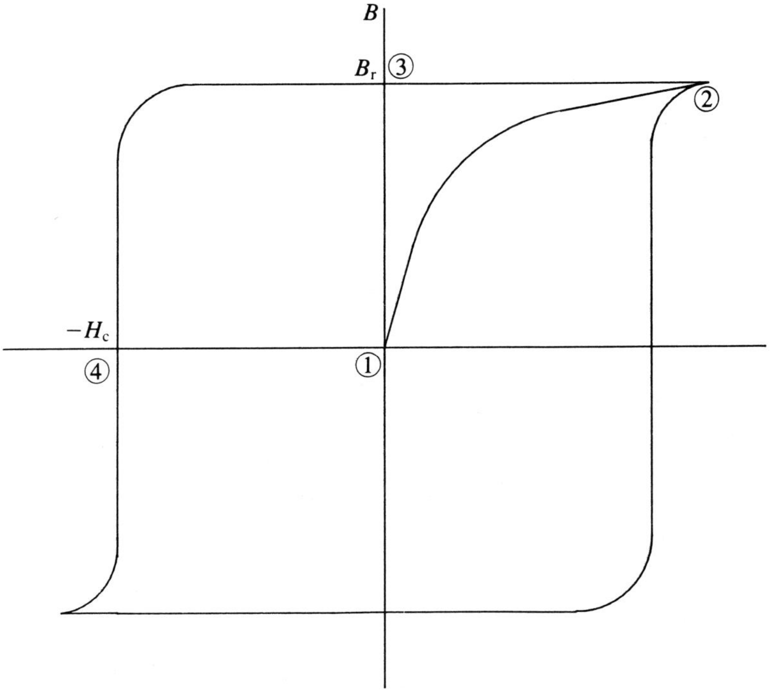

As can be seen the B–H curve shows a hysteresis effect. The plot never returns to the origin, point 1, unless external forces are supplied. To summarize, the reluctance of a magnetic material is *not* usually calculated from the formula

$$\mathscr{R} = \frac{l}{\mu A}$$

It is customary to calculate the reluctance in air, $\mathscr{R} = l/\mu_0 A$; from which we calculate the flux, Φ, assuming the magnetic potential, V_m, is known. Now if we assume no flux leakage, the same flux flows through the magnetic material. Knowing the dimensions we can then calculate the flux density, B. We follow with an example.

Example 4.14 Find the number of turns of wire carrying current $I = 25$ A in the coil shown in Figure 4.30 to provide a flux of 0.15 Webers

Figure 4.30. See example 4.14. $L_1 = L_3 = 2 \cdot 10^{-2}$ m, $L_2 = 3 \cdot 10^{-2}$ m, $L_4 = L_5 = 1.45 \cdot 10^{-2}$ m, Area $= 0.25 \cdot 10^{-4}$ m^2.

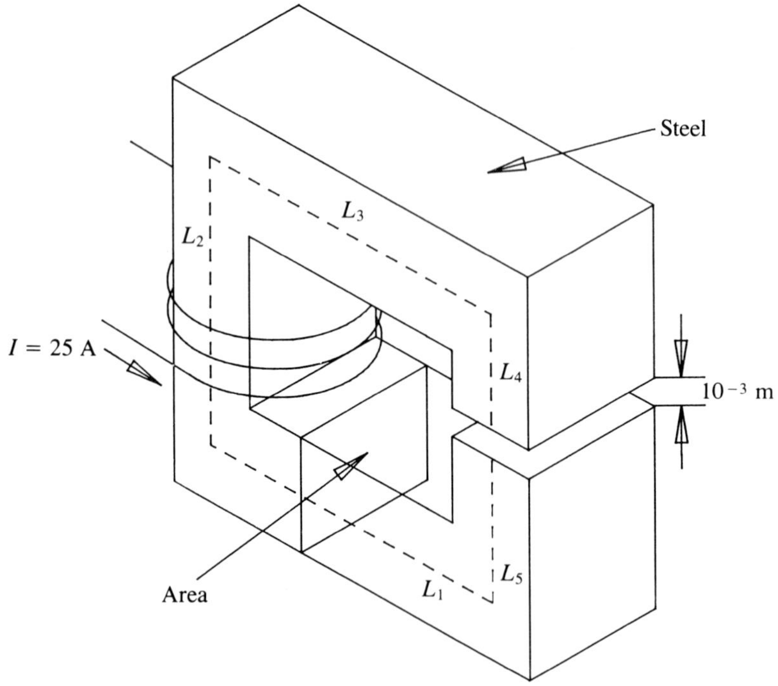

in the air gap. The B–H curve for the steel portion is shown in Figure 4.31.

$$V_m = \oint \mathbf{H} \cdot d\mathbf{l} = NI$$

$$= V_{m_{steel}} + V_{m_{air}}$$

$$V_{m_{air}} = \Phi \mathscr{R}$$

$$= 0.15 \frac{10^{-3}}{4\pi \cdot 10^{-7} \cdot 0.25 \cdot 10^{-2}} = 4.77 \cdot 10^4 \text{ A}$$

Also

$$B = \frac{\Phi}{\text{Area}} = \frac{0.15 \cdot 10^{-4}}{0.25 \cdot 10^{-4}}$$

$$= 0.6 \text{ T}$$

From Figure 4.31

$H = 80\,000$ A/m in Alnico 2 material

Figure 4.31. A typical *B–H* curve for Alnico 2 magnetic material.

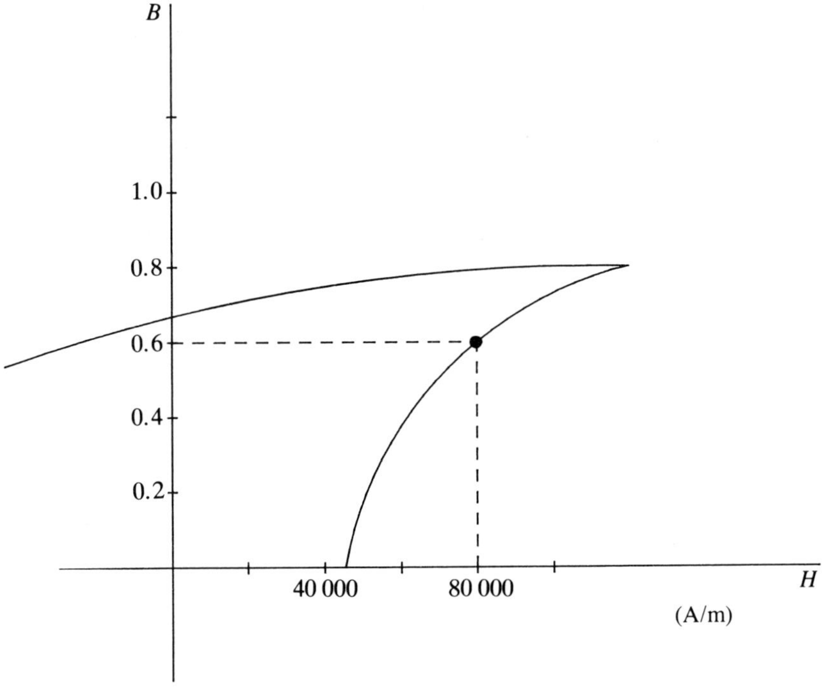

The flux will spread in the air gap but it is customary to assume as a good approximation that the flux in iron equals the flux in the gap. Therefore we can write

$$V_m = \oint \mathbf{H} \cdot d\mathbf{l} = NI$$

$$V_{m_{steel}} = H(L_1 + L_2 + L_3 + L_4 + L_5) = 79\,200 \text{ A-t}$$

$$V_m = 4.77 \cdot 10^4 + 7.92 \cdot 10^4 = 12.7 \cdot 10^4 \text{ A-t}$$

or

$$N = \frac{V_m}{I} = 5080 \text{ turns}$$

D4.12 A toroid is wound on a steel core with $\mu_r = 100$ and has an average radius of 10^{-1} m. The cross sectional area is $4.5 \cdot 10^{-4}$ m^2 and $N = 1500$ turns. An air gap is introduced with a length $l = 10^{-4}$ m. What current is required to provide a flux $\Phi = 5 \cdot 10^{-6}$ Webers?
Answer. 37.6 mA.

As indicated before, the transformer is the third type of inductor which is very useful. The transformer is used to either raise or lower voltages. If the secondary in Figure 4.24 has a higher voltage than the primary we call it a step up transformer. If the secondary has a lower voltage than the primary it is a step down transformer. In addition the current in the windings behaves opposite to the voltage; that is, for a step up transformer the current actually steps down from primary to secondary. Now the voltage varies linearly with the number of turns; that is

$$\frac{V_{pri}}{V_{sec}} = \frac{N_1}{N_2}$$

where N is the number of turns.

In addition to the (self) inductance of several turns of wire there exists a mutual inductance when two or more inductors are in close proximity, as in a transformer. This mutual inductance induces a voltage or voltages in the nearby inductors. Let us illustrate with an example:

Example 4.15 Write mesh equations for the circuit in Figure 4.32. For Mesh 1

$$v_s(t) = i_1(t)R_s + L_1\frac{di_1}{dt} - M\frac{di_2}{dt} \tag{4.32}$$

For Mesh 2

$$0 = i_2(t)R_L + L_2\frac{di_2}{dt} - M\frac{di_1}{dt} \tag{4.33}$$

In equations (4.32) and (4.33) the mutual inductance is 'M' in henries. The sign in front of M is determined by the dot and current directions. In Figure 4.32 since the currents in the primary and secondary are in the opposite directions, with reference to the dot, then M is minus. If either current direction changes or if either dot position changes the sign will be positive. If both change, the sign in front of M will remain negative.

There is more which could be said about transformers but we move on to magnetic dipoles. We use magnetic dipoles to model magnetic materials.

D4.13 A transformer has a turns ratio $N_1/N_2 = 10$. If the primary voltage is 50 V, what is the secondary voltage? Calculate the secondary current for every ampere of primary current.
Answer. 5 V, 10 A.

Figure 4.32. Circuit for example illustrating mutual inductance of transformer.

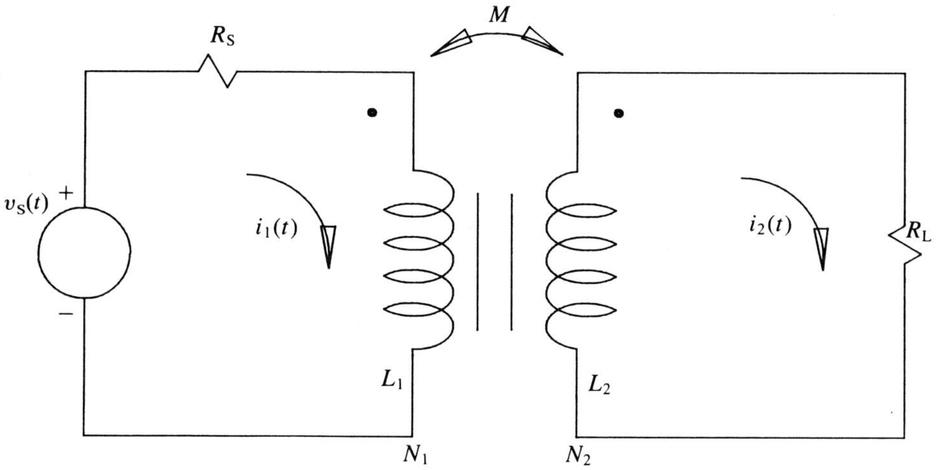

D4.14 Write the two mesh equations for Figure 4.32 (*a*) if the dot in the primary is at the bottom and (*b*) if both dots are at the bottom.

4.12 Magnetic dipoles

A dipole is more a convenient configuration for investigating fields than the single test charge. Whereas a single charge will be attracted to either pole along the lines of force (flux), a dipole will experience a twisting action – unless lined up with the flux lines. The dipole in electricity was a positive and a negative charge brought close together. In electricity we defined the electric dipole as

$$c \triangleq Qd$$

In magnetics the dipole is a single turn of wire with current I and we define the magnetic dipole as

$$\mathbf{m} \triangleq I\mathbf{A}$$

where I is current in amperes and $\mathbf{A}$ is the cross-sectional area of the loop in square meters.

For example, if we consider a loop of wire with current flowing in the wire as shown in Figure 4.33(*a*), the magnetic dipole is the product of the area enclosed by the wire and the current. If there is more than one turn as shown in Figure 4.33(*b*), then the dipole moment

$$m = NIA$$

where N is the number of turns.

Since a loop placed in a magnetic field will tend to turn (assuming current in loop) then a torque will exist. The 'force' – the force times the distance over which it is applied – is called torque. It is similar to a moment. If R is the distance from the center of a loop to one side, then

Figure 4.33. Magnetic dipoles, (*a*) single turn, (*b*) N turns.

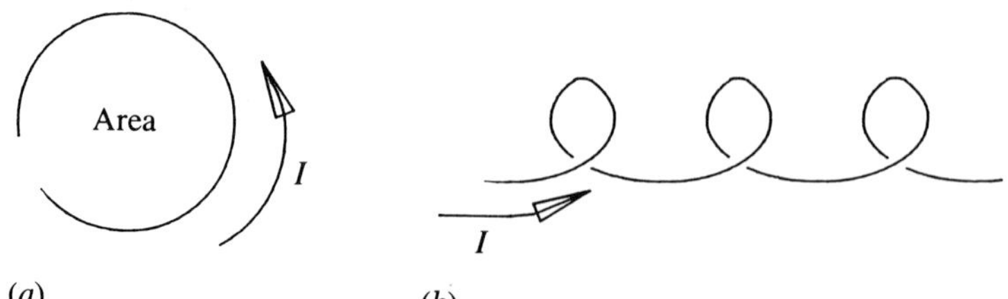

(*a*) (*b*)

the torque is the cross product of **R** and the force **F** on the side tending to rotate it.

$$\mathbf{T} = \mathbf{R} \times \mathbf{F} \quad \text{but} \quad \mathbf{F} = I\mathbf{L} \times \mathbf{B}$$

We now define torque. In differential form

$$\mathrm{d}\mathbf{T} \triangleq I\,\mathrm{d}\mathbf{A} \times \mathbf{B} = \mathrm{d}\mathbf{m} \times \mathbf{B}$$

Note that **B** is an external field and not the field resulting from the current *I*.

Example 4.16 Find the torque on a square loop of wire with current *I* in uniform **B** field as shown in Figure 4.34. Let us assume a differential current loop so that **B** is considered constant throughout and

$$\mathbf{T} = I\mathbf{A} \times \mathbf{B} = I(2a)a\hat{\mathbf{a}}_x \times (-B_y\hat{\mathbf{a}}_y) = -I2a^2\hat{\mathbf{a}}_z$$

The torque on the loop is in the direction of minimum force. The loop wants to line up such that the field it produces is in line with the external magnetic field. The reference for the torque may be anywhere; we choose the center of the loop. Hence, the torque is in the negative $\hat{\mathbf{a}}_z$-direction.

Figure 4.34. A loop of wire with current *I* in a uniform **B** field.

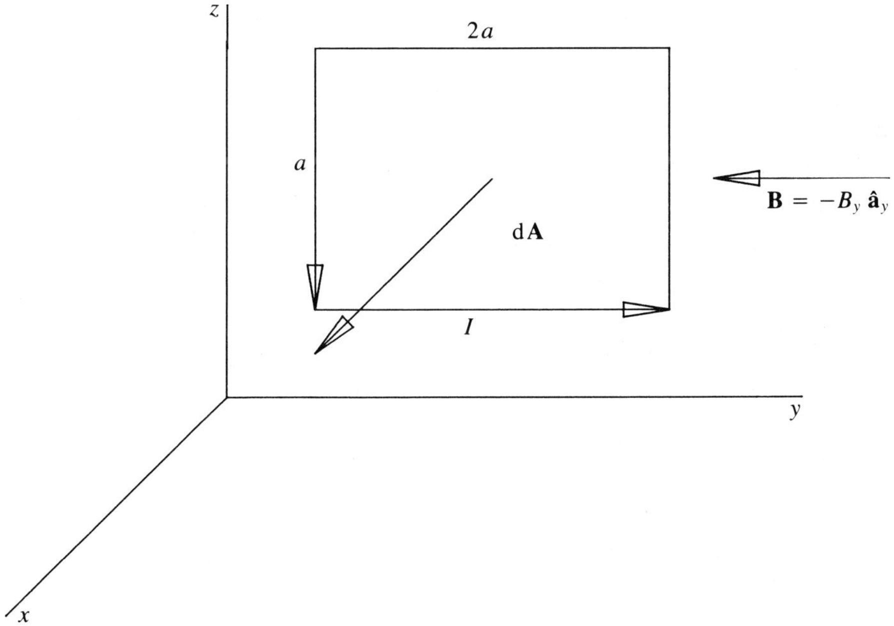

4.13 Energy density

In Chapter 3, Electrostatics, we determined the energy in an electric field. We found, from equation (3.84), that the electric energy stored in between the plates of a capacitor is

$$W_e = \int_{\text{vol}} \tfrac{1}{2}\bar{\mathbf{D}} \cdot \bar{\mathbf{E}} \, dv$$

By duality the energy stored in a magnetic field can be written

$$W_m = \int_{\text{vol}} \tfrac{1}{2}\bar{\mathbf{B}} \cdot \bar{\mathbf{H}} \, dv \tag{4.34}$$

If **B** and **H** are in the same direction we write

$$W_m = \int_{\text{vol}} \tfrac{1}{2}\mu H^2 \, dv \tag{4.35}$$

4.14 Magnetic vector potential

We have seen that it is convenient to define a potential in electrostatics and magnetostatics to describe flow in regions of 'charge' or flux. In both cases these are scalar potentials. In magnetics it is instructive to define a vector potential. From this vector potential we can find a magnetic field intensity. The operation is a differentiation; recall the previous methods – Ampere's law and Biot–Savart – were integrations. We begin with a vector identity

$$\nabla \cdot \nabla \times \mathbf{F} = 0 \tag{4.36}$$

From Maxwell's equation

$$\nabla \cdot \mathbf{B} = 0$$

or from equation (4.36) let

$$\mathbf{B} = \nabla \times \mathbf{F}$$

We now define the vector potential

$$\mathbf{B} \triangleq \nabla \times \mathbf{A} \tag{4.37}$$

from which

$$\mathbf{H} = \frac{1}{\mu} \nabla \times \mathbf{A}$$

Taking the curl of both sides

$$\nabla \times \mathbf{H} = \frac{1}{\mu} \nabla \times \nabla \times \mathbf{A} \tag{4.38}$$

A vector identity gives

$$\nabla \times \nabla \times \mathbf{A} = \nabla(\nabla \cdot \mathbf{A}) - \nabla^2 \mathbf{A}$$

Choose

$$\nabla \cdot \mathbf{A} = 0$$

or

$$\nabla \times \mathbf{H} = \frac{1}{\mu}(-\nabla^2 \mathbf{A}) \tag{4.39}$$

Recall Ampere's law (in point form)

$$\nabla \times \mathbf{H} = \mathbf{J}$$

Substituting into equation (4.39) we obtain

$$\nabla^2 \mathbf{A} = -\mu \mathbf{J} \tag{4.40}$$

Equation (4.40) is a vector Poisson's equation. If $\mathbf{J} = 0$ then

$$\nabla^2 \mathbf{A} = 0 \tag{4.41}$$

Equation (4.41) is a vector Laplace's equation. It can be shown that solutions to equation (4.41) can be written

$$\mathbf{A} = \oint \frac{\mu_0 I \, \mathrm{d}\mathbf{l}}{4\pi R} \tag{4.42}$$

or

$$\mathbf{A} = \iint \frac{\mu_0 K \, \mathrm{d}\mathbf{A}}{4\pi R} \tag{4.43}$$

or

$$\mathbf{A} = \iiint \frac{\mu_0 \mathbf{J} \, \mathrm{d}v}{4\pi R} \tag{4.44}$$

D4.15 (*a*) If $\mathbf{A} = 3x^2y\hat{\mathbf{a}}_x + 3xy^2\hat{\mathbf{a}}_y$ find **B**. (*b*) Calculate **B** at (1, 2, 1).
Answer. (*a*) $3(y^2 - x^2)\hat{\mathbf{a}}_z$. (*b*) $9\hat{\mathbf{a}}_z$ T.

This concludes the discussion of magnetics. We have seen that defining a scalar or vector potential is a convenient way of calculating field intensity or flux. If neither is available we can use Ampere's law or the Biot–Savart integral to find field intensity or magnetic flux. We now proceed to gravitational forces.

Problems

1. Calculate the magnetic field intensity and magnetic flux density at a radial distance of $r = 5$ m from a line filament carrying a current of 0.25 A.

2. Two infinitely long, thin conductors each with current $I/2\,A$ in the z-direction are at $y = \pm 2$ m. Find **H** along the z-axis.

3. Repeat problem 2 but reverse the direction of the current in one of the conductors.

4. An infinite conductor of radius $a(a > 1)$ meters has current (uniformly) flowing in the $-z$-direction. The conductor is along the z-axis. A one-inch diameter hole is drilled along the z-axis. Find **H** in the three regions.

5. Calculate the magnetic field intensity and magnetic flux density at a radial distance of $r = 5$ m from a cylindrical conductor carrying a uniformly distributed *surface* current of 0.05 A/m. The diameter of the conductor is 10^{-2} m.

6. Find the current density, $|\mathbf{J}|$, of a cylindrical conductor of radius $a = 1.5$ cm carrying current $I = 0.5$ A in the axial direction.

7. Calculate the current density at $r = 1$ cm of the conductor in problem 6 where the current I flows radially, in the $\hat{\mathbf{a}}_r$ direction, instead of axially, in the $\hat{\mathbf{a}}_z$ direction. The current is distributed from $z = 0$ to $z = 0.15$ m and from $\phi = 0$ to $\phi = \pi$.

8. Find **B** everywhere due to an infinite conductor along the z-axis carrying current I in the $-z$-direction.

9. Calculate the magnetic polarization **M** for the conductor in problem 8 which is surrounded by a magnetic material where $\mu_r = 2.5$.

10. An infinite sheet is in the xz $(y = 0)$ plane carrying surface current 0.03 A/m in the $\hat{\mathbf{a}}_z$ direction. Determine the magnetic field intensity at $y = 10$ m.

11. Determine for the configuration of problem 10, the total flux passing through a rectangle defined by $(0, 1, 0)$ $(0, 1.5, 0)$ $(0, 1, 2)$ and $(0, 1.5, 2)$.

12. Given two media where $\mu_{r_1} = 10.5$ for $z > 0$ and $\mu_{r_2} = 15$ for $z < 0$; if $\mathbf{H}_1 = 2.5\hat{\mathbf{a}}_x - 3.1\hat{\mathbf{a}}_y + \hat{\mathbf{a}}_z$ A/m, $z > 0$, find $\mathbf{B}$ in both regions.

13. Repeat problem 12 if $K = 1.2$ A/m at $z = 0$ the interface between the two regions.

14. Find the force $\mathbf{F}$ between two parallel conductors in the $x = 0$ plane each carrying current $I = 2$ A in the z-direction. The conductors are separated by a distance of 3 meters and are equidistant from the z-axis.

15. Repeat problem 14 if the current flows in the opposite direction in one of the conductors.

16. Inside a conductor $\mathbf{H}$ is known. Find $\mathbf{J}$ (a) $\mathbf{H} = 3xy\hat{\mathbf{a}}_x + 3xy\hat{\mathbf{a}}_y$ in rectangular coordinates, (b) $\mathbf{H} = r\hat{\mathbf{a}}_r + \sin\phi\hat{\mathbf{a}}_\phi$ in cylindrical coordinates, (c) $\mathbf{H} = \cos\theta\hat{\mathbf{a}}_\theta + r\sin\phi\hat{\mathbf{a}}_\phi$ in spherical coordinates.

17. For a coaxial conductor $a = 10^{-3}$ m, $b = 2 \cdot 10^{-3}$ m and $c = 2.5 \cdot 10^{-3}$ m. If $I = 0.25$ A find and plot $\mathbf{H}$ in all regions. See Fig. 4.7.

18. A current $I = 0.15$ A is flowing uniformly in a conductor lying along the z-axis. Find $\mathbf{J}$. The cross-sectional areas are shown in Figure 4.35.

19. Find $\mathbf{H}$ within the conductors shown in problem 18. Leave in integral form.

Figure 4.35. See problems 4.18 and 4.19.

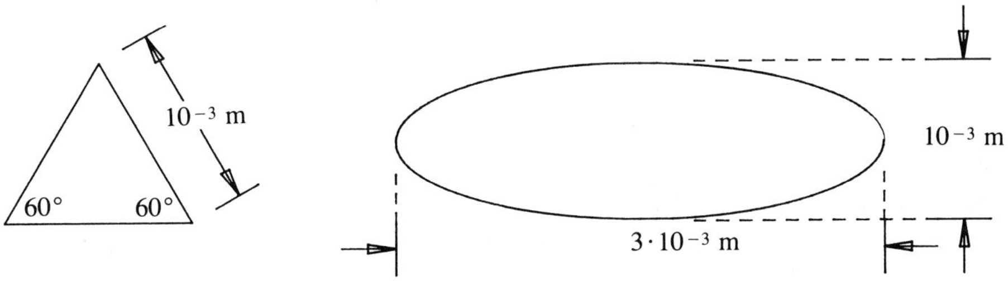

20. $\mathbf{B}_1 = \frac{1}{2}\hat{\mathbf{a}}_x - 2\hat{\mathbf{a}}_y - \frac{3}{2}\hat{\mathbf{a}}_z$ T in region 1 where $\mu_{r_1} = 10$. Find **B** and **H** in both regions where $z = 0$ is the boundary between regions. $\mu_{r_2} = 100$ in region 2.

21. In problem 20 assume a current sheet $K = 3$ A/m is flowing in the $\hat{\mathbf{a}}_y$ direction. Find **B** and **H** in both regions.

22. A sheet of current $K = 2$ A/m is flowing in the $\hat{\mathbf{a}}_y$ direction in the $z = 0$ plane. Find the scalar magnetic potential everywhere. Let $V_m = 0$ at $x = 0$.

23. In problem 22 a second parallel sheet is carrying $K = 2$ A/m in the $-\hat{\mathbf{a}}_y$ direction in the $z = 1$ plane. Find the scalar magnetic potential everywhere. Let $V_m = 0$ at $x = 0$.

24. Write loop equations for the primary and secondary circuits shown in Figure 4.36.

25. Write loop equations for the primary and secondary circuits shown in Figure 4.37.

26. Calculate the number of turns necessary in a toroid (air-filled) which has an average radius of $r_{\text{ave}} = 10^{-2}$ m and $r_{\text{turn}} = 2 \cdot 10^{-4}$ m to provide a uniform flux density $B = 10^{-3}$ T.

27. If we wind the toroid designed in problem 26 around a magnetic material (linear – a bad assumption normally) where $B = 300$ H, find the flux in the toroid.

Figure 4.36. See problem 4.24.

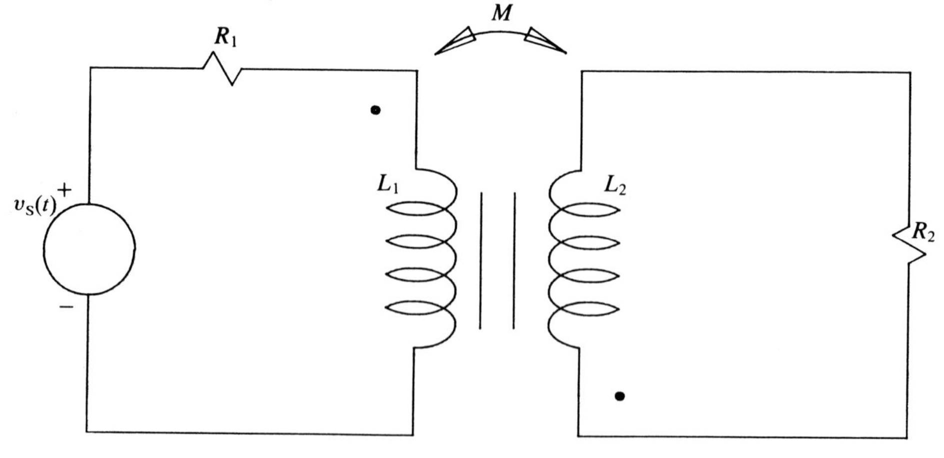

Figure 4.37. See problem 4.25.

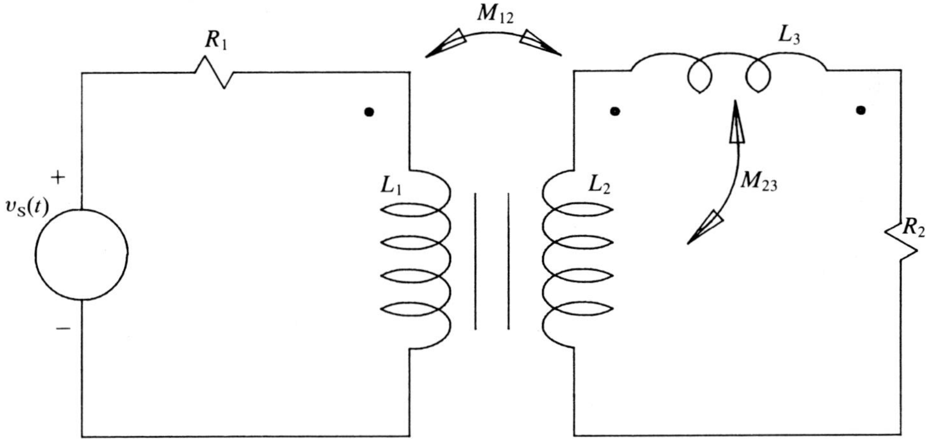

Figure 4.38. See problem 4.31.

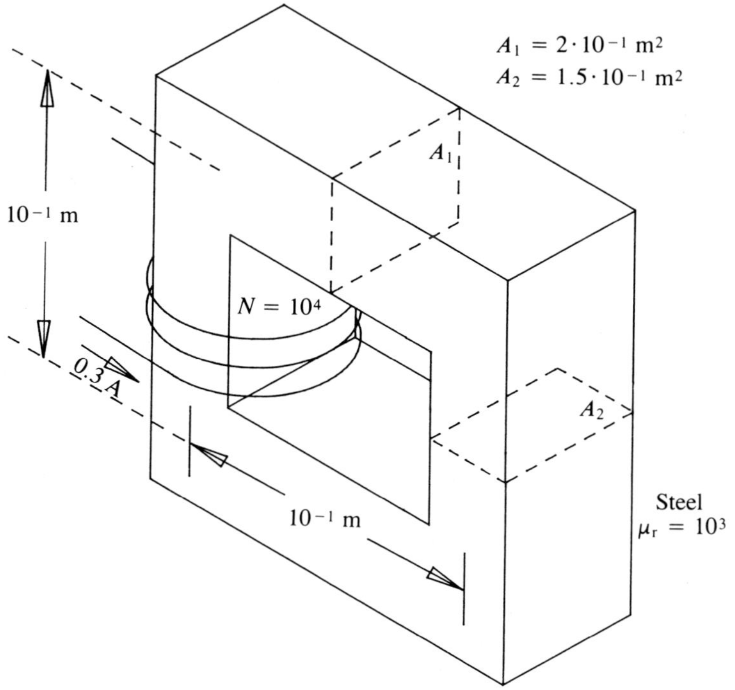

Figure 4.39. See problem 4.32.

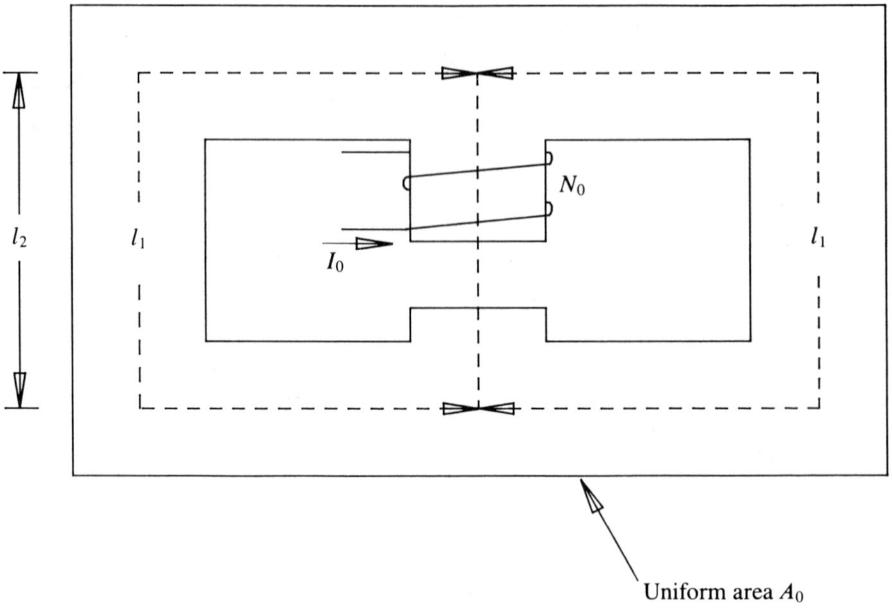

Figure 4.40. See problem 4.34.

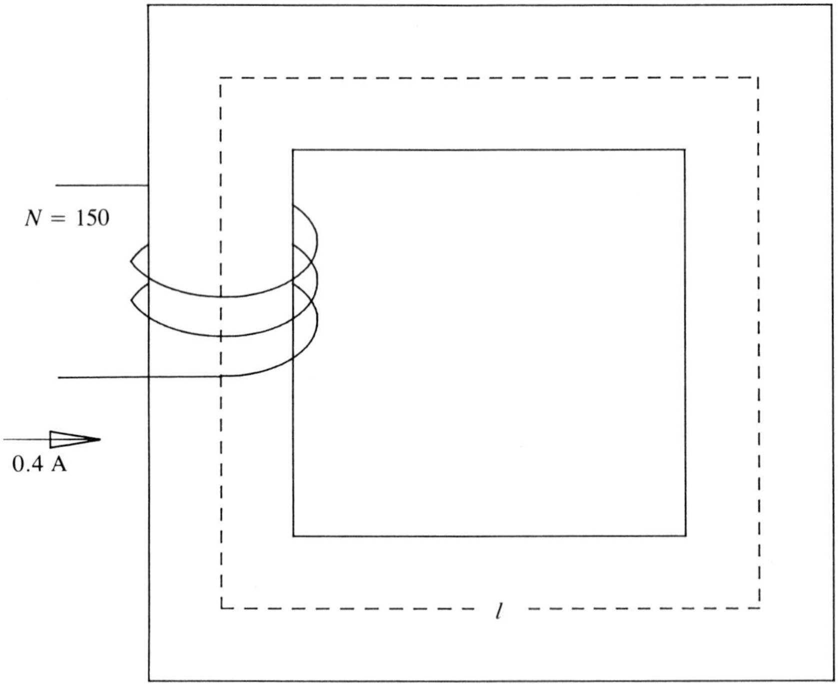

28. If we introduce an air gap $l = 10^{-6}$ m in the toroid of problem 27 what will the flux be in the gap? Assume uniformity.

29. Given a long solenoid with $N = 10$ with $r_{\text{turn}} = 10^{-3}$ m, $I = 0.25$ A in the z-direction. Find $\mathbf{m}$.

30. Find the torque on the solenoid of problem 29 from $(2, 3, 0)$ due to $\mathbf{B} = r\hat{\mathbf{a}}_r$.

31. (a) Calculate the flux within the steel form shown in Figure 4.38, (b) Calculate the flux density in the leg which has an area A_2.

32. Determine the flux in the air gap shown in Figure 4.39.

33. Three long straight parallel wires carry currents of $I/4$ A and are spaced at the corners of an equilateral triangle 1 m on a side. Calculate the magnitude of force on one wire due to the other two.

Figure 4.41. See problems 4.35 and 4.36.

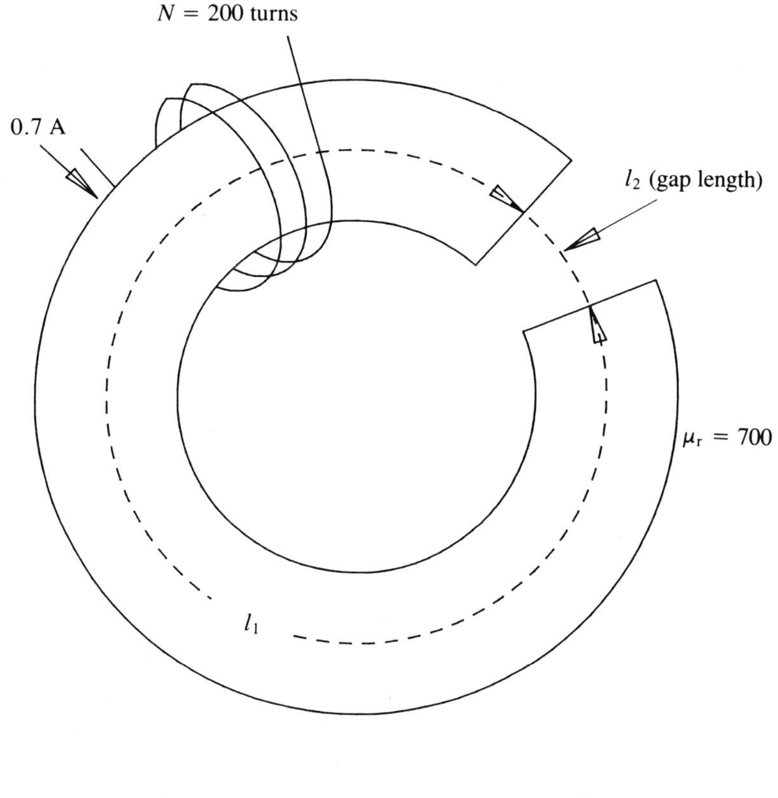

34. Calculate (*a*) the magnetic field intensity, **H**, if $l = 2 \cdot 10^{-1}$ m, (*b*) the magnetic flux density, **B**, if the medium is steel with $\mu_r = 450$ as shown in Figure 4.40.

35. Determine the flux density in the air gap of Figure 4.41 if $l_1 = 10^{-1}$ m and $l_2 = 10^{-3}$ m. Assume the metal is a linear material.

36. Calculate the number of turns required to provide a flux density of 3.5 T in the air gap of the toroid shown in Figure 4.41.

37. Calculate the vector magnetic potential for a line filament with current $I = 1.5$ A at a point $(1, 30°, z)$. The line is along the z-axis.

5

Gravitation

5.1 Introduction

The gravitational field is the third field we study in this text. It is one of the fundamental fields of nature and all matter is subject to it. One of the important equations in gravitational fields is the law of gravitational force deduced by Newton. Newton discovered the law following his study of Kepler's work. This law of gravitational force has the same form as the electrostatic force studied in Chapter 3. Recall that Coulomb discovered an inverse square relationship between two charges. Likewise Newton discovered an inverse square relationship between masses. It is stated below.

$$F \propto \frac{m_1 m_2}{R^2} \tag{5.1}$$

In this chapter we examine gravitational fields, define terms and compare with previous work.

5.2 The law of universal gravitation

Equation (5.1) can be written

$$\mathbf{F} = G\frac{m_1 m_2}{R^2}\hat{\mathbf{a}}_R \tag{5.2}$$

Equation (5.2) is expressed in SI units where $G = 6.67 \cdot 10^{-11}\,\mathrm{Nm^2/kg}$. Equation (5.2) is the law of universal gravitation.

D5.1 If m_1 $(=10^{-4}\,\mathrm{kg})$ is at the origin and m_2 $(=10^{-5}\,\mathrm{kg})$ is at $(10^{-2}, 10^{-2}, 10^{-2})$, calculate the force, $\mathbf{F}$, on m_2 caused by m_1.
Answer. $1.28 \cdot 10^{-16}\,[\hat{\mathbf{a}}_x + \hat{\mathbf{a}}_y + \hat{\mathbf{a}}_z]\,\mathrm{N}$.

We now define a gravitational field intensity by dividing equation (5.2) by m_1

$$\mathbf{g} \triangleq \frac{\mathbf{F}}{m_1} = G\frac{m_2}{R^2}\hat{\mathbf{a}}_R \tag{5.3}$$

The gravitational force acting on the mass m_1 is the gravitational field intensity times m_1. The field intensity $\mathbf{g}$ is analogous to the field intensities in Chapter 3, $\mathbf{E}$, and Chapter 4, $\mathbf{H}$.

D5.2 Calculate the gravitational field intensity at the origin caused by a mass, $m = 10^{-4}$ kg at $(0.5, 0.5, 1)$.
Answer. $3.63 \cdot 10^{-15} \left[\frac{1}{2}\hat{\mathbf{a}}_x + \frac{1}{2}\hat{\mathbf{a}}_y + \hat{\mathbf{a}}_z\right] \mathrm{m/s^2}$.

Equation (5.3) can be used to calculate the mass. Rearrange equation (5.3)

$$m_2 = \frac{gR^2}{G} \tag{5.4}$$

Example 5.1 Let's find the mass of the earth from equation (5.4). We first find g from

$$F = mg \tag{5.5}$$

For a unit mass we measure $g = F = 9.81$ N/kg. We are able to measure the radius of the earth. From equation (5.4)

$$\begin{aligned} m_{\text{earth}} = m_1 &= \frac{gR_{\text{earth}}^2}{G} \\ &= \frac{9.81(6.37 \cdot 10^6)^2}{6.67 \cdot 10^{-11}} \\ &= 5.97 \cdot 10^{24} \text{ kg} \end{aligned}$$

Equation (5.2) is for stationary masses. Since all bodies are in motion we now consider what happens if the masses are in rotation.

Consider a rotating ball as shown in Figure 5.1. If we assume it to have radius a and be rotating at a constant angular velocity ω, then the linear velocity at any point is $u = d\omega$. In particular, the velocity at any point on the surface of the ball is

$$u = a\omega \sin \alpha$$

where α is measured from the axis of rotation.

Now we turn the ball such that the axis of rotation is normal to the page as shown in Figure 5.2(a). Notice that the velocity vector is drawn tangentially. From Figure 5.2(b) we see that there is a component of this velocity in the radial direction pointing inward. Thus, there are actually two forces pointing toward the center – the gravitational force and the centripetal force. A simple freebody diagram in Figure 5.2(c)

shows us that the supportive weight of the ball equals the sum of the gravitational and centripetal forces. That is,

$$\mathbf{F} = \mathbf{F}_g + \mathbf{F}_c$$

or in scalar form

$$F = G\frac{m_1 m_2}{r^2} - \frac{m_2 u^2}{r} \tag{5.6}$$

Figure 5.1. Rotating sphere.

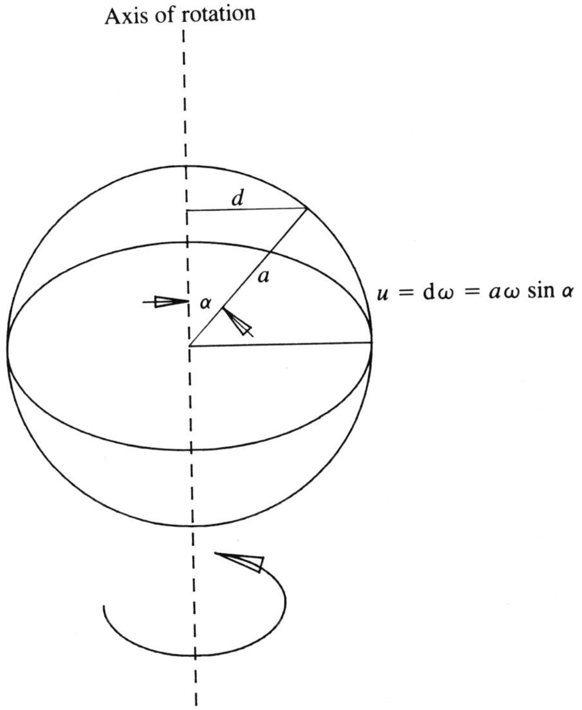

Figure 5.2. A rotating body showing the components of velocity. (*a*) Axis of rotation, (*b*) force due to rotation, (*c*) free body.

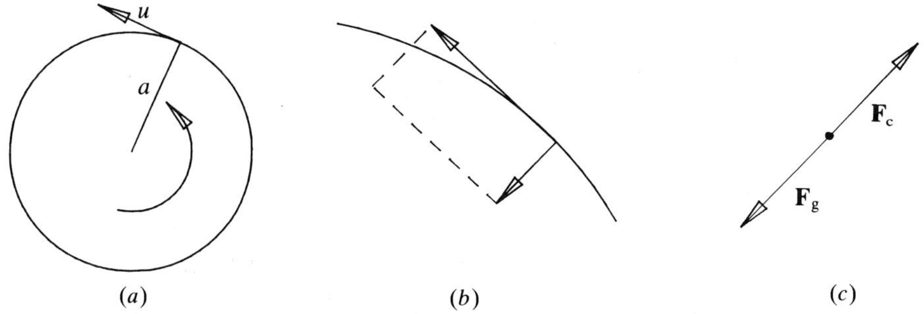

Let's check out the terms in equation (5.6) for a better understanding. The mass of the earth is m_1 and m_2 is the mass of the ball. The radius of the earth is $r = 6.37 \cdot 10^6$ m and u is the velocity of the ball in the direction of rotation due to the rotation of the earth. Note that the velocity is maximum; that is, we're at the equator with the ball.

The second term on the right of equation (5.6) is actually the correction to the gravitational constant due to the rotation of the earth. Plugging in appropriate values, we obtain

$$\frac{u^2}{r} = \frac{(r\omega)^2}{r} = \left(\frac{2\pi}{86400}\right)^2 6.37 \cdot 10^6$$
$$= 0.034 \text{ m/s}^2$$

This is negligible when compared to the gravitational acceleration of 9.8 m/s^2.

D5.3 If the rotating body in Figure 5.3 is the earth and if the other mass is a soccer ball, calculate the force using equation (5.6).
Answer. 24.62 N.

D5.4 Repeat D5.3 neglecting the second term on the right of equation (5.6).
Answer. 24.53 N.

It should be clear that the velocity vector and the acceleration vector are perpendicular. This is shown in Figure 5.3. Also notice the results of varying the velocity. For large velocities the mass m_2 escapes the pull of mass m_1. For small velocities mass m_2 collapses toward m_1. If the orbit is circular the velocity is constant. Let's do an example.

Example 5.2 What is the orbital speed of a mass, m_2, in circular orbit at an altitude of 500 km about the earth, m_1? For a circular orbit

$$F_{\text{acc}} = F_{\text{gravity}}$$

$$\frac{m_2 u^2}{R} = G\frac{m_1 m_2}{R^2}$$

$$u = \sqrt{\frac{Gm_1}{R}} \tag{5.7}$$

$$= 7.926 \cdot 10^3 \text{ m/s}$$

where $R = r_{\text{earth}} + \text{altitude} = 6.87 \cdot 10^6$ m and $m_1 = 6 \cdot 10^{24}$ kg.

D5.5 What is the orbital speed of a satellite about the earth at an altitude of $5 \cdot 10^6$ m.
Answer. $5.92 \cdot 10^4$ m/s.

D5.6 Calculate the period of the satellite's orbit of D5.5.
Answer. 3 hrs, 21 mins 12 s.

5.3 Flux density

We define flux density vector, as the flux divided by the area through which the flux is passing, or

$$|D_g| = \frac{\text{flux}}{\text{area}}$$

Recall Gauss' law

$$Q = \oint \mathbf{D} \cdot \mathrm{d}\mathbf{A}$$

Figure 5.3. A mass orbiting a large mass.

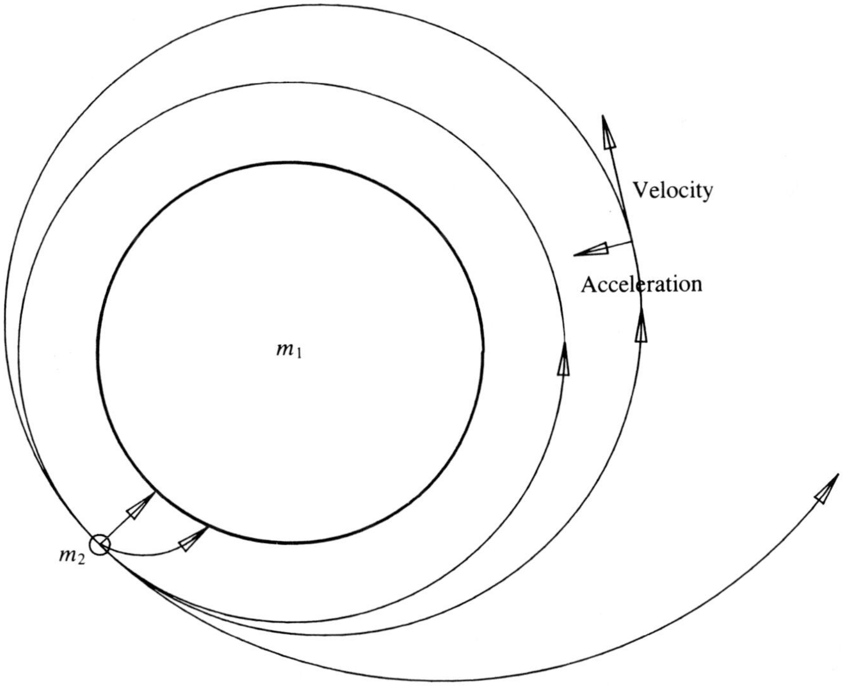

For Gauss' law in gravitational fields we write

$$\oint \mathbf{D}_g \cdot d\mathbf{A} = m \equiv \Phi \tag{5.8}$$

$$\iint D_g \hat{\mathbf{a}}_r \cdot r^2 \sin\theta \, d\theta \, d\phi \hat{\mathbf{a}}_r = \Phi \equiv m$$

where m is a spherical mass in kilograms.

Since m is a spherical mass, D_g is independent of the integral and we get

$$\mathbf{D}_g = \frac{m}{4\pi r^2} \hat{\mathbf{a}}_r \tag{5.9}$$

Compare equation (5.9) with the electric flux density for a spherical source in Chapter 3.

$$\mathbf{D} = \frac{Q}{4\pi r^2} \hat{\mathbf{a}}_r$$

Substituting equation (5.9) into equation (5.5), we obtain

$$\mathbf{F} = 4\pi G \mathbf{D}_g m \tag{5.10}$$

If we define a gravitational permittivity k where

$$k \triangleq \frac{1}{4\pi G} \tag{5.11}$$

we obtain

$$\mathbf{F} = \frac{\mathbf{D}_g m}{k} \tag{5.12}$$

We now can write

$$\mathbf{D}_g \triangleq k\mathbf{g} \tag{5.13}$$

Compare equation (5.13) with $\mathbf{D} = \epsilon \mathbf{E}$ in electrostatics and $\mathbf{B} = \mu \mathbf{H}$ in magnetostatics.

Force and field intensity are related by

$$\mathbf{F} = m\mathbf{g} \tag{5.14}$$

D5.7 Calculate the gravitational flux density at the point $(1, 2, 3)$ for a spherical mass $m = 0.2$ kg at $(0, 0, 0)$.
Answer. $1.14 \cdot 10^{-3} \hat{\mathbf{a}}_r$ kg/m^2.

Table 5.1 lists some of the analogous quantities we have covered thus far. Now we proceed to some more examples.

Example 5.3 Find the gravitational flux density as a function of radius in all regions for a sphere of radius a with uniform density ρ kg/m^3 as shown in Figure 5.4, where $\rho = m/(\frac{4}{3}\pi a^3)$.

Since we have a symmetric problem, we use Gauss' law again.

$r > a$

$$\oint \mathbf{D}_g \cdot d\mathbf{A} = \Phi = \iiint \rho \, dv = m$$

$$\iint D_g \hat{\mathbf{a}}_r \cdot r^2 \sin\theta \, d\theta \, d\phi \hat{\mathbf{a}}_r = \iiint \rho r^2 \sin\theta \, dr \, d\theta \, d\phi$$

$$D_g = \frac{\rho a^3}{3r^2}$$

$$\mathbf{D}_g = \frac{\rho a^3}{3r^2} \hat{\mathbf{a}}_r \tag{5.15}$$

Since the density is uniform

$$\rho = \frac{\Phi}{\frac{4}{3}\pi a^3} \tag{5.16}$$

we substitute equation (5.16) into equation (5.15) and obtain

$$\mathbf{D}_g = \frac{\Phi}{4\pi r^2} \hat{\mathbf{a}}_r \qquad r > a \tag{5.17}$$

This is the same density as we obtained for the point mass above in equation (5.9). In other words, for $r > a$ whether the spherical volume

Table 5.1. *Analogous quantities*

	Electricity	Magnetics	Gravity
Density	$\mathbf{D}$	$\mathbf{B}$	$\mathbf{D}_g$
Field intensity	$\mathbf{E}$	$\mathbf{H}$	$\mathbf{g}$
Constant	ϵ	μ	k
Point force	$\frac{Q_1 Q_2}{4\pi\epsilon R^2}$	none	$\frac{m_1 m_2}{4\pi k R^2}$

is small or large makes no difference. In fact, the density is identical to that of a point mass.

$r < a$ We now calculate the density for the Gaussian surface within the volume shown in Figure 5.4.

$$\iint \mathbf{D}_{\text{g}} \cdot \mathrm{d}\mathbf{A} = \Phi$$

where Φ = mass enclosed by Gaussian surface

$$\int_0^{2\pi} \int_0^{\pi} D_{\text{g}} \hat{\mathbf{a}}_r \cdot r^2 \sin\theta \, \mathrm{d}\theta \, \mathrm{d}\phi \hat{\mathbf{a}}_r = \int_0^{2\pi} \int_0^{\pi} \int_0^{r} \rho r'^2 \sin\theta \, \mathrm{d}r' \, \mathrm{d}\theta \, \mathrm{d}\phi$$

and

$$\mathbf{D}_{\text{g}} = \frac{\rho r}{3} \hat{\mathbf{a}}_r \qquad r < a \tag{5.18}$$

D5.8 Calculate the gravitational flux density of the earth at an altitude of 1000 m.

Answer. $1.17 \cdot 10^{10} \hat{\mathbf{a}}_r$ kg/m^2.

Figure 5.4. A uniform density sphere of radius *a*.

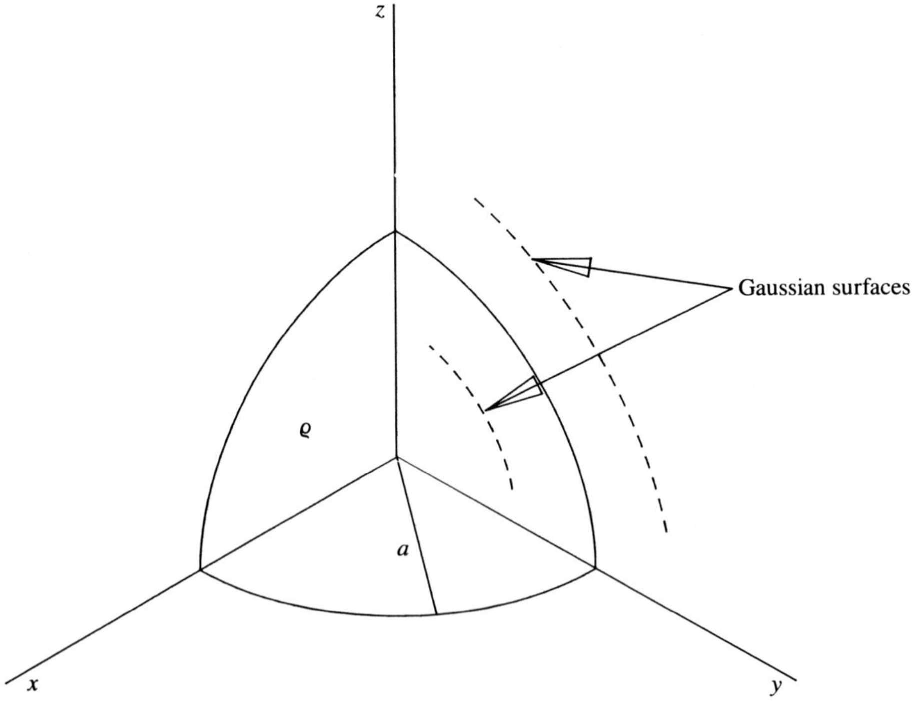

Returning to equation (5.8) we write

$$\oint \mathbf{D}_g \cdot d\mathbf{A} = m = \iiint \rho \, dv \tag{5.19}$$

Now use the divergence theorem on the left-hand side which gives

$$\oint \mathbf{D}_g \cdot d\mathbf{A} = \iiint \nabla \cdot \mathbf{D}_g \, dv \tag{5.20}$$

Combining equations (5.19) and (5.20) we obtain

$$\nabla \cdot \mathbf{D}_g = \rho \tag{5.21}$$

corresponding to Maxwell's first equation. Equation (5.21) can also be written

$$\nabla \cdot k\mathbf{g} = \rho$$

and

$$\nabla \cdot \mathbf{g} = \frac{\rho}{k} \tag{5.22}$$

5.4 Gravitational potential

We now define one more quantity – potential. Just as we defined electrical and magnetic potential, we define gravitational potential, V_g, as

$$V_g = V_B - V_A \triangleq -\int_A^B \mathbf{g} \cdot d\mathbf{l} \tag{5.23}$$

and

$$\mathbf{g} = -\nabla V_g \tag{5.24}$$

For conservative fields

$$\oint \mathbf{g} \cdot d\mathbf{l} = 0$$

Substituting equation (5.24) into (5.22) we obtain

$$\nabla \cdot (-\nabla V_g) = \frac{\rho}{k}$$

or

$$\nabla^2 V_g = -\frac{\rho}{k} \tag{5.25}$$

Equation (5.25) is Poisson's equation. If there is no mass in the region, equation (5.25) becomes

$$\nabla^2 V_g = 0$$

which is Laplace's equation. Using equation (5.23), let us calculate the potential for example 5.3.

For $r > a$

$$V_g = -\int \mathbf{g} \cdot d\mathbf{l}$$

and

$$\mathbf{D}_g = \frac{\Phi}{4\pi r^2}\,\hat{\mathbf{a}}_r$$

$$\mathbf{g} = \frac{\Phi}{4\pi k r^2}\,\hat{\mathbf{a}}_r$$

which gives

$$V_g = -\int \frac{\Phi}{4\pi k r^2}\,\hat{\mathbf{a}}_r \cdot dr\hat{\mathbf{a}}_r$$

$$= \frac{\Phi}{4\pi k r} \qquad r > a \tag{5.26}$$

For $r < a$

$$\mathbf{D}_g = \frac{\rho r}{3}\,\hat{\mathbf{a}}_r$$

and

$$\mathbf{g} = \frac{\rho r}{3k}\,\hat{\mathbf{a}}_r$$

$$V_g = -\int \frac{\rho r}{3k}\,\hat{\mathbf{a}}_r \cdot dr\hat{\mathbf{a}}_r$$

$$= -\frac{\rho r^2}{6k} \qquad r < a \tag{5.27}$$

D5.9 Calculate the potential due to a sphere of radius 10^{-1} m and uniform density $\rho = 3\pi/4\ \mathrm{kg/m^3}$ at the points $r = 10^{-2}$ m and $r = 0.15$ m. Let $V_g = 0$ at $r \to \infty$.
Answer. $3.3 \cdot 10^{-14}$, $4.39 \cdot 10^{-12}\ \mathrm{m^2/s^2}$.

Example 5.4 Find the gravitational field intensity due to an infinite cylindrical volume of radius a and uniform density ρ. Picture a pencil-shaped satellite whose radius is much less than its length.

We choose the cylindrical coordinate system and place the source (cylinder) along the z-axis as shown in Figure 5.5. Since our source is 'long', the flux is radial only or

$$\mathbf{D}_g = D_g \hat{\mathbf{a}}_r$$

Using Gauss' law

$$\oint D_g \hat{\mathbf{a}}_r \cdot r\, d\phi\, dz \hat{\mathbf{a}}_r = \int_0^L \int_0^{2\pi} \int_0^a \frac{m}{\pi a^2 L} r\, dr\, d\phi\, dz$$

$$D_g = \frac{m}{2\pi r L}$$

In vector form

$$\mathbf{D}_g = \frac{m}{2\pi r} \hat{\mathbf{a}}_r \qquad r > a$$

where $L = 1$ m and the gravitational field intensity is

$$\mathbf{g} = \frac{m}{2\pi k r} \hat{\mathbf{a}}_r \qquad r > a$$

Figure 5.5. A long pencil shaped satellite with radius a, (a) side view, (b) top view.

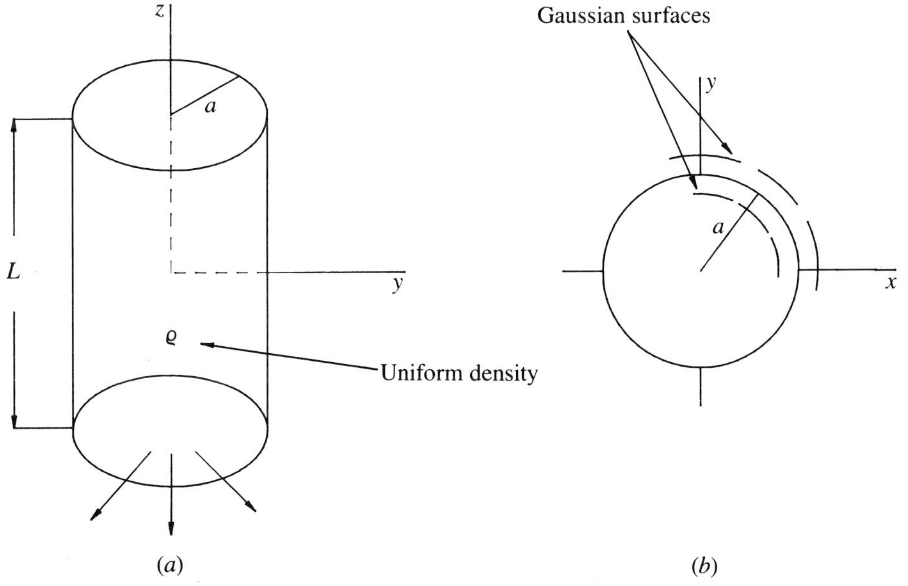

In addition

$$V_g = -\int \mathbf{g} \cdot d\mathbf{l}$$

$$= -\int \frac{m}{2\pi k r} dr$$

$$= \frac{m}{2\pi k} \ln \frac{r}{r_{ref}} \qquad r > a$$

$r < a$ Now we repeat the calculations for the region inside the cylinder, $r < a$, as shown in Figure 5.5(b).

$$\int_0^1 \int_0^{2\pi} D_g r \, d\phi \, dz = \int_0^1 \int_0^{2\pi} \int_0^r \frac{m}{\pi a^2(1)} r' \, dr' \, d\phi \, dz$$

$$D_g = \frac{mr}{2\pi a^2(1)}$$

or

$$\mathbf{D}_g = \frac{mr}{2\pi a^2} \hat{\mathbf{a}}_r \qquad r < a$$

and the gravitational field intensity is

$$\mathbf{g} = \frac{mr}{2\pi k a^2} \hat{\mathbf{a}}_r \qquad r < a$$

In addition,

$$V_g = -\int \mathbf{g} \cdot d\mathbf{l}$$

$$= -\int \frac{mr}{2\pi a^2 k} dr$$

$$= -\frac{mr^2}{4\pi a^2 k} \qquad r < a$$

D5.10 Given a small cylindrical satellite with radius $r = 10^{-1}$ m and length $L = 10$ m, if the density is $\rho = 10/\pi$ kg/m^3 calculate the gravitational potential at $r = 0.01$ m and $r = 1$ m. Choose a convenient reference. *Answer.* $3.07 \cdot 10^{-11}$, $6.67 \cdot 10^{-14}$ m^2/s^2. $r_{ref} = 0.1$ m.

We are now able to compare our results with some of those in electrostatics in Chapter 3.

From equation (3.23)

$$\mathbf{D} = \frac{Q}{4\pi r^2}\hat{\mathbf{a}}_r \qquad r > a$$

and from equation (5.9)

$$\mathbf{D}_g = \frac{m}{4\pi r^2}\hat{\mathbf{a}}_r \qquad r > a$$

for a uniformly 'charged' sphere.

From equation (3.26)

$$V = \frac{Q}{4\pi\epsilon_0 r} \qquad r > a$$

and from equation (5.26)

$$V_g = \frac{m}{4\pi k r} \qquad r > a$$

Figure 5.6. Cross section of a long pencil satellite with density ρ.

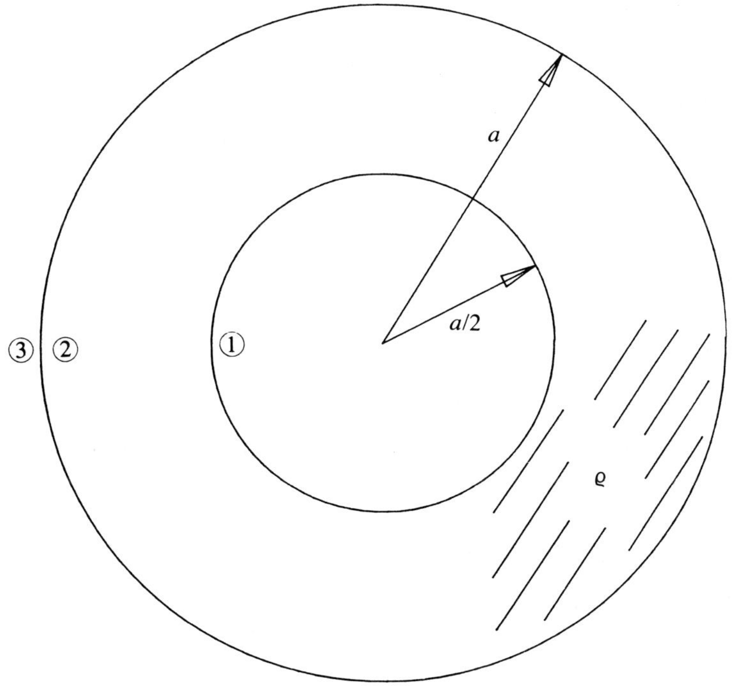

D5.11 Figure 5.6 is the cross section of a long satellite of radius a. The region $0<r<a/2$ is a vacuum. The region $a/2<r<a$ has a uniform density ρ. Calculate $\bar{\mathbf{D}}_g$ everywhere.

Answer. $0<r<\frac{a}{2}$; $\rho\left[\frac{r}{2}-\frac{a^2}{8r}\right]\hat{\mathbf{a}}_r$ $\quad \frac{a}{2}<r<a$; $\frac{3\rho a^2}{r}\hat{\mathbf{a}}_r$ $\quad r>a$.

This concludes the chapter on gravitational fields. We have seen many similarities with other fields – in particular electric fields. We now move on to fluid fields.

Problems

1. Two electrons are suspended on a small – mightly small – string with a knot at the bottom. Find the separation d.

2. A 2 kg soccer ball is tied to a 3 m rope. If one end of the rope is held while the sphere rotates at $\omega = 3$ rad/s, what is the force on the rope?

3. If the ball of problem 2 is released and goes into orbit around the earth, what is the orbital velocity for the sphere to 'fly' at 100 m?

4. What must the velocity of the soccer ball in problem 2 be to escape the earth's gravity?

5. What is the force on the soccer ball of problem 4 when it is half way between the earth and the moon? Assume uniform densities. Assume the density of the earth to be $5.5 \cdot 10^3$ kg/m^3 and the density of the moon to be $3.3 \cdot 10^3$ kg/m^3.

6

Fluid flow

6.1 Introduction

Another engineering field in which we use familiar mathematics is the field of fluid flow.

What is a fluid? A *fluid* is a material which has no resistance to a shear force. As an illustration, if we were to stick a fork into a piece of wood and twist, there would be resistance. On the other hand, if we were to stick the same fork into a jar of water, there would be no resistance. Thus we consider the wood a solid and the water a fluid. Further, two major divisions of fluids are laminar and turbulent. Laminar flow is fluid which can be described by streamlines and flow tubes. The fluid is pictured as flowing in layers. When the velocity of the fluid exceeds a certain maximum value, the flow becomes turbulent. In this text we confine ourselves to laminar flow. In addition, we can classify laminar flow into two types: rotational and irrotational. Mathematically, we define these two types with the closed contour integral

$$\oint \mathbf{N} \cdot \mathbf{dl}$$

In electrostatics this was written

$$\oint \mathbf{E} \cdot \mathbf{dl} = 0$$

In magnetostatics this was written

$$\oint \mathbf{H} \cdot \mathbf{dl} = I$$

In gravitation this was written

$$\oint \mathbf{g} \cdot \mathbf{dl} = 0$$

In fluids this is written

$$\oint \mathbf{u} \cdot d\mathbf{l} = k \tag{6.1}$$

where **u** is the velocity and k some constant which has the units m^2/s.

For irrotational flow $k = 0$; for rotational flow $k \neq 0$ and we say there is circulation. If we assume the fluid is incompressible, then irrotational fluids are governed by Laplace's equation

$$\nabla^2 V_f = 0 \tag{6.2}$$

where V_f is the fluid potential in m^2/s.

First consider static fluids, which are assumed incompressible and irrotational.

6.2 Plane source

Figure 6.1 is a picture of incompressible, uniform fluid flow. No fluid crosses the imaginary streamlines. The fluid flows in the x-direction. Also shown is one side of a Gaussian surface (closed surface). Repeating Gauss' law here

$$\Phi = \oint \mathbf{D} \cdot d\mathbf{A}$$

For ideal fluid flow we write in general

$$\Phi = \oint \mathbf{u} \cdot d\mathbf{A} \tag{6.3}$$

Thus, we see that Φ is a volume rate and has units of m^3/s in ideal fluid flow. For the *closed* surface depicted in Figure 6.1, (and no sources or

Figure 6.1. Uniform flow in the x-direction.

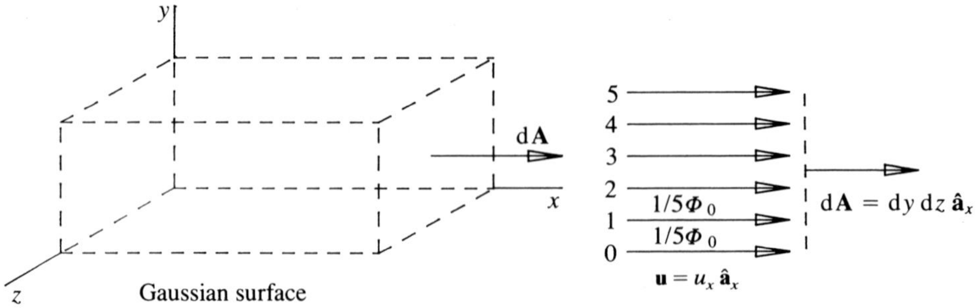

sinks within the surface), we write, for incompressible fluids,

$$\oint \mathbf{u} \cdot \mathrm{d}\mathbf{A} = 0$$

Notice that this is analogous to the magnetostatic situation,

$$\oint \mathbf{B} \cdot \mathrm{d}\mathbf{A} = 0$$

where we noted magnetic lines do not diverge. Similarly, in magnetics we wrote

$$\iint \mathbf{B} \cdot \mathrm{d}\mathbf{A} = \Phi$$

for an open surface integration. In fluids we write

$$\iint \mathbf{u} \cdot \mathrm{d}\mathbf{A} = \Phi \tag{6.4}$$

Since in Figure 6.1

$$\mathbf{u} = u_x \hat{\mathbf{a}}_x$$

and

$$\mathrm{d}\mathbf{A} = \mathrm{d}y\, \mathrm{d}z \hat{\mathbf{a}}_x$$

equation (6.4) becomes

$$\int_0^1 \int_0^y u_x \hat{\mathbf{a}}_x \cdot \mathrm{d}y\, \mathrm{d}z \hat{\mathbf{a}}_x = \Phi$$

and

$$u_x = \frac{\Phi}{y(1)} \tag{6.5}$$

Remember Φ is a volume of fluid per second or a volume rate of flow. Dividing by '$y(1)$' we obtain a velocity. That is, standing at some point on the x-axis in Figure 6.1 we observe a certain volume of fluid passing every second. We simplify somewhat by considering our volume to have a depth of one meter in the z-direction. What can we say about the y-direction? Equation (6.5) states that for every y value we define a streamline. Between these streamlines the same amount of fluid flows. Fluid will not cross the streamlines.

For example, in Figure 6.1 there are six streamlines and five flux tubes. We arbitrarily label each stream line starting from the bottom: $y = 0, 1, 2, \cdots, 5$. So each flux tube represents $\frac{1}{5}\Phi_0$.

Having said this about flux, can we imagine a potential – as we have in other fields? Let's take a glimpse ahead. From equation (6.1)

$$\oint \mathbf{u} \cdot \mathrm{d}\mathbf{l} = k \tag{6.1}$$

For an open integral

$$V_{\mathrm{f}} = -\int \mathbf{u} \cdot \mathrm{d}\mathbf{l} \tag{6.6}$$

Specifically in Figure 6.1

$$\begin{aligned} V_{\mathrm{f}} &= -\int u_x \, \mathrm{d}x \\ V_{\mathrm{f}} &= -u_x x + \text{constant} \end{aligned} \tag{6.7}$$

That is, the 'equipotentials' are orthogonal to the streamlines. Does that sound familiar? More about potential later.

The plane source in electrostatics was an infinite plane with uniform charges distributed uniformly. The charges were the source of flux lines. In fluids the charge is the volume rate of fluid. Therefore the flux is the result of this volume rate. For an infinite plane source the fluid behaves like a wall advancing uniformly. This is shown in Figure 6.1. The flow is uniform in the x-direction. The wall extends from $z = -\infty$ to $z = \infty$ and from $y = -\infty$ to $y = \infty$. In other words we will consider uniform flow as the result of a plane (wall) source.

D6.1 Given the velocity of a fluid is $\mathbf{u} = 2(\hat{\mathbf{a}}_x + \hat{\mathbf{a}}_y)$ m/s, sketch several streamlines in the $z = 0$ plane. Label the flux tubes.

D6.2 Find the potential function for D6.1 and sketch several equipotentials. *Answer.* $-2x - 2y$ m^2/s.

6.3 Point source

Next consider a point source. For compressible fluids we use equation (6.3), Gauss' law for fluids, and write

$$\Phi = \oint \mathbf{u} \cdot \mathrm{d}\mathbf{A} \tag{6.8}$$

or

$$\mathbf{u} = \frac{\Phi}{4\pi r^2}\hat{\mathbf{a}}_r \tag{6.9}$$

D6.3 The volume rate of fluid from a point source is $0.25\,\text{m}^3/\text{s}$. Calculate the velocity at $r = 0.65\,\text{m}$.
Answer. $4.71 \cdot 10^{-2}\,\text{m/s}$.

From equation (6.6) we can determine the fluid potential

$$V_\text{f} = -\int \mathbf{u} \cdot \text{d}\mathbf{l} \tag{6.6}$$

Combining equation (6.6) and equation (6.9)

$$V_\text{f} = -\int \frac{\Phi}{4\pi r^2}\,\text{d}r$$

$$= \frac{\Phi}{4\pi r} \tag{6.10}$$

where we assume $V_\text{f} = 0$ at $r = \infty$.

D6.4 The volume rate of fluid from a point source is $0.3\,\text{m}^3/\text{s}$. Calculate the fluid potential at $r = 0.2\,\text{m}$.
Answer. $0.119\,\text{m}^2/\text{s}$.

Equations (6.9) and (6.10) should look familiar to us after the first several chapters. From Chapters 3 and 5 the field intensity for a point source is inversely proportional to r^2. Similarly the potential is inversely proportional to r. Let us proceed to a line source.

6.4 Line source

Consider an infinite line source from which the fluid is flowing radially as shown in Figure 6.2. Again, using Gauss' law, we write

$$\Phi = \oint \mathbf{u} \cdot \text{d}\mathbf{A} \tag{6.3}$$

$$= \int_0^1 \int_0^{2\pi} u_r\hat{\mathbf{a}}_r \cdot r\,\text{d}\phi\,\text{d}z\hat{\mathbf{a}}_r$$

That is,

$$u_r = \frac{\Phi}{2\pi r(1)}$$

and

$$\mathbf{u} = \frac{\Phi}{2\pi r}\,\hat{\mathbf{a}}_r \tag{6.11}$$

D6.5 The volume rate of fluid from a long, thin cylindrical source is $0.15\ \mathrm{m^3/s}$. Calculate the velocity at the point $(0.2, 10°, 5)$.
Answer. $0.119\ \mathrm{m/s}$.

Again we determine the fluid potential using equation (6.6)

$$V_{\mathrm{f}} = -\int \mathbf{u}\cdot \mathrm{d}\mathbf{l}$$

$$= -\int \frac{\Phi}{2\pi r}\,\hat{\mathbf{a}}_r \cdot \mathrm{d}r\hat{\mathbf{a}}_r$$

$$= \frac{\Phi}{2\pi}\ln\frac{r}{r_{\mathrm{ref}}} \tag{6.12}$$

Figure 6.2. A long cylindrical source along the z-axis.

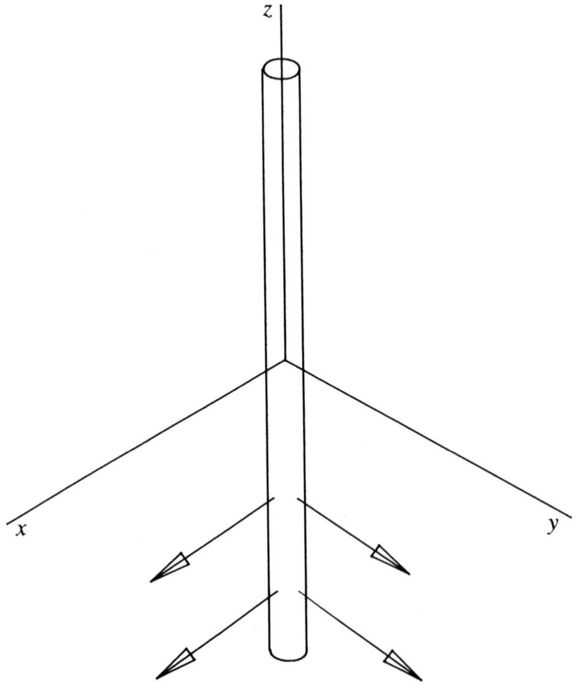

Equation (6.12) requires us to choose a reference potential in any particular problem.

We see that the three situations – uniform, point, and line sources – have forms which are similar to those same configurations in Chapters 3 and 5. The difference is that the 'charge', Φ, has the units of m^3/s. After this familiar introduction, let us deal with fluids in more detail. First let us examine the static situation.

6.5 Flow combinations

One of the instructive properties of the flow function is the superposition property. We can model more complicated flows by combining several different functions. If each function satisfies Laplace's equation then the sum will also satisfy Laplace's equation.

1. Line source/line sink – radial flow

The first combination is the line source and line sink with radial flow. The velocity function for the line source was derived above.

$$\mathbf{u} = \frac{\Phi}{2\pi r}\hat{\mathbf{a}}_r \tag{6.11}$$

The flow function is obtained from

$$\begin{aligned}\Phi &= \iint \mathbf{u} \cdot \mathrm{d}\mathbf{A} \\ &= \int_0^1 \int_0^\phi \frac{\Phi_0}{2\pi r}\hat{\mathbf{a}}_r \cdot r\,\mathrm{d}\phi\,\mathrm{d}z\hat{\mathbf{a}}_r\end{aligned}$$

$$\Phi = \frac{+\Phi_0\phi}{2\pi} \tag{6.13}$$

For a line sink $+\Phi_0$ is replaced by $-\Phi_0$. For a line sink

$$\Phi = -\frac{\Phi_0\phi}{2\pi} \tag{6.14}$$

The geometry for a line source and line sink of values Φ_0 and $-\Phi_0$ respectively, is shown in Figure 6.3. Summing the two flux functions, equations (6.13) and (6.14), we obtain

$$\begin{aligned}\Phi &= \Phi_{\text{source}} + \Phi_{\text{sink}} \\ &= \frac{\Phi_0}{2\pi}(\phi_1 - \phi_2) \\ &= -\frac{\Phi_0\alpha}{2\pi}\end{aligned} \tag{6.15}$$

Figure 6.3. A source–sink combination.

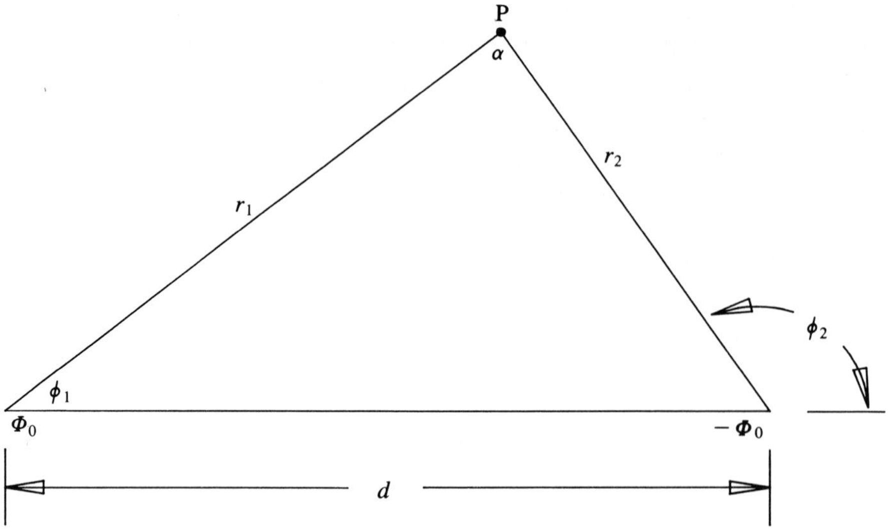

Figure 6.4. A source–sink combination observed from a distant point *P*.

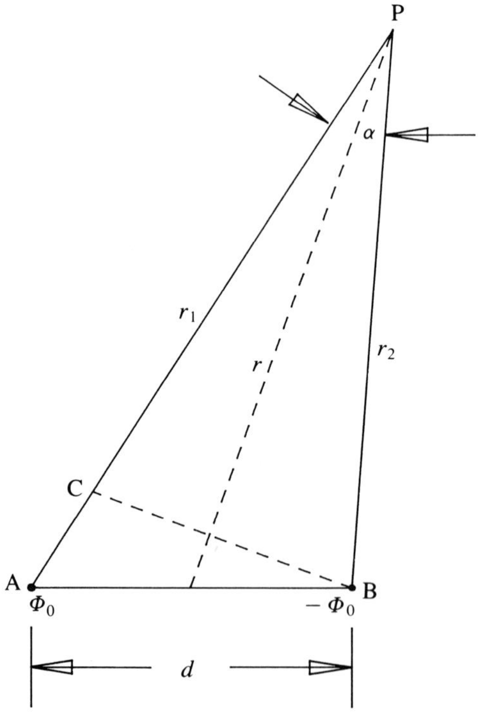

Figure 6.4 shows the source–sink combination as viewed from a distant point P. As r increases (or d decreases) the three lines r, r_1 and r_2 become nearly parallel. As $r \to \infty$ there is cancellation of the two 'charges' – the source and sink.

Figure 6.5 shows the flux function for two line functions; Figure 6.5(*a*) is a plot for opposite polarities, while Figure 6.5(*b*) is a plot for the same polarities. As one would expect opposites attract. The flow from the source terminates at the sink as shown in Figure 6.5(*a*). Conversely likes repel. Whether or not both likes are sources or sinks the flow lines oppose each other as shown in Figure 6.5(*b*). If both are sources the flow lines terminate at infinity. If both are sinks, the flow lines begin at infinity and terminate at the sources.

D6.6 Sketch the flow lines for the combination of a double line source, $2\Phi_0$, and a line sink, $-\Phi_0$, separated by a distance, d.

---

2. Doublet

From our definition of a doublet in electrostatics

$$\Phi_0 d = \text{constant}$$

From Figure 6.4

$$BC = d \sin \phi_1 \approx r\alpha$$

or

$$\alpha = \frac{d \sin \phi_1}{r}$$

Substitute into equation (6.15)

$$\Phi = -\frac{\Phi_0 d \sin \phi_1}{2\pi r} \tag{6.16}$$

Figure 6.5. Flux functions for two parallel, infinite line 'charges' of (*a*) opposite and (*b*) same polarities.

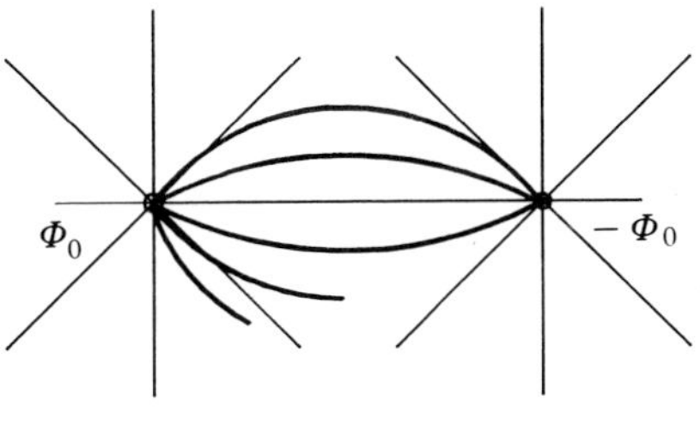

(*a*)

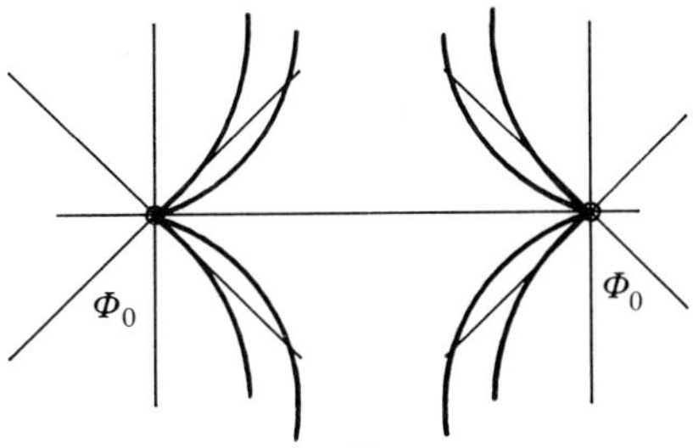

(*b*)

From Figure 6.6, we write

$$\Phi = -\frac{\Phi_0 d \dfrac{y}{\sqrt{x^2+y^2}}}{2\pi\sqrt{x^2+y^2}}$$

$$= \frac{-\Phi_0 d\, y}{2\pi(x^2+y^2)}$$

$$2\pi\Phi(x^2+y^2) + \Phi_0 d\, y = 0$$

$$x^2 + y^2 + \frac{\Phi_0 d\, y}{2\pi\Phi} + \left(\frac{\Phi_0 d}{4\pi\Phi}\right)^2 = \left(\frac{\Phi_0 d}{4\pi\Phi}\right)^2$$

$$x^2 + \left(y + \frac{1}{b}\right)^2 = \left(\frac{1}{b}\right)^2 \tag{6.17}$$

where

$$b = \frac{4\pi\Phi}{d\Phi_0}$$

Equation (6.17) is the equation of a circle with the origin at $(0, -1/b)$ and a radius of $1/b$ for $b > 0$. The flow for the doublet combination is shown in Figure 6.7.

Figure 6.6. A doublet observed at point P in rectangular and polar coordinates.

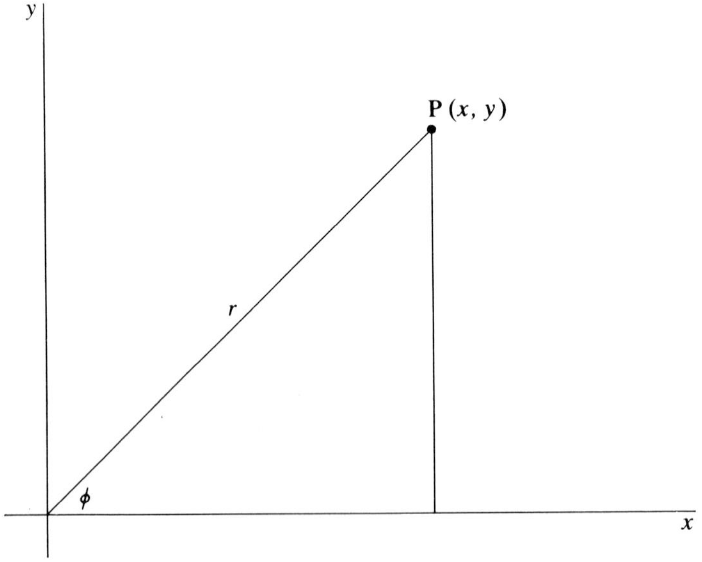

3. The line source and uniform field

The third useful combination is the line source (or sink) and the uniform field. (The streamlines are the same for a line source and uniform and a line sink and uniform.) This combination can model the flow around a half body in an infinite fluid. Rewriting equation (6.5) for a uniform field we have

$$\Phi = u_x y \equiv \Phi_{\text{uniform}}$$

or

$$\Phi_{\text{uniform}} = u_x r \sin \phi$$

Also from equation (6.13) we have

$$\Phi_{\text{source}} = \frac{\Phi_0 \phi}{2\pi}$$

Thus

$$\begin{aligned}\Phi &= \Phi_{\text{uniform}} + \Phi_{\text{source}} \\ &= u_x r \sin \phi + \frac{\Phi_0 \phi}{2\pi} \qquad (6.18)\end{aligned}$$

Equation (6.18) is plotted in Figure 6.8. Note that for $\phi = 0$, $\Phi = 0$ and

Figure 6.7. Flow streamlines for the doublet combination along the x-axis.

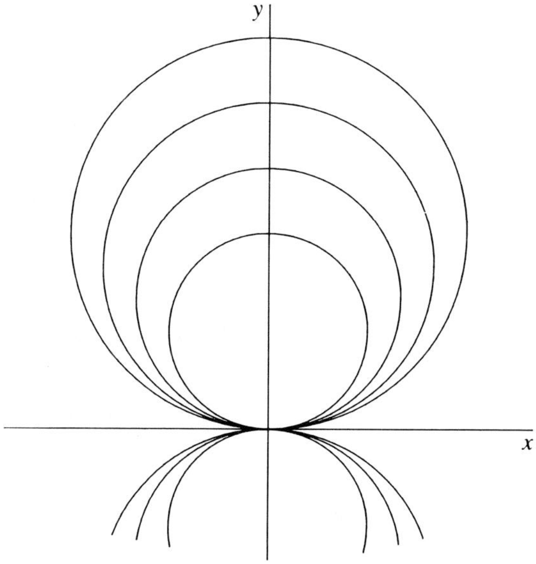

$\phi = \pi$, $\Phi = \Phi_0/2$. As will be seen below, the velocity components of the resulting flow field are obtainable from $u_r = (1/r)\, \partial\Phi/\partial\phi$ and $u_\phi = -\partial\Phi/\partial r$.

$$\frac{\partial\Phi}{\partial\phi} = u_x r \cos\phi + \frac{\Phi_0}{2\pi}$$

For $u_r = 0$ and $\phi = \pi$ we obtain

$$r = \frac{\Phi_0}{2\pi u_x} \qquad (6.19)$$

Equation (6.19) expresses the distance to the *stagnation* point shown in Figure 6.8. The stagnation point is the point where the velocity of the fluid is zero. The contour line passing through the stagnation point extends on the right to infinity and outlines a half body.

Figure 6.8. The line source and uniform field combination.

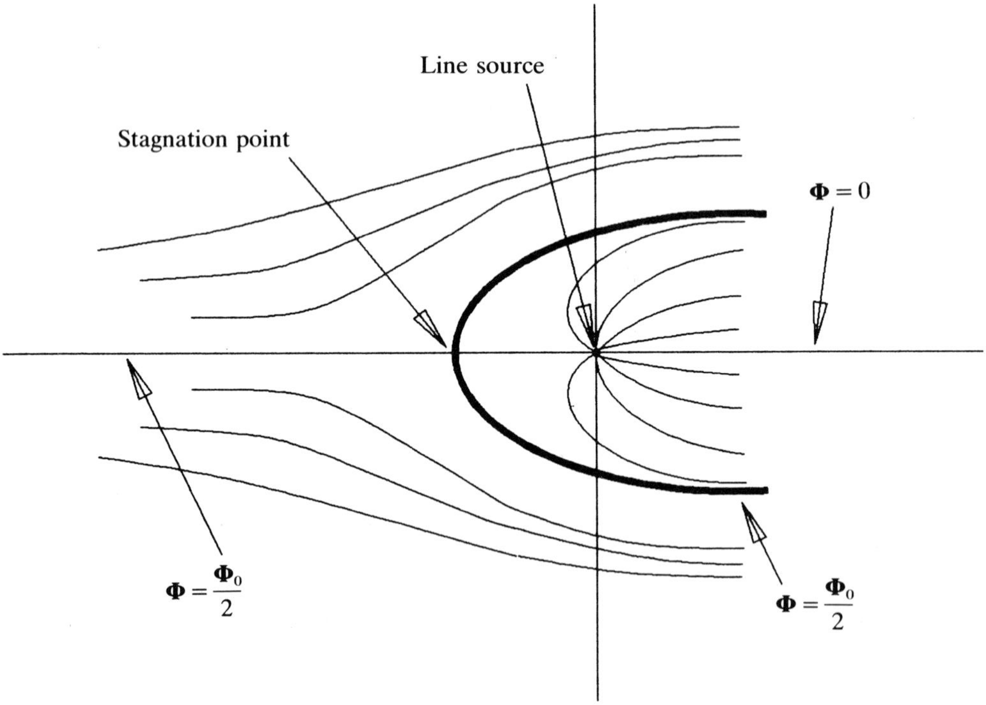

D6.7 Construct a table for the location of the stagnation point using equation (6.19). Use several values for Φ_0 and u_x; e.g., $\Phi_0 = u_x$, $\Phi_0 = 3u_x$, $\Phi_0 = \frac{1}{3}u_x$ and so forth.

---

4. The doublet and uniform field

The next useful combination is the doublet and the uniform field.

$$\Phi = \Phi_{\text{doublet}} + \Phi_{\text{uniform}}$$

From equation (6.16) we have

$$\Phi_{\text{doublet}} = -\frac{\Phi_0 d \sin \phi}{2\pi r}$$

From section 3 above

$$\Phi_{\text{uniform}} = u_x r \sin \phi$$

Thus

$$\Phi = -\frac{\Phi_0 d \sin \phi}{2\pi r} + u_x r \sin \phi$$

$$= \frac{u_x \sin \phi}{r}\left(r^2 - \frac{\Phi_0 d}{2\pi u_x}\right) \qquad (6.20)$$

Equation (6.20) is plotted in Figure 6.9. For $r = \sqrt{(\Phi_0 d / 2\pi u_x)}$, $\Phi = 0$ and there exist two stagnation points. Figure 6.9 can be used to model fluid flow around a cylinder.

Let's go on to the concept of potential.

D6.8 Construct a table for the location of the stagnation points using equation (6.20). Use several values for Φ_0 and u_x. Let $d = 10^{-1}$, 10^{-3} m.

6.6 Potential

In order to derive the potentials of interest, we recall electrostatics once again. In Chapter 3 we found two relationships between potential and field intensity, that is, $\mathbf{E} = -\nabla V$ and $V = -\int \mathbf{E} \cdot d\mathbf{l}$. We used the gradient equation to find field intensity when potential was known. Similar relationships in fluids are

$$\mathbf{u} = -\nabla V_f \tag{6.21}$$

and from above

$$V_f = -\int \mathbf{u} \cdot d\mathbf{l}$$

Important assumptions made for these two equations are that the fluid is ideal and has zero viscosity. The second assumption implies the fluid has no resistance, therefore, no pressure is necessary to cause the flow. Later we will assume a viscosity so that the potential, V_f, will *be* the pressure.

As an example of the use of equation (6.21), let us assume a fluid field

$$V_f = (x - 2)y$$

then

$$\nabla V_f = y\hat{\mathbf{a}}_x + (x - 2)\hat{\mathbf{a}}_y = -\mathbf{u}$$

Clearly this is a vector field for the velocity of the fluid.

Figure 6.9. The doublet and the uniform field.

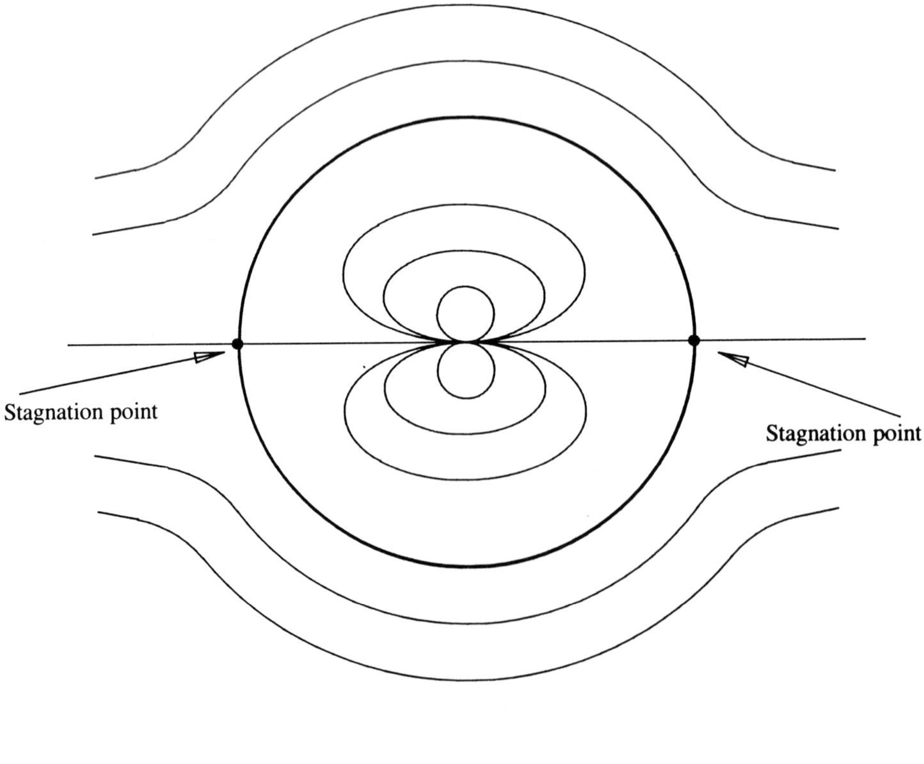

Example 6.1 As an example of the use of equation (6.6), let us find the potential for a line source where the flow is radial. From equation (6.11)

$$\mathbf{u} = \frac{\Phi}{2\pi r}\,\hat{\mathbf{a}}_r$$

Substituting this into equation (6.6), we obtain

$$\begin{aligned} V_\mathrm{f} &= -\int \frac{\Phi}{2\pi r}\,\hat{\mathbf{a}}_r \cdot \mathrm{d}r\hat{\mathbf{a}}_r \\ &= -\frac{\Phi}{2\pi}\ln r \end{aligned}$$

where we have assumed $V_\mathrm{f} = 0$ at some arbitrary reference.

Figure 6.10 shows several equipotentials (circles). The radii are shown to the right ($+x$-axis), while the natural log values are shown to the left ($-x$-axis). The velocity vectors are shown as radial vectors. We see that the velocity vectors (field intensity or flux) are orthogonal to the equipotentials (surfaces in three dimensions). We have assumed $V_\mathrm{f} = 0$ at $r = 1$ m.

Figure 6.10. Streamlines and equipotentials for a line source.

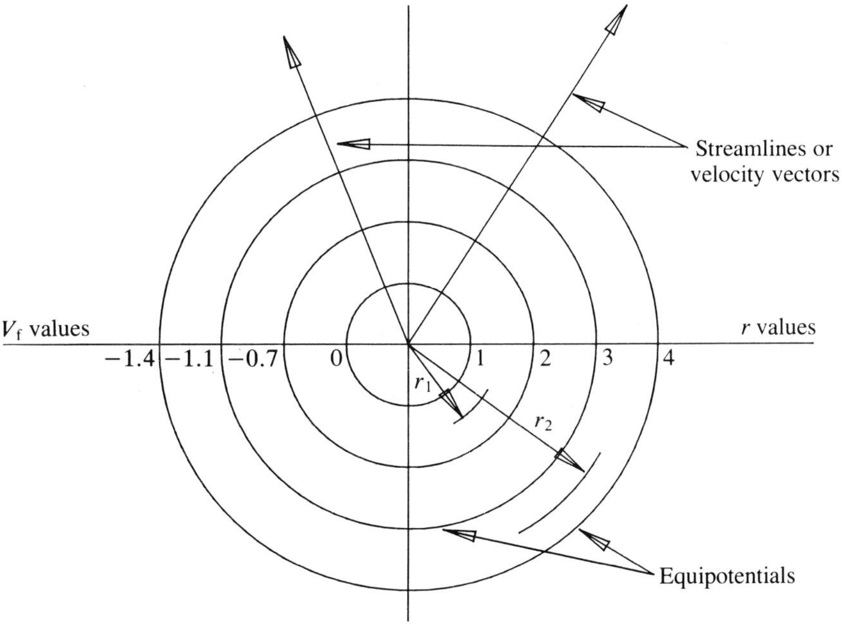

Further, just as we portrayed combination flow graphically in Figure 6.5, we can portray combination potentials graphically. From equation (6.12) we see that the equipotentials are circles which vary as the natural log. To obtain the plot for any combination, we simply (maybe not so simply) add the separate potentials. As an illustration consider the combination of two lines 'charges'. The charges can be the same or opposite polarities. We consider the two cases together

$$V_f = V_{f_1} \pm V_{f_2}$$

where V_{f_1} and V_{f_2} represent the potential functions of sources or sinks. Substituting appropriate values, we obtain

$$V_f = \frac{\Phi}{2\pi}(\ln r_1 \pm \ln r_2) \tag{6.22}$$

where r_1 and r_2 are shown in Figure 6.10. If the lines are of opposite polarity, we have

$$V_f = \frac{\Phi}{2\pi}\ln\left(\frac{r_1}{r_2}\right) \tag{6.23}$$

and if the lines are of the same polarity, we write

$$V_f = \frac{\Phi}{2\pi}\ln r_1 r_2 \tag{6.24}$$

Equation (6.23) is plotted in Figure 6.11; equation (6.24) is plotted in Figure 6.12. Figure 6.11 is a sketch in which the charges are of opposite polarity; that is a source and a sink. Figure 6.12 is a sketch in which both are of the same sign – either both are sources or both are sinks. Notice that the circles are spaced equally according to the natural log. The values assigned to the circles are arbitrary and do not affect the shapes. The only requirements are that the strengths are equal in magnitude and that the numbers ascend or descend in proper sequence.

D6.9 Given two infinite parallel line charges $\Phi_1 = 2\ \text{m}^3/\text{s}$ at $(0.5, 0)$ and $\Phi_2 = \pm 2\ \text{m}^3/\text{s}$ at $(-0.5, 0)$, calculate the potential at $(9, 9, 0)$ if (*a*) $\Phi_2 = 2\ \text{m}^3/\text{s}$ and if (*b*) $\Phi_2 = -2\ \text{m}^3/\text{s}$.
Answer. (*a*) $1.67\ \text{m}^2/\text{s}$ (*b*) $1.77 \cdot 10^{-2}\ \text{m}^2/\text{s}$.

6.7 The line source and uniform flow

Now we repeat the line source and uniform flow but use another method. That is, we find the potential, V_f, and then take the gradient.

Figure 6.11. Equipotential surfaces for two parallel, infinite line 'charges'.

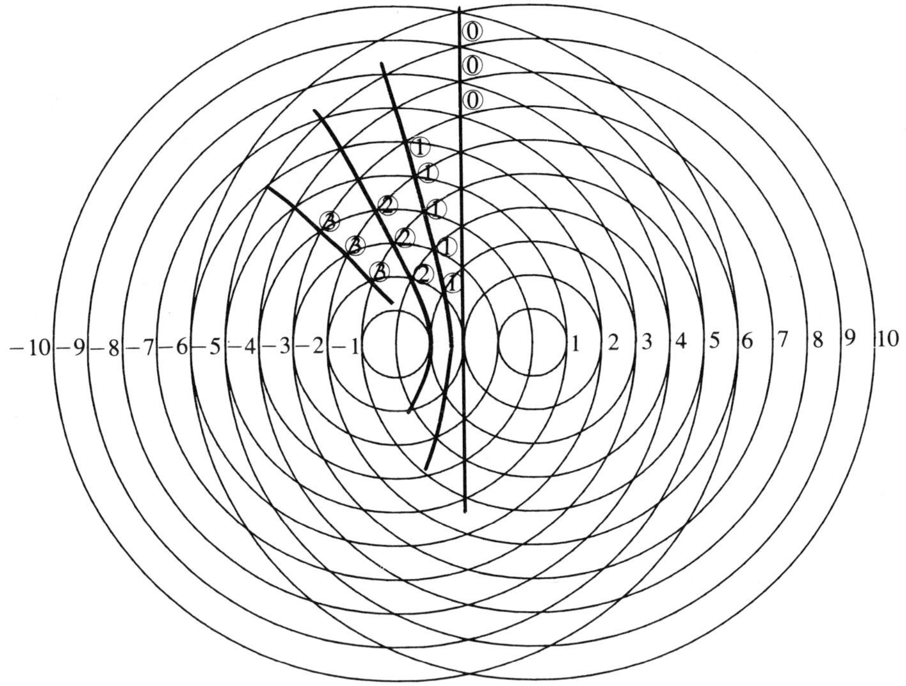

Figure 6.12. Equipotential surfaces for two parallel, infinite line 'charges'.

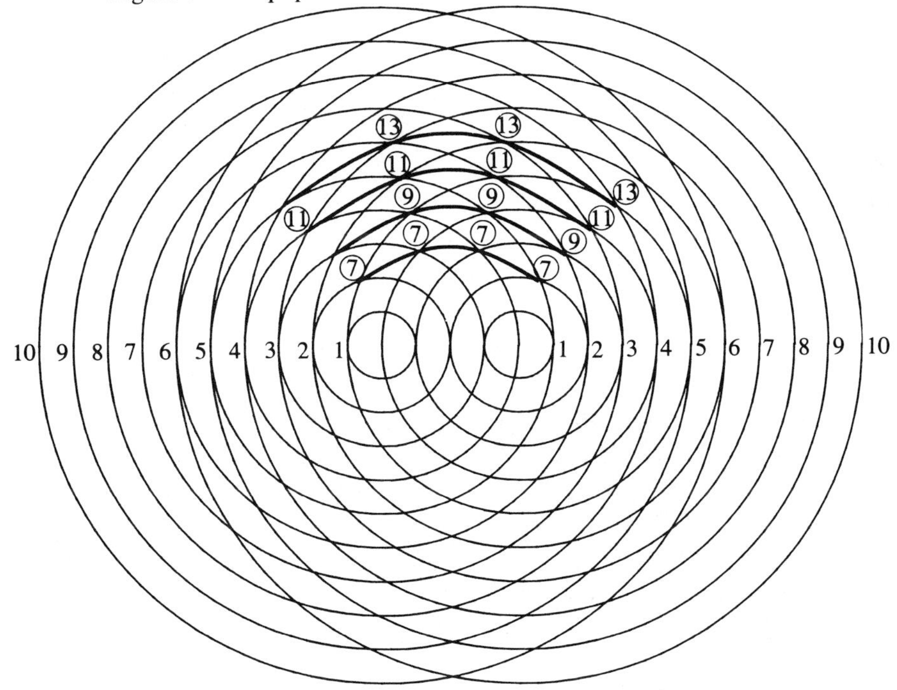

From Figure 6.1

$$V_{\mathrm{f}} = -u_0 r \cos\phi$$

From equation (6.12)

$$V_{\mathrm{f}} = -\frac{\Phi_0}{2\pi}\ln r$$

Using superposition

$$V_{\mathrm{f}} = -\frac{\Phi_0}{2\pi}\ln r - u_0 r \cos\phi$$

Now taking the gradient, we obtain

$$\mathbf{u} = -\nabla V_{\mathrm{f}}$$

$$= \left(\frac{\Phi_0}{2\pi r} + u_0 \cos\phi\right)\hat{\mathbf{a}}_r - u_0 \sin\phi\hat{\mathbf{a}}_\phi \tag{6.25}$$

Equation (6.25) is plotted in Figure 6.13. Notice the stagnation point; that is the point at which no radial flow exists. From equation (6.25) this implies

$$u_0 \cos\phi + \frac{\Phi_0}{2\pi r} = 0$$

Figure 6.13. The line source and uniform fiow.

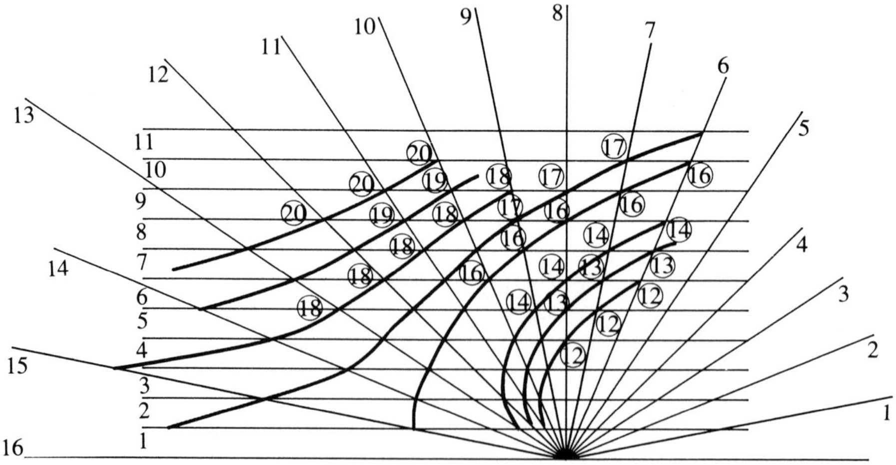

At the stagnation point $\phi = \pi$ and

$$\frac{\Phi_0}{2\pi r} = u_0$$

or

$$r = \frac{\Phi_0}{2\pi u_0} \tag{6.19}$$

where r is the distance from the line source to the stagnation point. As the velocity, u_0, increases, the distance decreases. As the strength of the line source increases, the distance to the stagnation point increases. Any of the streamlines in Figure 6.13 can model a body. This will prove useful later. Also note from equation (6.25) that the u_ϕ component is zero also, that is,

$$u_\phi = u_0 \sin \pi = 0$$

D6.10 Locate the stagnation point for the combination uniform flow and line source (radial) flow. (*a*) $u_0 = 1.5\ \text{m/s}$; $\Phi_0 = 10\ \text{m}^3/\text{s}$, (*b*) repeat for $u_0 = 10\ \text{m/s}$ and $\Phi_0 = 1.5\ \text{m}^3/\text{s}$.
Answer. (*a*) 1.06 m (*b*) $2.39 \cdot 10^{-2}$ m.

D6.11 Calculate the velocity at the point $(-10, 10, 0)$ for both parts of D6.10.
Answer. (*a*) $-0.948\hat{\mathbf{a}}_r + 1.06\hat{\mathbf{a}}_\phi$ (*b*) $-7.05\hat{\mathbf{a}}_r + 7.07\hat{\mathbf{a}}_\phi$ m/s.

6.8 Line vortex

For a line charge there can be, in addition to radial flow, another type – circumferential flow. Take, for example, the fluid in a washing machine (without clothes). The fluid flows in the $\hat{\mathbf{a}}_\phi$ direction as the tub rotates. This flow can be subdivided further into two kinds – forced vortex and free vortex. As the tub spins, there is a velocity at which the fluid turns as one body – with the same velocity as the tub. This is a forced vortex. Now if the tub stops spinning, the water continues in the $\hat{\mathbf{a}}_\phi$ direction. Then if the water is drained from the center, we note that the velocity increases as it approaches the center hole.

It is convenient to point out the duality of the line source/sink and the line vortex, that is, the flux lines and the equipotentials are interchangeable as shown in Figure 6.14. For the line source in which the flow is radial (Figure 6.14(*a*)) the equipotentials are circumferential; for the line source in which the flow is circumferential (Figure

Figure 6.14. The duality (*a*) radial and (*b*) circumferential flows.

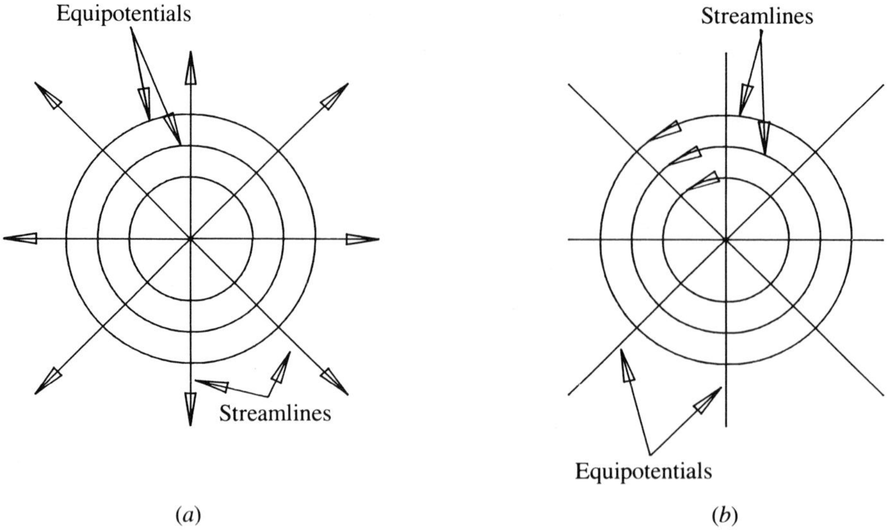

Figure 6.15. A differential area of fluid flow in the *xy* plane.

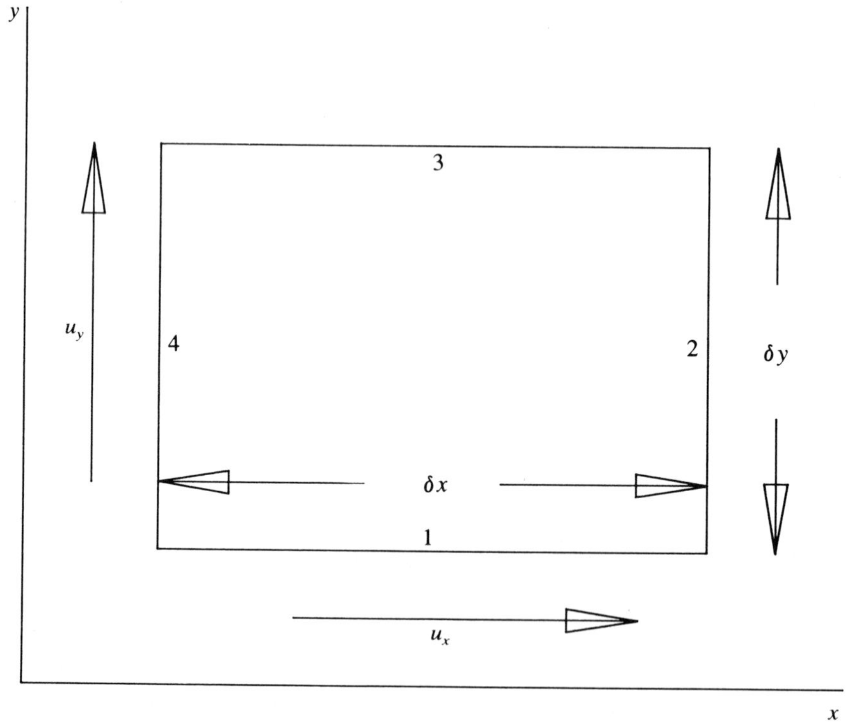

6.14(*b*)), the equipotentials are radial. These will be considered in more detail later.

First, let us return to flow in rectangular coordinates. Figure 6.15 shows a fluid flowing in the xy plane with the velocity varying in the x- and y-directions. Using equation (6.1) we write

$$\oint \mathbf{u} \cdot d\mathbf{l} = \begin{cases} 0 & \text{no rotation} \\ k & \text{rotation} \end{cases}$$

or integrating around the path 1-2-3-4

$$\oint \mathbf{u} \cdot d\mathbf{l} = \int_1 + \int_2 + \int_3 + \int_4 = k$$

we obtain

$$\oint \mathbf{u} \cdot d\mathbf{l} \approx u_x \delta x + \left(u_y + \frac{\partial u_y}{\partial x} \delta x\right) \delta y - \left(u_x + \frac{\partial u_x}{\partial y} \delta y\right) \delta x - u_y \, \delta y$$

or

$$\frac{\partial u_y}{\partial x} \delta x \, \delta y - \frac{\partial u_x}{\partial y} \delta x \, \delta y = k$$

Dividing both sides by the area, $\delta x \, \delta y$, we define vorticity

$$\frac{\partial u_y}{\partial x} - \frac{\partial u_x}{\partial y} = \frac{k}{\delta x \, \delta y} \triangleq \text{vorticity} \tag{6.26}$$

In other words, circulation per area is defined as vorticity. Clearly if the circulation is zero, the vorticity is zero. Furthermore, if $k = 0$ then

$$\frac{\partial u_y}{\partial x} = \frac{\partial u_x}{\partial y}$$

Consider an example.

Example 6.2 Determine the fluid velocity in the vicinity of a boundary. Figure 6.16 is a plot representing fluid flowing in the x-direction with the velocity, u_x, varying in the y-direction. The region is divided into laminar and boundary layer flows. For the laminar section, $y > y_0$, there is no rotation, vorticity is zero. For the boundary layer, $y < y_0$, the velocity is zero at the boundary ($y = 0$) and increases nonlinearly to the laminar flow velocity at $y = y_0$. For the laminar section we write

$$u_y = 0$$

and

$$\frac{\partial u_x}{\partial y} = 0 \qquad (k = 0)$$

Integrating, we obtain

$$u_x = \text{constant}$$

For the boundary layer we write

$$u_y = 0$$

and

$$\frac{\partial u_x}{\partial y} = \frac{k}{\delta x\, \delta y}$$

with the boundary conditions $u_x = 0$ for $y = 0$ and $u_x = u_0$ at $y = y_0$. Now consider a two-dimensional flow in cylindrical coordinates. Vortex flow

Figure 6.16. Fluid flow along a boundary.

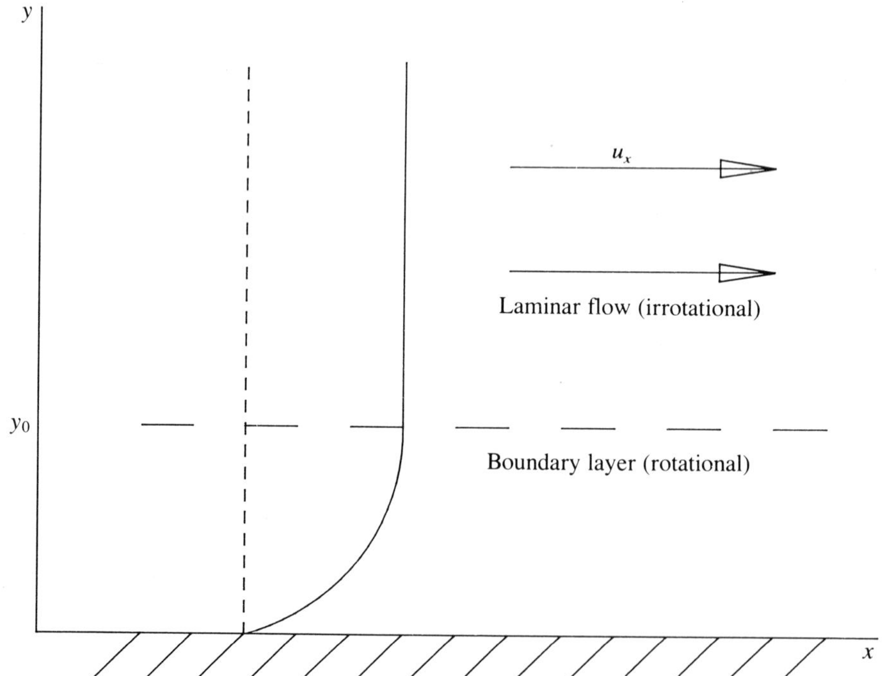

is shown in Figure 6.17. Remember vortex flow is of two types – free and forced.

D6.12 Determine if the fluid is rotational for (*a*) $\mathbf{u} = x\hat{\mathbf{a}}_x + y\hat{\mathbf{a}}_y$, (*b*) $\mathbf{u} = y\hat{\mathbf{a}}_x + x\hat{\mathbf{a}}_y$, (*c*) $\mathbf{u} = 2y\hat{\mathbf{a}}_x + x\hat{\mathbf{a}}_y$.
Answer. (*a*) irrotational, (*b*) rotational, (*c*) rotational.

6.9 Free vortex

As we divided the flow in the previous section into irrotational and rotational, we divide cylindrical flow similarly. The free vortex is an irrotational flow where the circumferential flow is constant and the radial flow velocity is not. Approximations to this are hurricanes, tornadoes, and fluid emptying from a funnel. This is the type of flow we have when water leaves a central hole in a container. Let us assume the fluid is rotating tangentially. We still assume the radial velocity is zero. Consider the small element of fluid as shown in Figure 6.18.

Figure 6.17. Free vortex flow in two dimensions.

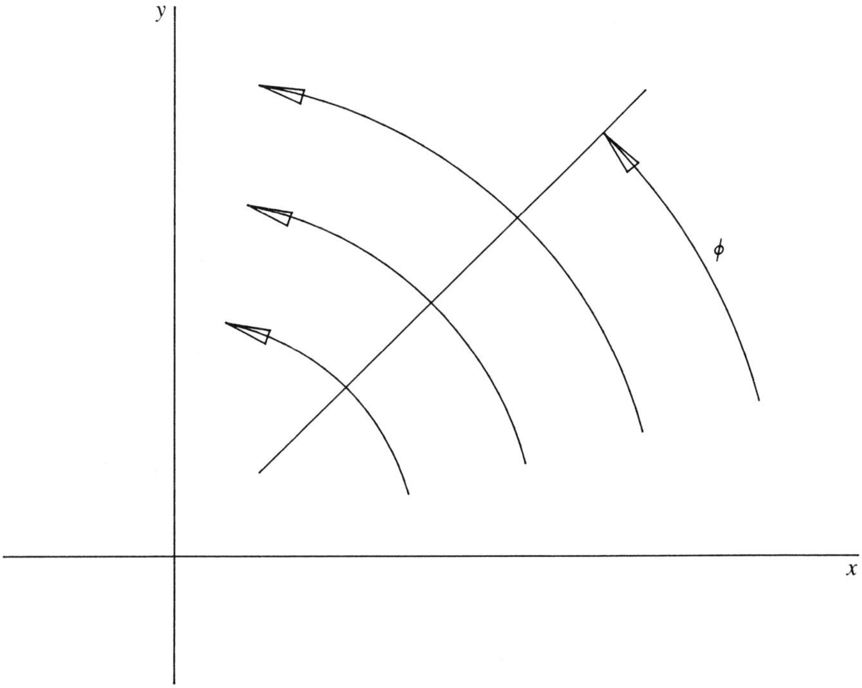

We integrate around the contour of the element of fluid in the counter-clockwise direction.

$$\oint \mathbf{u}\cdot d\mathbf{l} = \int_1 + \int_2 + \int_3 + \int_4$$

$$= \int \left[u_r\, \delta r + \left(u_\phi + \frac{\partial u_\phi}{\partial r}\,\delta r\right)(r+\delta r)\,\delta\phi \right.$$

$$\left. - u_r\,\delta r - u_\phi r\,\delta\phi \right]$$

$$\approx 0 + \left(u_\phi + \frac{\partial u_\phi}{\partial r}\,\delta r\right)(r+\delta r)\,\delta\phi + 0 - u_\phi r\,\delta\phi$$

If we neglect the second order term, we obtain

$$u_\phi\,\delta r\,\delta\phi + r\frac{\partial u_\phi}{\partial r}\,\delta r\,\delta\phi = k$$

$$\frac{u_\phi}{r} + \frac{\partial u_\phi}{\partial r} = \frac{k}{r\,\delta r\,\delta\phi} = \frac{k}{\text{Area}} \tag{6.27}$$

which is the vorticity.

Figure 6.18. A differential area of fluid flow in the $(r,\ \phi)$ plane.

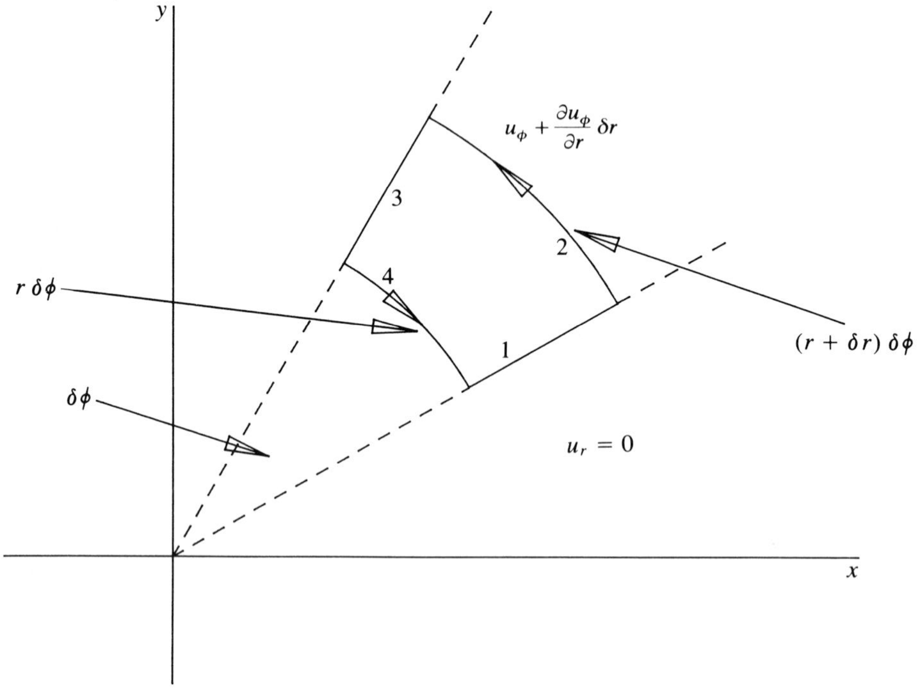

D6.13 Determine if the fluid is rotational for (*a*) $\mathbf{u} = \sin\phi\hat{\mathbf{a}}_\phi$, (*b*) $\mathbf{u} = r\sin\phi\hat{\mathbf{a}}_\phi$, (*c*) $\mathbf{u} = 1/r\sin\phi\hat{\mathbf{a}}_\phi$.
Answer. (*a*) rotational, (*b*) rotational, (*c*) irrotational.

Now if the flow is irrotational or $k = 0$, then equation (6.27) becomes

$$\frac{u_\phi}{r} + \frac{\partial u_\phi}{\partial r} = 0 \tag{6.28}$$

Rewriting, we obtain

$$\frac{\partial r}{r} = -\frac{\partial u_\phi}{u_\phi}$$

and integrating, we get

$$\ln r = -\ln u_\phi + c_1$$

$$= \ln\frac{1}{c_2 u_\phi}$$

$$r = \frac{1}{c_2 u_\phi}$$

or

$$u_\phi = \frac{c_3}{r} \tag{6.29}$$

We see that the velocity tends toward infinity at the center of the field. Since this is an impossibility, we exclude the origin; we allow vorticity in the center region. Consider the vortex flow in Figure 6.19. Model vortex flow with the two regions shown. We now define forced vortex.

6.10 Forced vortex

A forced vortex is defined as a fluid which moves similar to a solid body. In Figure 6.19 the fluid in region 1 is defined as rotating in the $\hat{\mathbf{a}}_\phi$ direction with no translation. So that in region 1

$$u_{\phi_1} = r\omega$$

and in region 2

$$u_{\phi_2} = \frac{c_3}{r} \tag{6.29}$$

At the boundary $r = a$

$$u_{\phi_1} = u_{\phi_2}$$

and

$$a\omega = \frac{c_3}{a}$$

In other words

$$u_{\phi_2} = \frac{a^2\omega}{r} \qquad r > a \tag{6.30}$$

Now let us develop some general equations for forced vortex flow as we did for the free vortex flow. Consider Figure 6.20 where the fluid is rotating in the $\hat{\mathbf{a}}_\phi$ direction with a constant angular velocity ω. The fluid

Figure 6.19. Rotational and irrotational flow in vortex flow.

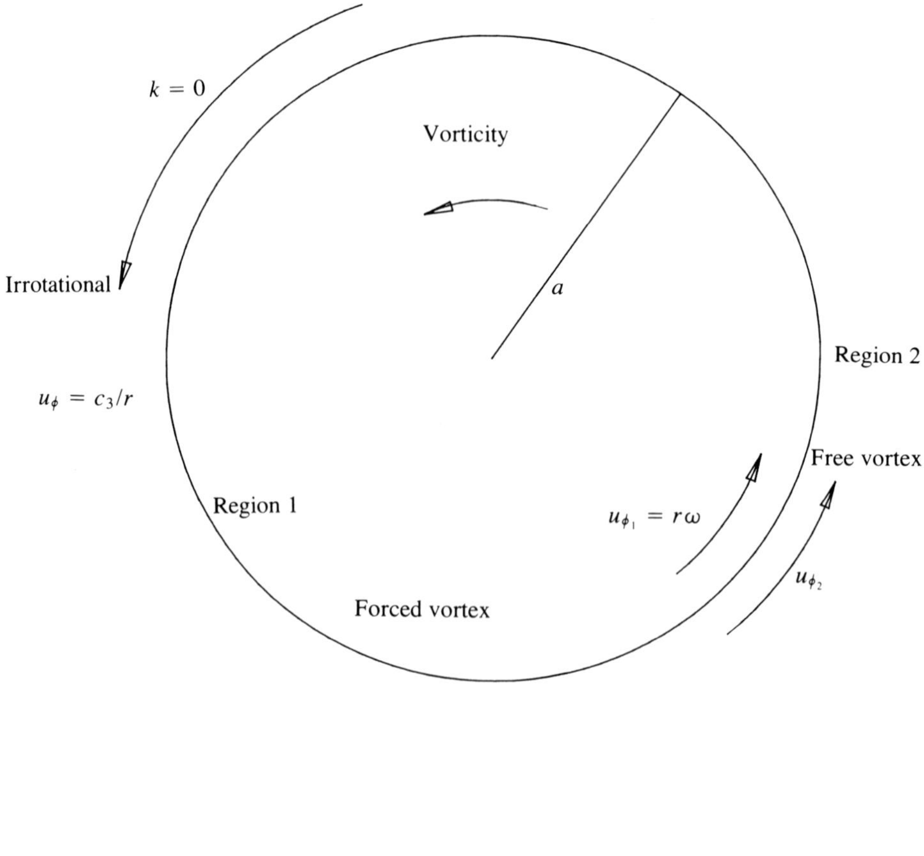

rotates as a mass – like cold honey! From equation (6.1) and Figure 6.20

$$\oint \mathbf{u} \cdot \mathrm{d}\mathbf{l} = k$$

$$\oint \mathbf{u} \cdot \mathrm{d}\mathbf{l} = \int_1 r\omega(-r\,\mathrm{d}\phi) + \int_2 u_r\,\mathrm{d}r$$

$$+ \int_3 (r + \delta r)\omega(r + \delta r)\,\mathrm{d}\phi - \int_4 u_r\,\mathrm{d}r$$

$$\approx -r^2\omega\,\delta\phi + (r^2 + (\delta r)^2 + 2r\,\delta r)\omega\,\delta\phi$$

$$= (\delta r)^2\omega\,\delta\phi + 2r\omega\,\delta r\,\delta\phi$$

$$= 2\omega\tfrac{1}{2}\,\delta r\,\delta\phi(\delta r + 2r)$$

$$\frac{k}{\frac{1}{2}\,\delta r\,\delta\phi(\delta r + 2r)} = 2\omega \tag{6.31}$$

where $\frac{1}{2}\,\delta r\,\delta\phi(2r + \delta r)$ is the area of the curvilinear rectangle shown in

Figure 6.20. Fluid flow in a forced vortex.

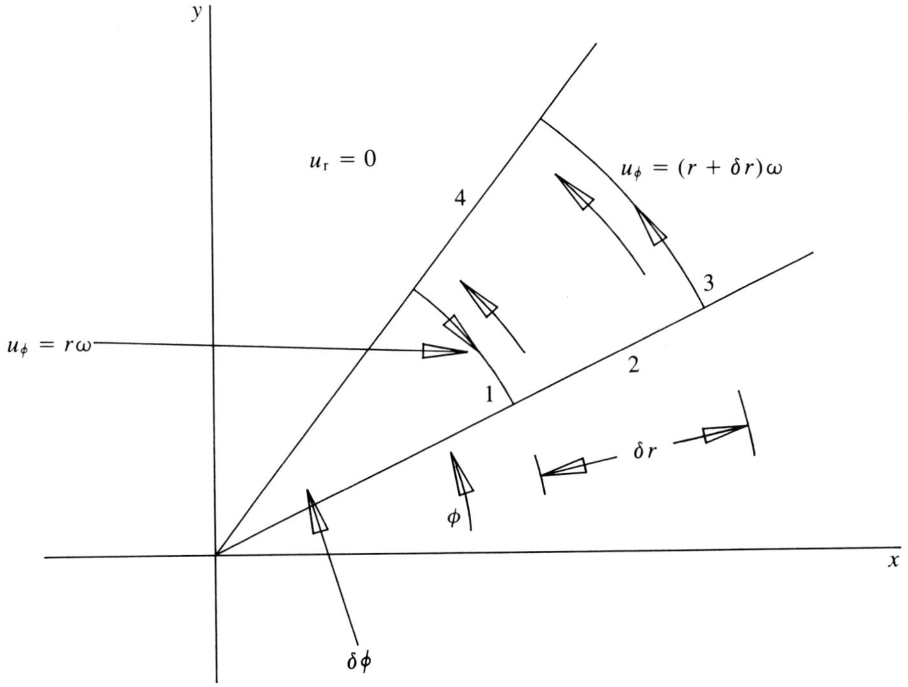

Figure 6.20. Thus, we write equation (6.31) as

$$\text{vorticity} = \frac{k}{\text{Area}} = 2\omega \tag{6.32}$$

where 2ω is the vorticity of the forced vortex. Understand that for a fluid with zero viscosity this condition would never arise. That is, with no friction in the fluid, the angular motion could not be transferred radially.

Example 6.3 One method of removing honey from the comb is by spinning it. What is the vorticity of honey rotating at 100 rpm just before it leaves the comb?

$$\text{vorticity} = 2\omega \approx 21 \text{ rad/s}.$$

D6.14 The eye of a tornado has a radius of 10 m. If the velocity is 40 m/s at 5 m, calculate the velocity at $r = 10$ m, 20 m.
Answer. 80 m/s, 40 m/s.

6.11 The potential due to a line vortex

From the definition of potential in an ideal fluid, we write

$$V_{\mathrm{f}} = -\int \mathbf{u} \cdot \mathrm{d}\mathbf{l}$$

Recall for a free vortex

$$u_\phi = \frac{c_3}{r}$$

or

$$\mathbf{u} = \frac{c_3}{r}\hat{\mathbf{a}}_\phi \tag{6.33}$$

The equipotential lines are along any radius for the streamlines to be orthogonal. Therefore,

$$\mathrm{d}\mathbf{l} = r\,\mathrm{d}\phi\hat{\mathbf{a}}_\phi$$

Integrating, we obtain

$$V_{\mathrm{f}} = -\int_0^\phi \frac{c_3}{r} r\,\mathrm{d}\phi$$

$$= -c_3\phi + c_4$$

From the boundary condition

$$V_f(0) = 0 \qquad c_4 = 0$$

and from the condition

$$V_f(2\pi) = 2\pi$$

the constant can be written

$$c_3 = \frac{V_{f_0}}{2\pi}$$

so that

$$V_f = -\frac{V_{f_0}}{2\pi}\phi \tag{6.34}$$

6.12 The flux due to a line vortex

From Gauss' law

$$\Phi = \oint \mathbf{u} \cdot d\mathbf{A}$$

For an open surface

$$\Phi = \iint \mathbf{u} \cdot d\mathbf{A} \tag{6.35}$$

Once again for a free vortex from equation (6.34) and $\mathbf{u} = -\nabla V_f$ we obtain

$$\mathbf{u} = \frac{c_2}{2\pi r}\hat{\mathbf{a}}_\phi$$

and from Figure 6.21

$$d\mathbf{A} = dr\, dz\hat{\mathbf{a}}_\phi$$

Integrating equation (6.35) we obtain

$$\Phi = \int_0^1 \int_{r_{ref}}^{r} \frac{c_2}{2\pi r}\, dr\, dz$$

$$= \frac{c_2}{2\pi} \ln \frac{r}{r_{ref}} \tag{6.36}$$

where $\Phi = 0$ at $r = r_{ref}$, an arbitrary reference.

Figure 6.21. The differential area for circumferential flow.

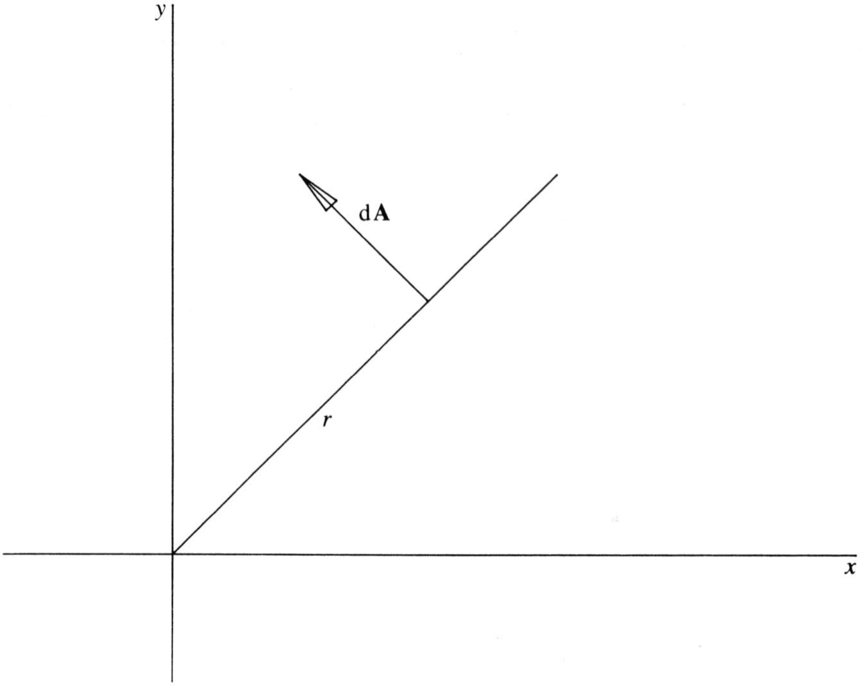

Figure 6.22. Fluid flow around a cylindrical radiator.

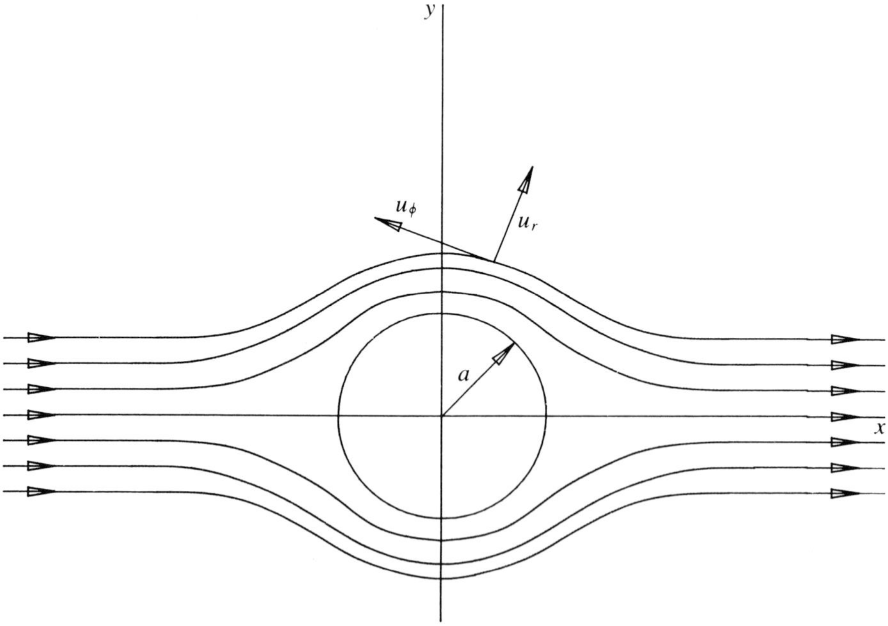

6.13 Application

In this section we model a real problem with a combination of flow. Fluid flow around a cylinder such as in most heat transfer devices can be modelled using a uniform and line source/sink combination. This combination was discussed on pages 187 and 188 and is shown in Figure 6.9. It is repeated above in Figure 6.22. The doublet inside the cylinder is not shown since we are interested in the streamlines around the outside of the cylinder. The cylinder represents the radiator which transfers heat to the fluid moving past, if the fluid within the cylinder is at a higher temperature than the fluid outside. If the fluid inside is at a lower temperature than the outside fluid, the heat transfer is in the opposite direction.

Note from Figure 6.22 that the velocity vectors are streamlines and are related by the following equations

$$u_r = \frac{\partial \Phi}{r\,\partial \phi} \tag{6.37}$$

$$u_\phi = -\frac{\partial \Phi}{\partial r} \tag{6.38}$$

The streamline function for Figure 6.22 is given by equation (6.20) page 187

$$\Phi = \frac{u_0 \sin \phi}{r}\left(r^2 - \frac{\Phi_0 d}{2\pi u_0}\right) \tag{6.20}$$

Combining equations (6.37), (6.38) and (6.20) we obtain

$$\mathbf{u} = u_0 \cos \phi\left(1 - \frac{\Phi_0 d}{2\pi u_0 r^2}\right)\hat{\mathbf{a}}_r - u_0 \sin \phi\left(1 + \frac{\Phi_0 d}{2\pi u_0 r^2}\right)\hat{\mathbf{a}}_\phi \tag{6.39}$$

From the boundary condition $u_r(a) = 0$

$$1 - \frac{\Phi_0 d}{2\pi u_0 a^2} = 0$$

$$a = \sqrt{\frac{\Phi_0 d}{2\pi u_0}}$$

Equation (6.39) becomes

$$\mathbf{u} = u_0 \cos \phi\left(1 - \frac{a^2}{r^2}\right)\hat{\mathbf{a}}_r - u_0 \sin \phi\left(1 + \frac{a^2}{r^2}\right)\hat{\mathbf{a}}_\phi \tag{6.40}$$

We can determine the potential function from

$$V_f = -\int \mathbf{u} \cdot \mathbf{dl}$$

where $\mathbf{dl} = \mathrm{d}r\hat{\mathbf{a}}_r + r\,\mathrm{d}\phi\hat{\mathbf{a}}_\phi$ in two dimensions. From equation (6.40) we obtain for the potential function

$$V_f = -u_0 \cos\phi\left(r + \frac{a^2}{r}\right) + \text{constant}$$

If $V_f(a) = 0$ then

$$V_f = u_0 \cos\phi\left(r + \frac{a^2}{r} - 2a\right) \tag{6.41}$$

Equations (6.40) and (6.41) are sketched in Figure 6.23. Streamlines and potential lines are orthogonal. Also note the duality; streamlines and potential lines may be interchanged. Again,

$$u_r(a) = 0$$

the radial velocity is zero on the circumference of the cylinder. Also

$$u_\phi(a) = -2u_0 \sin\phi$$

the tangential velocity is a function of ϕ, on the circumference of the cylinder. At the stagnation points, $(a, 0)$ and (a, π), both $u_r = 0$ and $u_\phi = 0$.

Figure 6.23. The orthogonality of potential and streamlines.

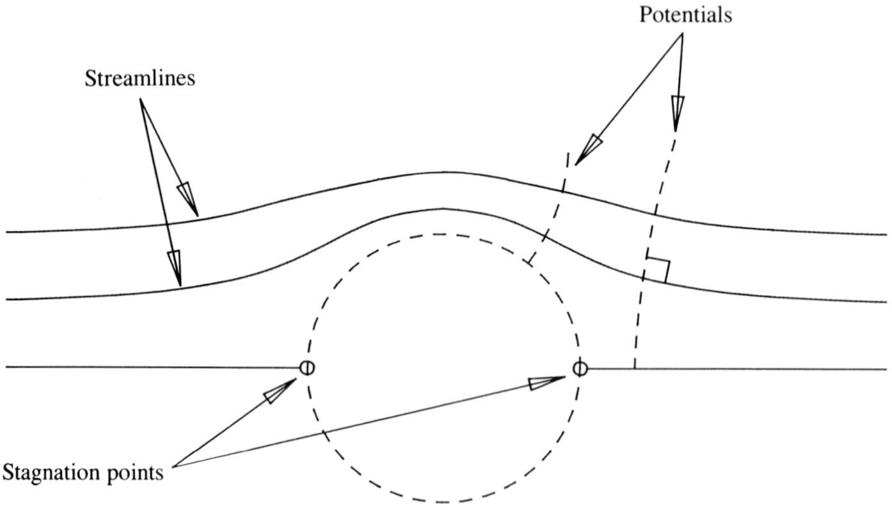

Finally we see that for $r \gg a$ we have

$$\mathbf{u} = u_0 \cos \phi \hat{\mathbf{a}}_r - u_0 \sin \phi \hat{\mathbf{a}}_\phi \tag{6.42}$$

For $\phi = 0$ the velocity is radial and for $\phi = \pi/2$, the velocity is in the $-\hat{\mathbf{a}}_\phi$ direction. For ϕ small $\cos \phi \approx 1$ and $\sin \phi \approx \phi$ equation (6.42) becomes

$$\mathbf{u} = u_0 \hat{\mathbf{a}}_r - u_0 \phi \hat{\mathbf{a}}_\phi \tag{6.43}$$

D6.15 Design one radiator of a cooling system such that the fluid flow is laminar as shown in Figure 6.22. Assume a uniform air flow velocity of 20 m/s.

D6.16 Plot the variation of the radius of the cylinder in D6.15 as a function of (*a*) $\Phi_0 d$ (the doublet) and (*b*) u_0 (uniform flow).

6.14 The hydrostatic equation

Consider a small volume of fluid in the shape of a disc as shown in Figure 6.24. The mass of the disc is the product of the density and the volume of the disc, that is

$$\text{mass} = \rho(\text{Area})\,\Delta y$$

where ρ is the density of the fluid in kg/m^3. The weight of the disc is

Figure 6.24. A differential disc of fluid.

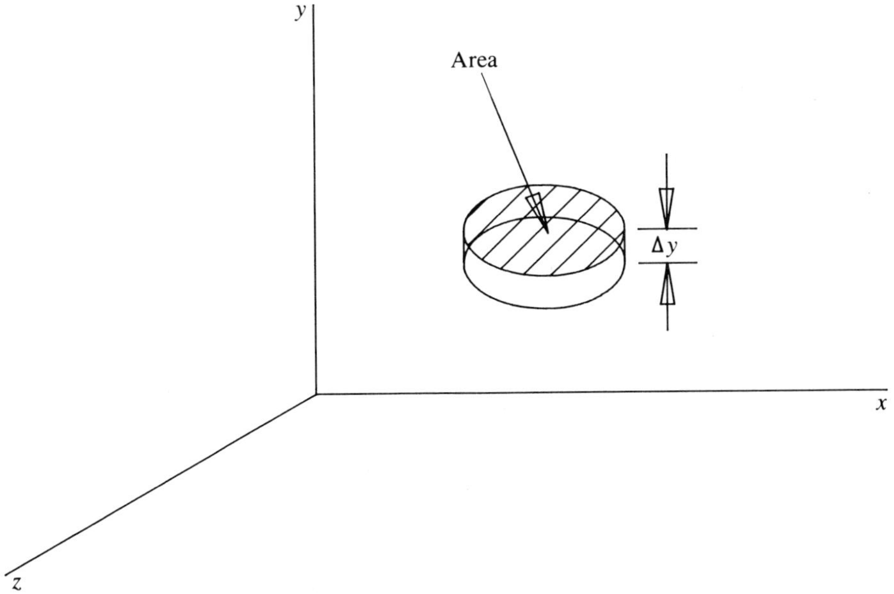

the product of the mass and the gravitational effect, g_0, or

$$\text{weight} = \rho(\text{Area})\,\Delta y\, g_0$$

Pressure is defined as force per unit area, or

$$p \triangleq \frac{\text{force}}{\text{area}}$$

The forces on the disc are the sum of the products of the respective pressures and areas.

Now consider a two-dimensional view of the disc as shown in Figure 6.25. As a result of gravity, the pressure differs as a function of height. In addition the disc has weight. Summing forces on the disc in the vertical direction, we obtain

$$\sum F_y = 0$$

Figure 6.25. A two-dimensional view of a disc of fluid.

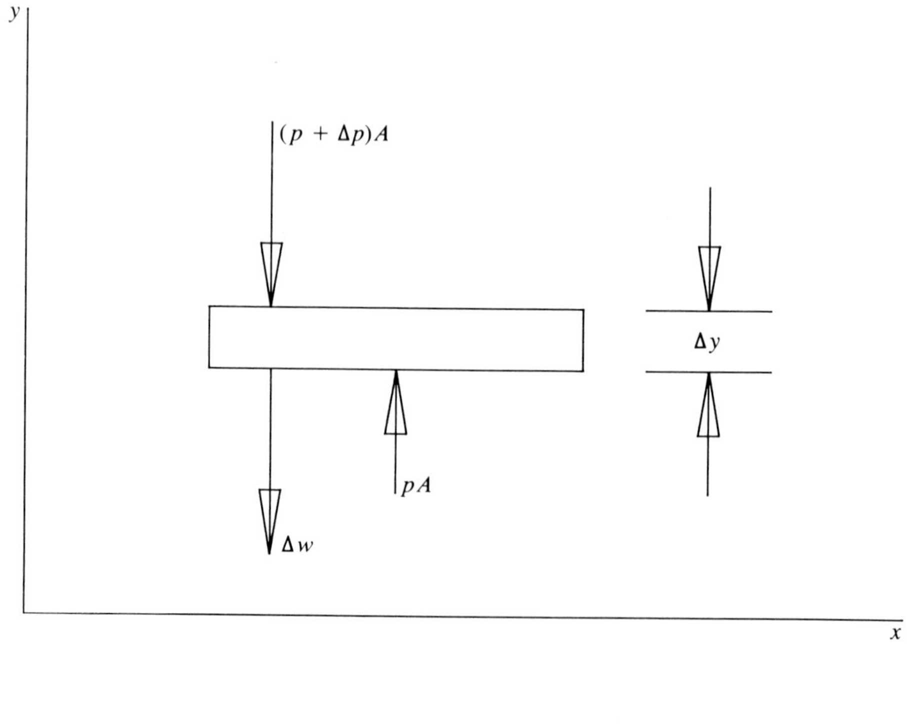

or

$$F_{up} - F_{down} = 0$$

$$pA - (p + \Delta p)A - \Delta w = 0$$

$$-\Delta pA - \Delta w = 0$$

$$-\Delta pA = \Delta w$$

$$= \Delta y \rho g_0 A$$

$$-\Delta p = \Delta y \rho g_0$$

$$-(p_2 - p_1) = (y_2 - y_1)\rho g_0$$

where ρ and g_0 are constants defined above

$$p_1 = p_2 + (y_2 - y_1)\rho g_0 \quad (6.44)$$

Equation (6.44) is sometimes written

$$p = p_0 - \rho g_0 h \quad (6.45)$$

where p_0 is the initial pressure and h is the height. This tells us that $p + \rho g_0 h$ is constant in static fluid. Another way of saying this is that the sum of the change in pressure and the change in potential is zero. This is called the hydrostatic equation. Rewriting equation (6.44) we get

$$\frac{p_1 - p_2}{y_2 - y_1} = \rho g_0$$

or

$$\frac{\Delta p}{\Delta y} = \rho g_0$$

In the limit

$$\frac{\mathrm{d}p}{\mathrm{d}y} = \rho g_0$$

In general

$$\nabla p = \rho g_0 \hat{\mathbf{a}}_A \quad (6.46)$$

Equation (6.46) is the vector form of the hydrostatic equation. This is in equilibrium. If the pressure or density varies in the fluid, convection currents result.

6.15 Fluids in motion/Bernoulli's equation

Rewriting equation (6.45) we have

$$p + \rho g_0 h = \text{constant}$$

If the density is constant (a good assumption), then p = constant; if the density is not constant, the fluid is compressible. Let us compress a volume of fluid then release it. Now as the fluid flows away from a point

$$\nabla \cdot (\rho \mathbf{u}) = -\frac{\partial \rho}{\partial t} \quad \text{conservation of mass} \tag{6.47}$$

Equation (6.47) is the continuity equation. Compare this with

$$\nabla \cdot \mathbf{J} = -\frac{\partial \rho}{\partial t}$$

in time varying electric fields. In other words, the divergence of the current density equals the decrease in the charge density rate.

We conclude this chapter with a brief discussion of an important theorem – Bernoulli's theorem. Most problems dealing with liquids in motion use this equation.

$$\frac{p}{\rho} + g_0 h + \tfrac{1}{2}u^2 = \text{constant} \tag{6.48}$$

Figure 6.26 illustrates Bernoulli's theorem in two dimensions. Consider a particle moving along a streamline. As the particle moves from point 1 to point 2, any or all of the three variables – p, h, or u – in equation (6.48) may vary. The sum is constant. For example if the three vary, equation (6.48) becomes

$$\frac{1}{\rho}(p_1 - p_2) + g_0(h_1 - h_2) + \tfrac{1}{2}(u_1{}^2 - u_2{}^2) = \text{constant}$$

We illustrate the theorem with an example. Figure 6.27 shows a container full of fluid with an opening near the bottom. We will assume the pressure is constant. In addition, we assume the velocity of the fluid level at the top is zero. Substituting into equation (6.48), we obtain

$$-\tfrac{1}{2}\rho u^2 + (p_1 - p_1) + \rho g_0 h = 0$$

$$u = \sqrt{2g_0 h}$$

This is the velocity of the emerging fluid, assuming the pressure is constant throughout (no gravitational effects) and the velocity of the fluid at the top (lowering of fluid level) is negligible.

Figure 6.26. An illustration of the Bernoulli theorem.

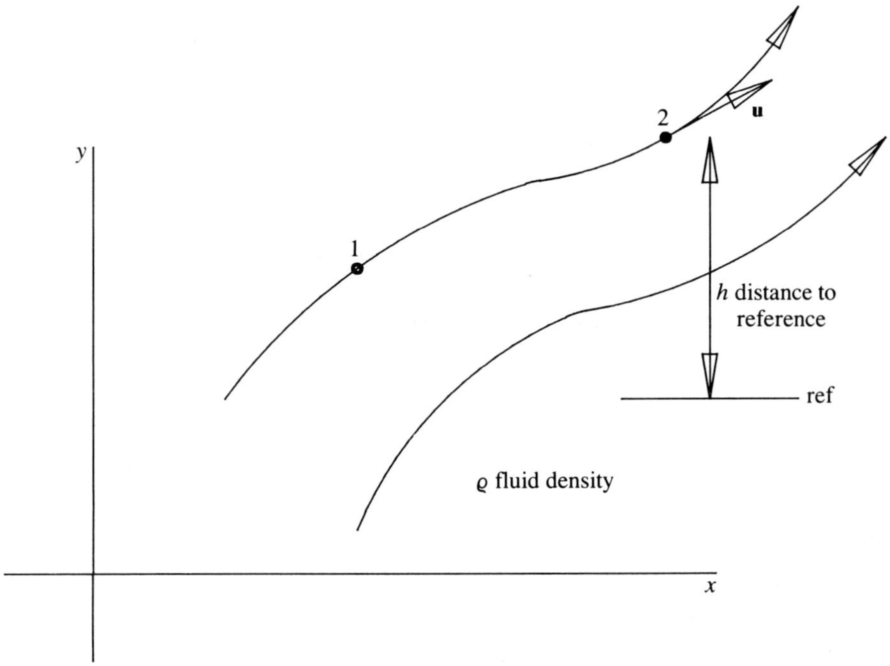

Figure 6.27. Container of fluid.

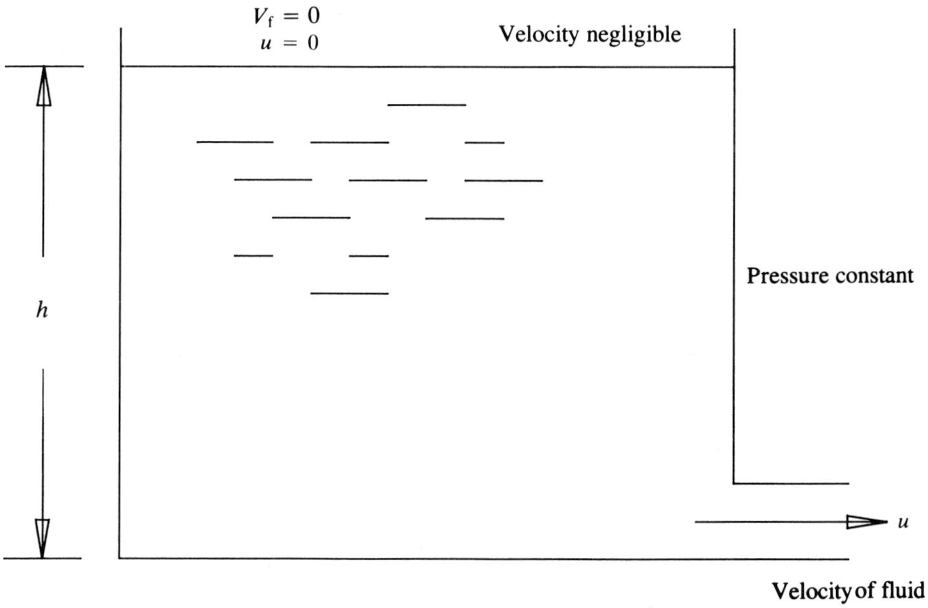

Problems

1. The sandy ground beneath a flooded field is absorbing water at the rate of $10^{-6}\,\mathrm{m^3/s}$. Determine the velocity vector as a function of distance. Assume the field is in the xy plane.

2. Determine the potential function for problem 1. Assume $V_f(0, 0, 0) = 0$.

3. There are two point sources each of value $0.25\,\mathrm{m^3/s}$, one at $(0, 0, 0)$ and a second at $(0, 1, 0)$. Calculate the velocity at $(0, 1, 1)$.

4. Plot the fluid potential for a spherical fluid source as a function of radius if there is an air filled balloon of radius $r = 0.05\,\mathrm{m}$ at the center of the fluid source. Choose $V_f = 0$ at $r \to \infty$ and $\Phi = 0.4\,\mathrm{m^3/s}$.

5. Two long cylindrical sources of value $\Phi_1 = 0.2\,\mathrm{m^3/s}$ at $x = 0.5$ and $\Phi_2 = 0.2\,\mathrm{m^3/s}$ at $x = -0.5$ are parallel to the z-axis. Calculate the velocity at $(5, 7, 1)$.

6. Determine the potential of a long, thin cylindrical source along the z-axis at $r = 8.5\,\mathrm{m}$ if $V_f = 0$ at (a) $r = 10\,\mathrm{m}$ and (b) $r = 5\,\mathrm{m}$. Assume $\Phi = 0.4\,\mathrm{m^3/s}$.

7. Sketch the streamlines for the combination of three parallel, infinitely long, line (filament) sources located at the vertices of an equilateral triangle – side length 1 m. Each source has value $\Phi = \Phi_0$.

8. Plot the streamlines for the combination of a uniform flow and one line source and two line sinks. The uniform flow is in the x-direction and $u_x = 1\,\mathrm{m/s}$. The source is at $(0, 0)$ and has value $\Phi_0 = 2\pi\,\mathrm{m^3/s}$. Each sink has the value $\Phi = -\pi\,\mathrm{m^3/s}$; one is at $(1, 0)$ and the second is at $(2, 0)$.

9. Repeat problem 8 with the following changes: 1. add a third line sink at $(3, 0)$ and 2. each sink has the value $\Phi = -2\pi/3\,\mathrm{m^3/s}$.

   The *continuity equation* for two-dimensional inviscid, incompressible flow is

$$\frac{\partial u_x}{\partial x} + \frac{\partial u_y}{\partial y} = 0$$

In problems 10–12 determine if the flow satisfies the continuity equation.

10. $u_x = u_y = x^2y^2$

11. $u_x = \dfrac{x}{x^2 + y^2}$

$u_y = \dfrac{y}{x^2 + y^2}$

12. $\mathbf{u} = r\cos\phi\hat{\mathbf{a}}_r + r\sin\phi\hat{\mathbf{a}}_\phi$

13. In vortex flow the potential is a function of angle only, that is $V_f = -C_1\phi$. Show that this satisfies Laplace's equation.

14. For the region $r > 2$ in cylindrical coordinates the velocity is

$$\mathbf{u} = \frac{C}{r}\hat{\mathbf{a}}_\phi$$

If $|\mathbf{u}| = 10$ m/s at $r = 10$ m determine C.

15. The region $r \leq 2$ in cylindrical coordinates is rotational and the fluid is rotating in the ϕ-direction at $\omega = 23$ rad/s. Determine the vorticity and the circulation.

16. If the region $r \leq 3$ in cylindrical coordinates is a forced vortex and the region $r > 3$ is a free vortex sketch the velocity distribution as a function of radius. At $r = 3$,

$$\mathbf{u} = 10\hat{\mathbf{a}}_\phi \text{ m/s}.$$

17. For a fluid (water) rotating at 16 rad/s there is a forced vortex for $r \leq 4$ m. At $r = 4$,

$$\mathbf{u} = -5\hat{\mathbf{a}}_\phi \text{ m/s}.$$

Determine the pressure difference outside the forced vortex (the eye of a tornado, for example) relative to the pressure at a distance.

For problems 18–20 calculate the force at the lower opening of Figure 6.28 subject to the following conditions:

18. The lower opening is a circle: $\alpha = 90°$

Figure 6.28. See problems 18–20.

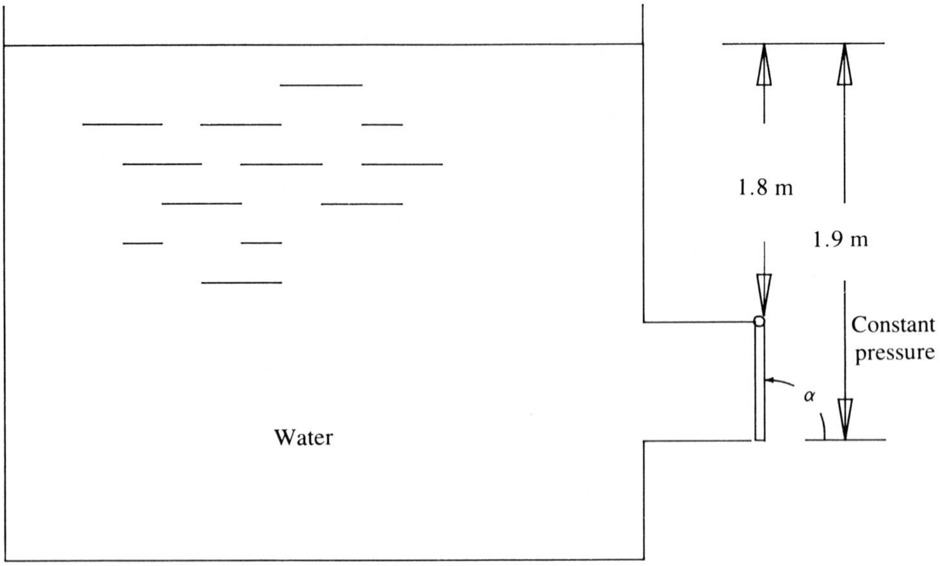

Figure 6.29. See problem 21.

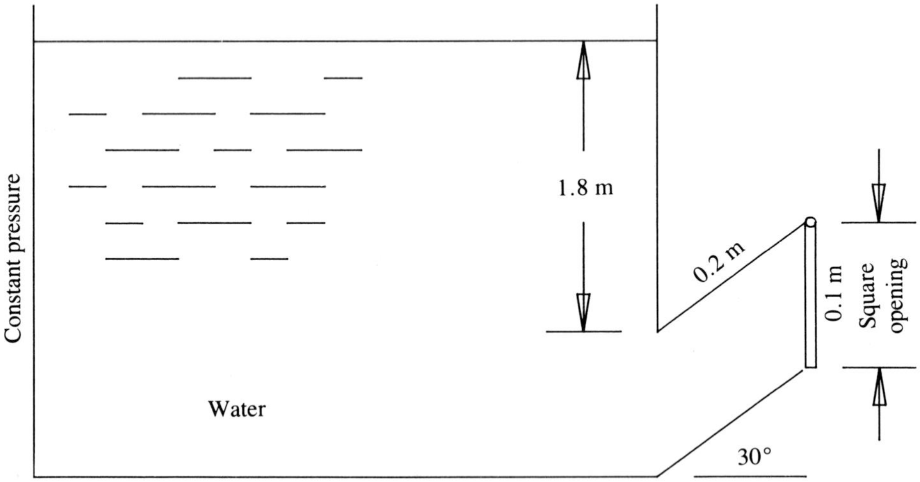

19. The lower opening is a square: $\alpha = 90°$

20. The lower opening is a rectangle of width 0.3 m and $\alpha = 45°$

21. Calculate the force on the cover at the lower opening of Figure 6.29.

22. In Figure 6.30 the pressure difference between points 'a' and 'b' is $12\ N/m^2$. If the velocity of the water at 'a' is 1.5 m/s calculate the velocity at 'b'.

23. In Figure 6.30 if the velocity at 'a' is 2.5 m/s calculate the volumetric flow rate.

24. In Figure 6.30 the velocity at 'a' is 1.7 m/s and the velocity at 'b' is 3.2 m/s. Calculate the pressure difference between points 'a' and 'b'.

25. If the volumetric flow rate in Figure 6.30 is $3 \cdot 10^{-4}\ m^3/s$ calculate (*a*) the velocity at 'a', (*b*) the velocity at 'b' and (*c*) the pressure difference.

Figure 6.30. See problems 22–25.

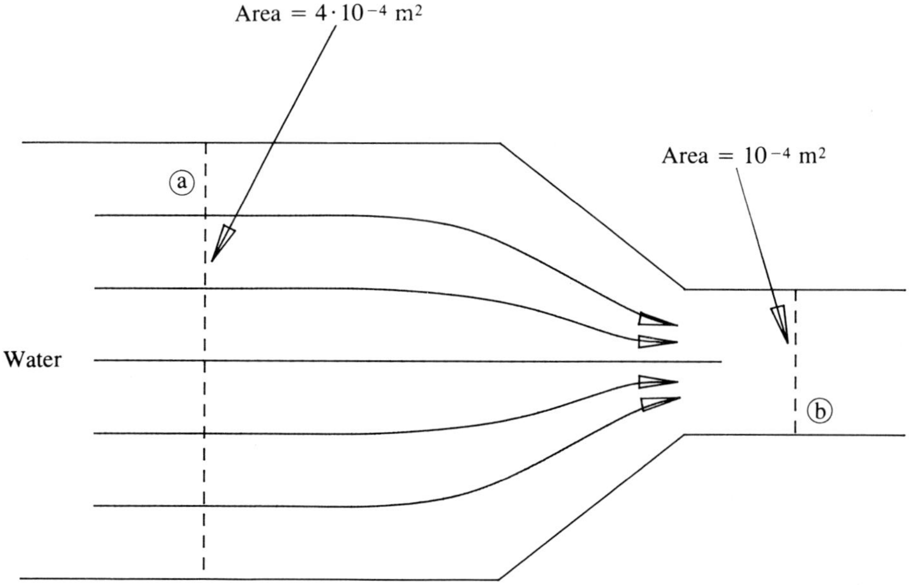

7

Thermal fields

7.1 Introduction

In this chapter we consider the methods by which heat is transferred. How is one body affected by a nearby body which is at a different temperature? We know that the difference in energies causes heat to flow from the higher energy (higher temperature) body to the lower energy body. This transfer takes place through conduction, convection, and radiation. Factors which influence the flow are temperature, absorptivity, emissivity, thermal conductivity, and heat transfer coefficient, among other things. We will discuss conduction and convection in more detail than heat transfer by radiation. Let's look at conduction first.

7.2 Conduction

Conduction is the main method of heat transfer for solids. It is governed by the Fourier law which says that power flow is proportional to the gradient of temperature

$$\mathbf{Q}_{\mathrm{h}} = -kA\,\nabla T = -kA\,\frac{\Delta T}{\Delta l}\,\hat{\mathbf{a}}_{\mathrm{l}} \tag{7.1}$$

where k is thermal conductivity in W/mK, A is the cross-sectional area in m^2, T is temperature in K, l is length in m and Q_{h} is heat flow in W.

Rearrange equation (7.1) as

$$Q = -\frac{\Delta T}{\Delta l/kA}$$

Here is an equation analogous to Ohm's law in electrostatics

$$\Delta I = \frac{\Delta V}{R}$$

where

$$R = \frac{l}{\sigma A}$$

l being the length, A the cross-sectional area, and σ the conductivity of the resistance.

Dividing both sides of equation (7.1) by the area, we obtain a density

$$D_h = \frac{Q_h}{A} = -k\frac{\Delta T}{\Delta l} \tag{7.2}$$

Recall Gauss' law

$$Q = \oint \mathbf{D} \cdot d\mathbf{A}$$

Dividing both sides by area, we obtain

$$\frac{Q}{A} = \frac{\oint \mathbf{D} \cdot d\mathbf{A}}{A} \Rightarrow \text{density}$$

This is the same form as equation (7.2). We are able to apply Gauss' law to conductive heat transfer. Rewriting the law in terms of a heat density, $\mathbf{D}_h$, we have, for a small volume,

$$\Phi = \oint \mathbf{D}_h \cdot d\mathbf{A} = -kA\frac{\Delta T}{\Delta l} \tag{7.3}$$

Before we apply equation (7.3) to a spherical source, let's define a field intensity and a potential. From equation (7.2)

$$D_h = -k\frac{\Delta T}{\Delta l} \tag{7.2}$$

we define field intensity

$$E_h = \frac{D_h}{k} \tag{7.4}$$

so that

$$\mathbf{D}_h = k\mathbf{E}_h \tag{7.5}$$

In addition from equation (7.2)

$$\Delta T = -\frac{1}{k} D_h \, \Delta l$$

or

$$T_2 - T_1 = -\frac{1}{k}\int \mathbf{D}_h \cdot d\mathbf{l}$$

$$\text{potential} \triangleq T_2 - T_1 \tag{7.6}$$

The difference in temperature is considered a potential. For a conservative field

$$\frac{1}{k}\oint \mathbf{D}_\mathrm{h} \cdot \mathrm{d}\mathbf{l} = 0$$

From equation (7.3)

$$\Phi = \oint \mathbf{D}_\mathrm{h} \cdot \mathrm{d}\mathbf{A} = \int_v \rho \, \mathrm{d}v$$

where ρ is the rate of heat flow per unit volume in $\mathrm{W/m^3}$ and the area encloses the heat source. From the Divergence theorem

$$\oint \mathbf{D}_\mathrm{h} \cdot \mathrm{d}\mathbf{A} = \int_v \nabla \cdot \mathbf{D}_\mathrm{h} \, \mathrm{d}v \tag{7.7}$$

From the right sides of equations (7.3) and (7.7) we obtain

$$\nabla \cdot \mathbf{D}_\mathrm{h} = \rho \tag{7.8}$$

Now from equation (7.2) we write

$$\mathbf{D}_\mathrm{h} = -k \, \nabla T \tag{7.9}$$

or

$$\mathbf{E}_\mathrm{h} = \frac{\mathbf{D}_\mathrm{h}}{k} = -\nabla T \tag{7.10}$$

Substituting equation (7.9) into equation (7.8) we obtain Poisson's equation

$$\nabla \cdot (-k \, \nabla T) = \rho$$

$$\nabla^2 T = -\frac{\rho}{k} \tag{7.11}$$

If $\rho = 0$ equation (7.11) becomes Laplace's equation

$$\nabla^2 T = 0 \tag{7.12}$$

7.3 Spherical source

In many circumstances a spherical body is the source of heat energy. The sun and earth are examples of spherical sources. The sun radiates heat energy to the earth. The earth absorbs the energy by a complicated process. Let's model the earth as a sphere. The sun is a conducting source; the earth is a conducting sink. The earth receives the energy by radiation; it redistributes the energy by re-radiation, conduction and to a lesser extent by convection.

Consider a sphere of radius a, with a strength of Q watts as shown in Figure 7.1. The volume density is $Q_h/\frac{4}{3}\pi a^3$. Apply Gauss' law

$$\Phi = Q_h = \oint \mathbf{D}_h \cdot d\mathbf{A} \tag{7.13}$$

We know in spherical coordinates

$$d\mathbf{A} = r^2 \sin\theta \, d\theta \, d\phi \hat{\mathbf{a}}_r$$

We've seen this integration before. The result is

$$D_h = \frac{Q_h}{4\pi r^2} \qquad r > a \tag{7.14}$$

For $r < a$

$$\int_v \rho \, dv = \oint \mathbf{D}_h \cdot d\mathbf{A}$$

$$\int_0^{2\pi} \int_0^{\pi} \int_0^{r} \frac{Q_h}{\frac{4}{3}\pi a^3} r^2 \sin\theta \, dr \, d\theta \, d\phi = \oint D_h r^2 \sin\theta \, d\theta \, d\phi$$

$$D_h = \frac{Q_h}{\frac{4}{3}\pi a^3} \frac{r}{3} \qquad r < a \tag{7.15}$$

Figure 7.1. A spherical source of Q_h joules/second.

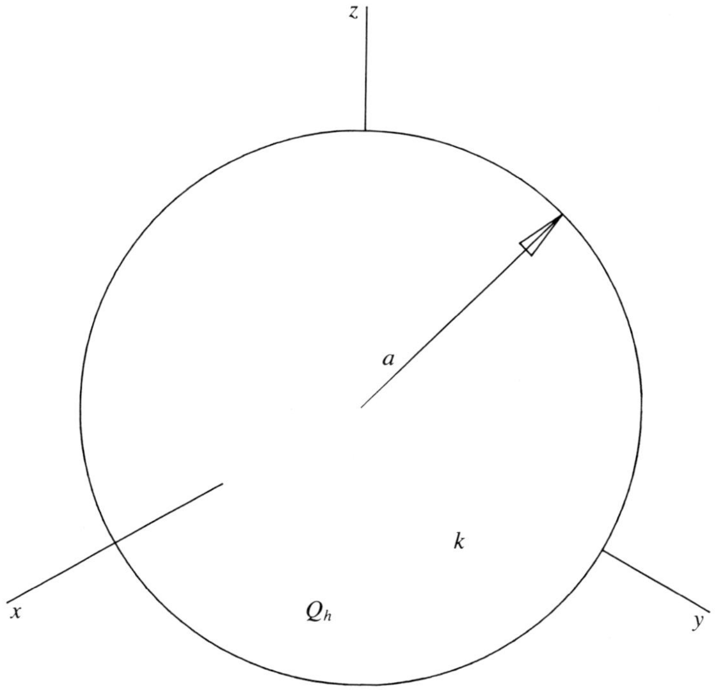

For field intensity equation (7.14) becomes

$$E_h = \frac{Q_h}{4\pi k_0 r^2} \qquad r > a$$

where k_0 is the effective thermal conductivity of the air surrounding the earth, and equation (7.15) becomes

$$E_h = \frac{Q_h}{\frac{4}{3}\pi a^3 k_i} \frac{r}{3} \qquad r < a$$

where k_i is the thermal conductivity of the earth. The potential is written

$$T_2 - T_1 = -\frac{1}{k} \int \mathbf{D}_h \cdot \mathbf{dl}$$

where

$$\mathbf{dl} = \mathrm{d}r\hat{\mathbf{a}}_r$$

$$T_2 - T_1 = -\frac{1}{k} \int \frac{Q_h}{4\pi r^2} \mathrm{d}r = \frac{Q_h}{4\pi k_0 r} \qquad r > a$$

$$T_2 - T_1 = -\frac{1}{k} \int \frac{Q_h}{\frac{4}{3}\pi a^3} \frac{r}{3} \mathrm{d}r = \frac{Q_h}{\frac{4}{3}\pi a^3 k_i} \frac{r^2}{6} \qquad r < a$$

Example 7.1 A forged ball bearing, radius 10^{-3} m, is heated at a rate of 100 W. Calculate the heat flux density at $r = 0.8 \cdot 10^{-3}$ m. From equation (7.15)

$$D_h = \frac{Q_h}{\frac{4}{3}\pi a^3} \frac{r}{3} = \frac{100}{\frac{4}{3}\pi (10^{-3})^3} \frac{0.8 \cdot 10^{-3}}{3} = 6.37 \cdot 10^6 \text{ W/m}^2$$

If the ball bearing is made of carbon steel $k = 54$ W/m °C the field intensity is

$$E_h = \frac{D_h}{k} = 14.156 \cdot 10^4 \text{ °C/m}$$

D7.1 Equation (7.1) is the Fourier law. Write ∇T in (*a*) Cartesian, (*b*) cylindrical, (*c*) spherical coordinates.

Answer.

(*a*) $\frac{\partial T}{\partial x}\hat{\mathbf{a}}_x + \frac{\partial T}{\partial y}\hat{\mathbf{a}}_y + \frac{\partial T}{\partial z}\hat{\mathbf{a}}_z$

(*b*) $\frac{\partial T}{\partial r}\hat{\mathbf{a}}_r + \frac{1}{r}\frac{\partial T}{\partial \phi}\hat{\mathbf{a}}_\phi + \frac{\partial T}{\partial z}\hat{\mathbf{a}}_z$

(*c*) $\dfrac{\partial T}{\partial r}\hat{\mathbf{a}}_r + \dfrac{1}{r}\dfrac{\partial T}{\partial \theta}\hat{\mathbf{a}}_\theta + \dfrac{1}{r\sin\theta}\dfrac{\partial T}{\partial \phi}\hat{\mathbf{a}}_\phi$

D7.2 Let the sphere in Figure 7.1 be a conductor of radius 10^{-2} m. It is coated with a dielectric of thickness 10^{-2} m in a still medium. This spherical resistor is dissipating 10 W. Calculate D_h and E_h at $1.5 \cdot 10^{-2}$ m.

Answer. 3.54 [kW/m²]; $\dfrac{3536}{k}$ [°C/m]

7.4 Cylindrical source

Many problems lend themselves to solutions in cylindrical coordinates. Heat transfer can occur radially, circumferentially, and axially or any combination of the three. Let's consider them in that order. If we assume uniform symmetric flow we can use Gauss' theorem.

First consider an infinite cylindrical rod, part of which is shown in Figure 7.2, which is transferring energy radially, due to a uniform

Figure 7.2. A portion of an infinite cylindrical heat source.

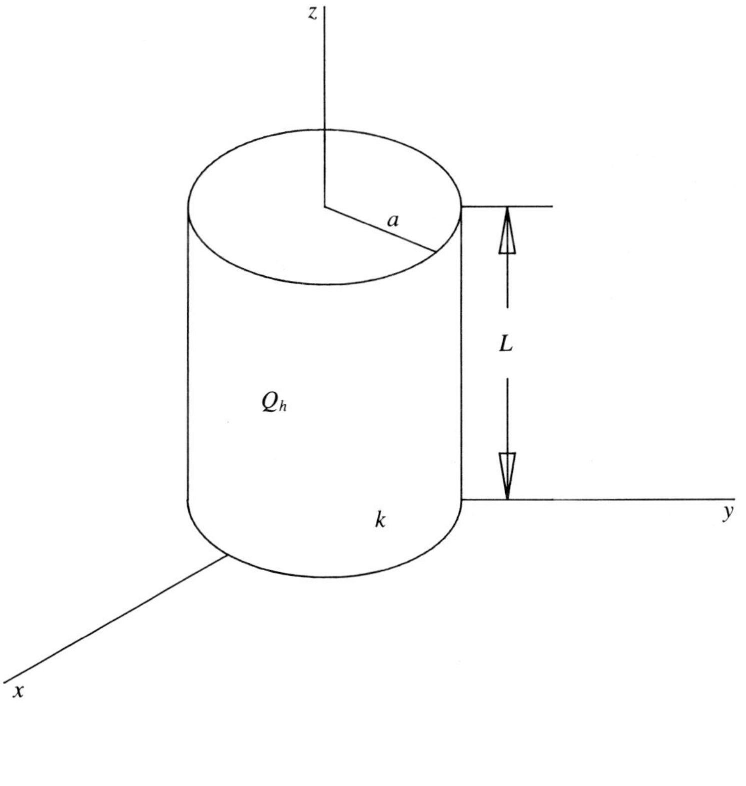

source of heat within it. From Gauss' theorem

$$\Phi \equiv Q_\text{h} = \iint \mathbf{D}_\text{h} \cdot \text{d}\mathbf{A}$$

$$= D_\text{h} \int_0^L \int_0^{2\pi} r \, \text{d}\phi \, \text{d}z$$

$$\mathbf{D}_\text{h} = \frac{Q_\text{h}}{2\pi r L} \hat{\mathbf{a}}_r \qquad r > a \tag{7.16}$$

Equation (7.16) is the heat flux density function for $r > a$. For $r < a$

$$\oint \mathbf{D}_\text{h} \cdot \text{d}\mathbf{A} = \int_0^1 \int_0^{2\pi} \int_0^r \frac{Q_\text{h}}{\pi a^2 L} r \, \text{d}r \, \text{d}\phi \, \text{d}z$$

$$D_\text{h} 2\pi r(1) = \frac{Q_\text{h}}{\pi a^2 L} \frac{r^2}{2} 2\pi(1)$$

$$\mathbf{D}_\text{h} = \frac{Q_\text{h} r}{2\pi a^2 L} \hat{\mathbf{a}}_r \qquad r < a \tag{7.17}$$

Equation (7.17) is the heat flux density function for $r < a$.

D7.3 If the center of the rod in Figure 7.2 is heated to 125 °C, while the surface temperature at $r = a$ is 25 °C, calculate the heat flux density at $r = a$. The rod is made of copper, $k = 390$ W/m °C; its diameter is 3 cm.
Answer. 2.6 MW/m^2.

Circumferential flow

Now let's assume that heat is transferred in the $\hat{\mathbf{a}}_\phi$ direction. This means that the streamlines are circumferential. Therefore the isotherms (equipotentials) are radial. This calls for using cylindrical coordinates and we note that $T = f(\phi)$. Instead of using Gauss' theorem we can use Laplace's equation to find the temperature function. Then we can take the gradient to find the density or intensity. Let's illustrate this with an example.

Example 7.2 A walkway is heated during inclement weather as shown in Figure 7.3. The heat flows in the $\hat{\mathbf{a}}_\phi$ direction; no heat flows either radially or in the $\hat{\mathbf{a}}_z$ direction. Calculate the heat flux density and the heat field intensity at the point $r = 30$ m, $\phi = \pi/4$. Let $k =$ 0.25 W/m °C.

From Laplace's equation

$$\nabla^2 T = 0 \tag{7.12}$$

In cylindrical coordinates

$$\nabla^2 T = \frac{1}{r}\frac{\partial}{\partial r}\left(r\frac{\partial T}{\partial r}\right) + \frac{1}{r^2}\frac{\partial^2 T}{\partial \phi^2} + \frac{\partial^2 T}{\partial z^2}$$

Since the temperature varies in the $\hat{\mathbf{a}}_\phi$ direction, equation (7.12) becomes

$$\frac{1}{r^2}\frac{\partial^2 T}{\partial \phi^2} = 0$$

Integrating twice we obtain

$$T = A\phi + B \tag{7.18}$$

From the boundary conditions given in Figure 7.3 equation (7.18)

Figure 7.3. Conduction heat transfer circumferentially from example 7.2.

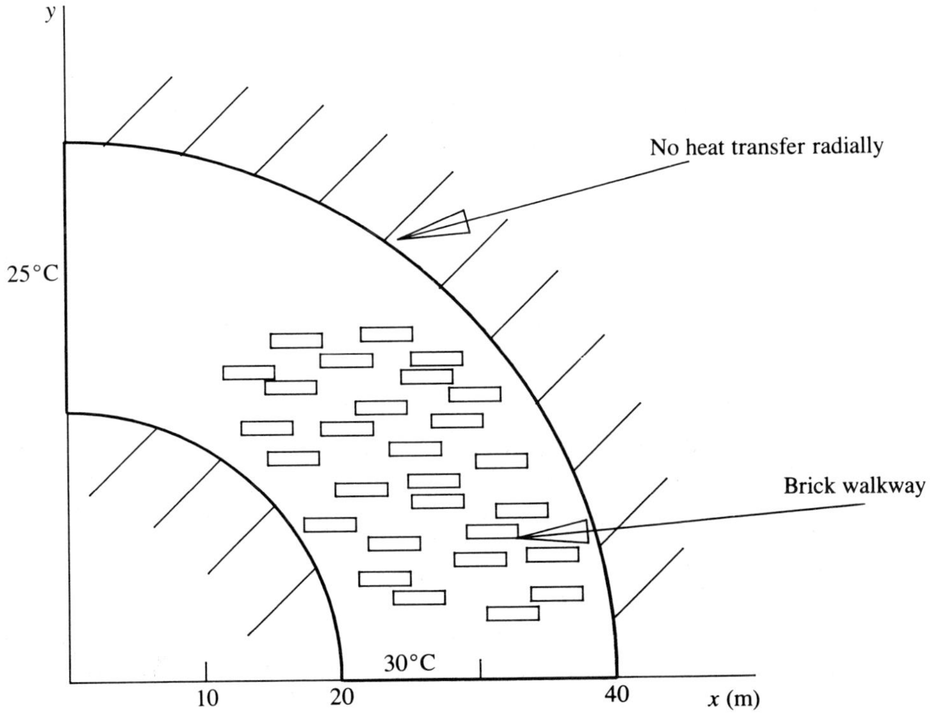

becomes

$$T = -\frac{10}{\pi}\phi + 30°$$

Now we take the gradient

$$\nabla T = -\frac{10}{\pi r}\hat{\mathbf{a}}_\phi$$

and calculate $\mathbf{D}_\mathrm{h}$

$$\mathbf{D}_\mathrm{h} = -k\,\nabla T \tag{7.9}$$

$$= \frac{10k}{\pi r}\hat{\mathbf{a}}_\phi \tag{7.19}$$

Substituting for k and r (note that D_h is ϕ independent) we obtain

$$\mathbf{D}_\mathrm{h} = \frac{10(0.25)}{\pi 30}\hat{\mathbf{a}}_\phi$$

$$= 0.0265\hat{\mathbf{a}}_\phi \ \mathrm{W/m^2}$$

and

$$\mathbf{E}_\mathrm{h} = \frac{\mathbf{D}_\mathrm{h}}{k} = 0.106\hat{\mathbf{a}}_\phi \ \mathrm{m/°C}$$

Equation (7.19) shows that $\mathbf{D}_\mathrm{h}$ (and therefore $\mathbf{E}_\mathrm{h}$) is inversely proportional to radius. Also from Figure 7.3 one would expect more heat to be transferred along the shorter arc lengths than the longer.

D7.4 If $T = \frac{5}{\pi}\phi - 10$ °C calculate the temperature and heat flux density at $r = 0.5$, $\phi = 90°$ where $k = 10$ W/m °C.
Answer. −7.5 °C, −31.8 W/m².

Axial flow

Since axial flow in the cylindrical configuration is in the z-direction let's analyze axial flow when we analyze planar sources next. In plane source analysis we use rectangular coordinates.

7.5 Plane source (or sink)

Consider a planar source as shown in Figure 7.4. Since the source is in the plane and the heat is transferred in the x-direction we have unidirectional flow or axial flow. Axial flow was discussed in section 7.4. Let the source be at a temperature T_0 and let the temperature at $x = x_1$ be $T = T_1$ where $T_0 > T_1$. From equation (7.12)

$$\nabla^2 T = 0 \tag{7.12}$$

or

$$\frac{d^2T}{dx^2} = 0$$

Integrating twice we obtain

$$T = Ax + B \tag{7.20}$$

Let's use equation (7.20) in two examples.

Example 7.3 Consider a plane source at $x = 0$ as shown in Figure 7.4. If the temperature at $x = 0$ is $T_0 = 40$ °C and the temperature at $x_1 = 5$ m is $T_1 = 20$ °C find the temperature as a function of distance. From equation (7.20)

$$T = 40° = A(0) + B$$

and

$$T = 20° = A(5) + B$$

Figure 7.4. A planar heat source

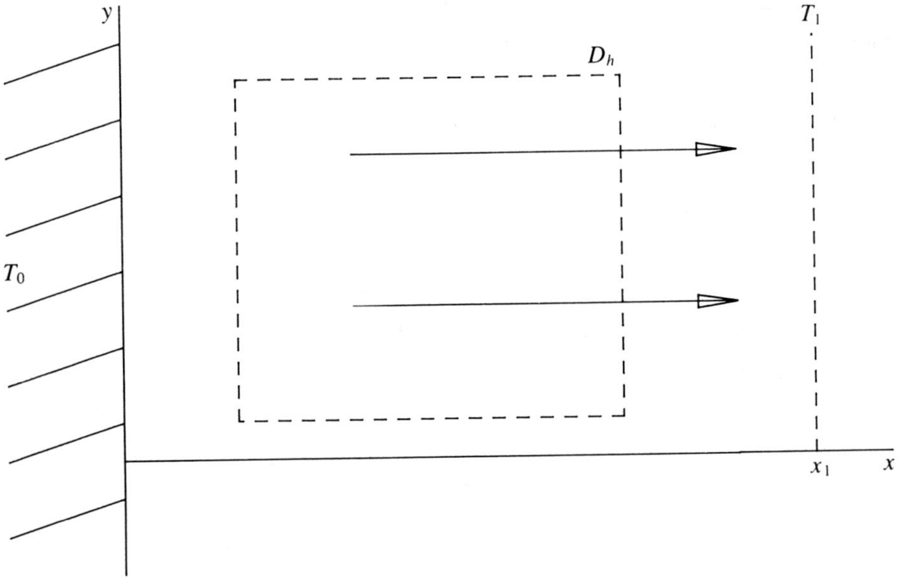

The temperature function is

$$T = -4x + 40$$

Example 7.4 Determine the heat flux density function from the results of example 7.3. From equation (7.9)

$$\mathbf{D}_{\mathrm{h}} = -k\,\nabla T$$

$$\nabla T = \frac{\partial T}{\partial x}\hat{\mathbf{a}}_x + \frac{\partial T}{\partial y}\hat{\mathbf{a}}_y + \frac{\partial T}{\partial z}\hat{\mathbf{a}}_z$$

so that

$$\mathbf{D}_h = -k\frac{\partial T}{\partial x}\hat{\mathbf{a}}_x$$

$$= 4k\hat{\mathbf{a}}_x \ \mathrm{W/m^2}$$

There is a relationship in heat transfer which reminds us of resistance from electric fields. The thermal resistance depends on length, area, and a constant. Recall electrical resistance

$$R = \frac{l}{\sigma A}$$

where σ is conductivity. The thermal resistance has the same form. We show by example.

Example 7.5 Calculate the thermal resistance of the brick shown in Figure 7.5. Note that the brick can be utilized three ways. We calculate

Figure 7.5. See example7.5.

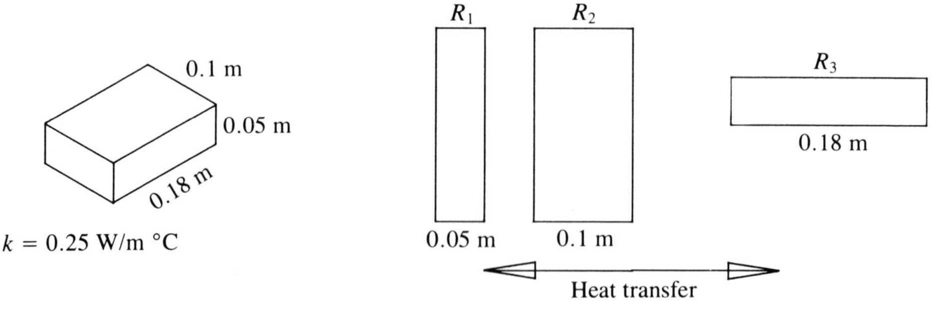

all three resistances.

$$R_1 = \frac{l_1}{kA_1} = \frac{0.05}{0.25(0.1)(0.18)} = 11.11$$

$$R_2 = \frac{l_2}{kA_2} = \frac{0.1}{0.25(0.05)(0.18)} = 44.44$$

$$R_3 = \frac{l_3}{kA_3} = \frac{0.18}{0.25(0.1)(0.05)} = 144.0\ °\mathrm{C/W}$$

Notice that R_3 has the highest thermal resistance.

D7.5 Two different metals are used in the block of an engine as shown in Figure 7.6. Calculate the thermal resistance per unit area. Assume a depth of one meter into the page.
Answer. $3.18 \cdot 10^{-2}$ °C/W.

D7.6 Two different metals are used in the block of an engine as shown in Figure 7.7. Calculate the thermal resistance per unit area. Assume a depth of one meter into the page.
Answer. $1.19 \cdot 10^{-2}$ °C/W

Figure 7.6. See **D7.5.**

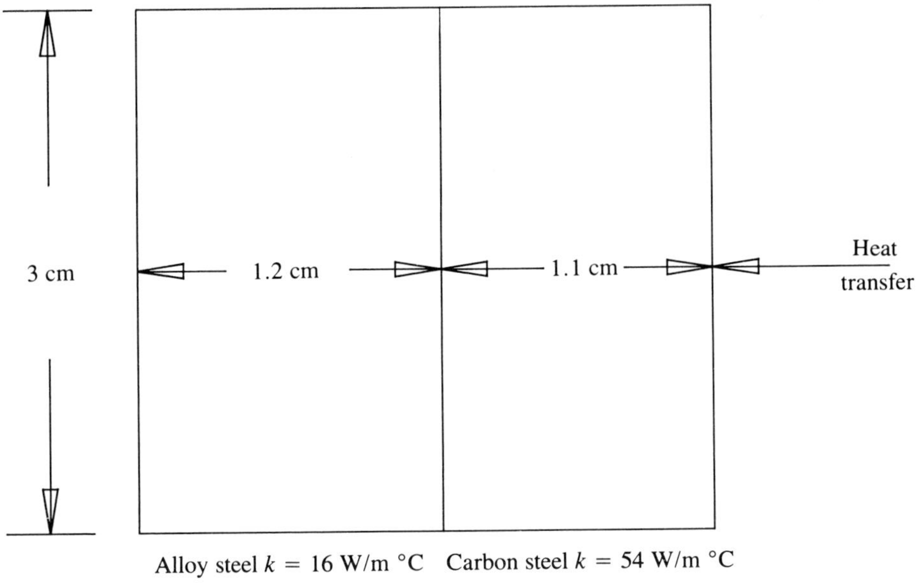

To conclude the conduction part of this chapter we refer back to Figure 7.4. The dashed square represents a Gaussian surface. If there is no source or sink within the surface Gauss' law becomes

$$0 = \oint \mathbf{D}_\mathrm{h} \cdot \mathrm{d}\mathbf{A}$$

Since $\mathbf{D}_\mathrm{h}$ is in the x-direction

$$\mathrm{d}\mathbf{A} = \pm \mathrm{d}y\ \mathrm{d}z\hat{\mathbf{a}}_x$$

Considering only $\mathrm{d}y\ \mathrm{d}z\hat{\mathbf{a}}_x$

$$\int_0^z \int_0^y D_\mathrm{h}\hat{\mathbf{a}}_x \cdot \mathrm{d}y\ \mathrm{d}z\hat{\mathbf{a}}_x = Q_\mathrm{h}$$

or

$$\mathbf{D}_\mathrm{h} = \frac{Q_\mathrm{h}}{yz}\hat{\mathbf{a}}_x \tag{7.21}$$

where Q_h represents the heat leaving the boundary at $x = 0$. Now let's consider heat transfer by convection.

Figure 7.7. See **D7.6.**

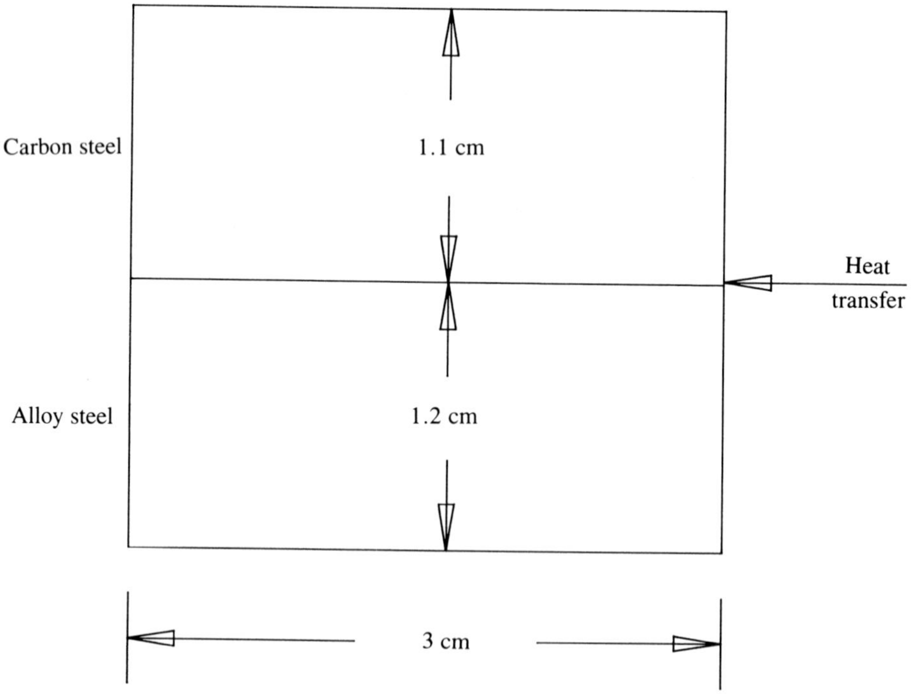

7.6 Convection

Convection is the second category of heat transfer in this chapter. Convection is concerned with heat transfer in fluids; in addition convection is divided into two categories – forced and natural convection. We have experienced heat transfer by convection. In natural convection the gravitational effects cause the heavier fluid to replace the lighter. In forced convection an external source, for example, a fan, forces air (fluid) across a heated surface to move the body of air. Examples are the heat transfer inside an oven or the cooling of an air-cooled engine. Natural convection also occurs inside a baking oven. The more dense, cooler air continually displaces the warmer air which rises to facilitate baking. Forced convection occurs in cooling an air-cooled engine for example. The moving cool air removes the heat from the surface of the engine. The heat transfer equation is given by

$$Q_{\mathrm{h}} = hA(T_2 - T_1) \tag{7.22}$$

where Q_{h} is the heat transfer rate in J/s, A is the surface area in m^2, $T_2 - T_1$ is the difference in temperature in °C, and h is the convective heat transfer coefficient in $\mathrm{W/m^2\,°C}$. We define a density, $\mathbf{D}_{\mathrm{h}}$, by dividing by area

$$\mathbf{D}_{\mathrm{h}} = \frac{Q_{\mathrm{h}}}{A}\hat{\mathbf{a}}_D \triangleq h(T_2 - T_1)\hat{\mathbf{a}}_D \tag{7.23}$$

where $\hat{\mathbf{a}}_D$ is in the direction of $\mathbf{D}_{\mathrm{h}}$.

Consider a simple example of forced convection. A spherical balloon is filled with hot air. As more air is forced into the balloon it continues to expand. At some unknown air volume the balloon will explode. By forced convection the hot air will mix with the surrounding cooler air. Neglect gravitational effect. Let the balloon explode at a radius of a meters. The temperature of the hot air is T_2; the external temperature is T_1. Assume the balloon material is removed; only the expanding gas is left.

Example 7.6 Find the heat flow per second due to the hot air, shown in Figure 7.8. Using Gauss' law, $Q_{\mathrm{h}} = \oint \mathbf{D}_{\mathrm{h}} \cdot \mathrm{d}\mathbf{A}$, and Figure 7.8 we obtain the heat flux density at $r = a$

$$D_{\mathrm{h}} = \frac{Q_{\mathrm{h}}}{4\pi a^2} \tag{7.24}$$

also

$$D_h = h(T_1 - T_2) \tag{7.25}$$

Combining equations (7.24) and (7.25) we obtain

$$Q_h = 4\pi a^2 h(T_1 - T_2)$$

D7.7 A balloon is filled with hot air to a bursting temperature of $T = 35\,°C$ and a radius of 10 cm. The outside temperature is 28 °C. The convective heat transfer coefficient of the outside air is $h = 8\ W/m^2\,°C$. Calculate the quantity of heat and the heat flux density at $r = 10$ cm. *Answer.* $7.04 \cdot 10^{-2}$ W, 56.02 W/m².

7.7 Cylindrical source

A more common configuration in heat transfer is the cylinder shown in Figure 7.9. Examples are steam pipes and cooling tubes. Consider a heated rod shown in Figure 7.9(b). The free air surrounding the rod will remove the heat by convection.

$$\begin{aligned} Q_h &= hA\,\Delta T \\ &= h2\pi a(1)(T_0 - T_{air}) \end{aligned} \tag{7.26}$$

Figure 7.8. An expanding balloon. See example 7.6.

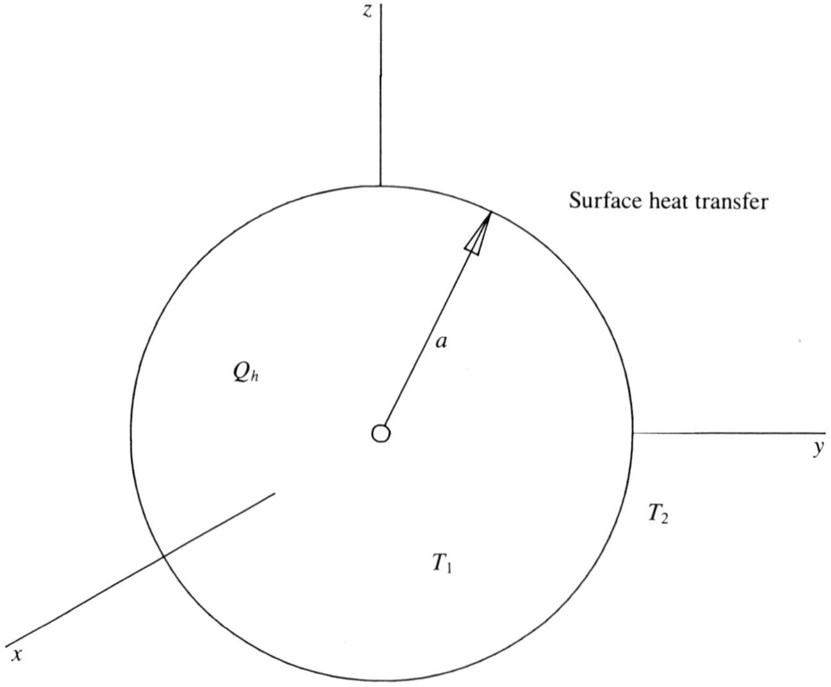

Figure 7.9. The cylindrical configuration, (*a*) orientation of cylinder, (*b*) heated rod.

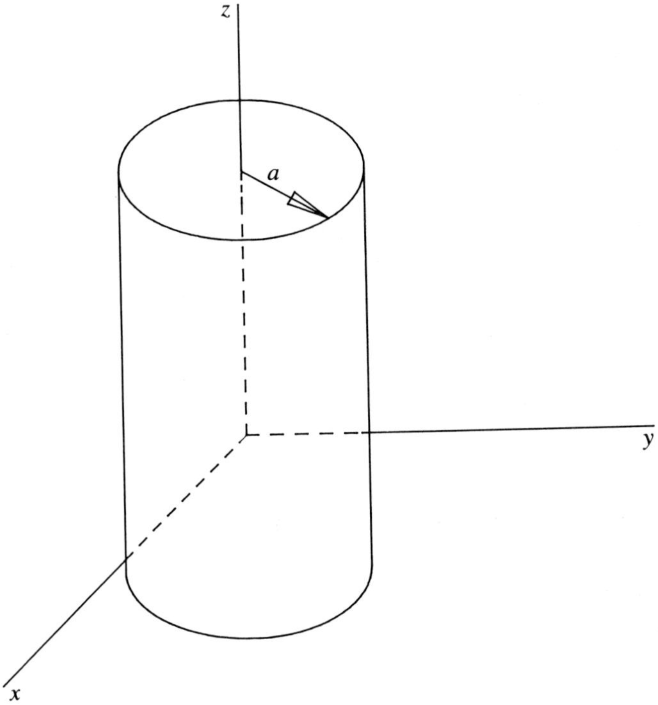

(*a*)

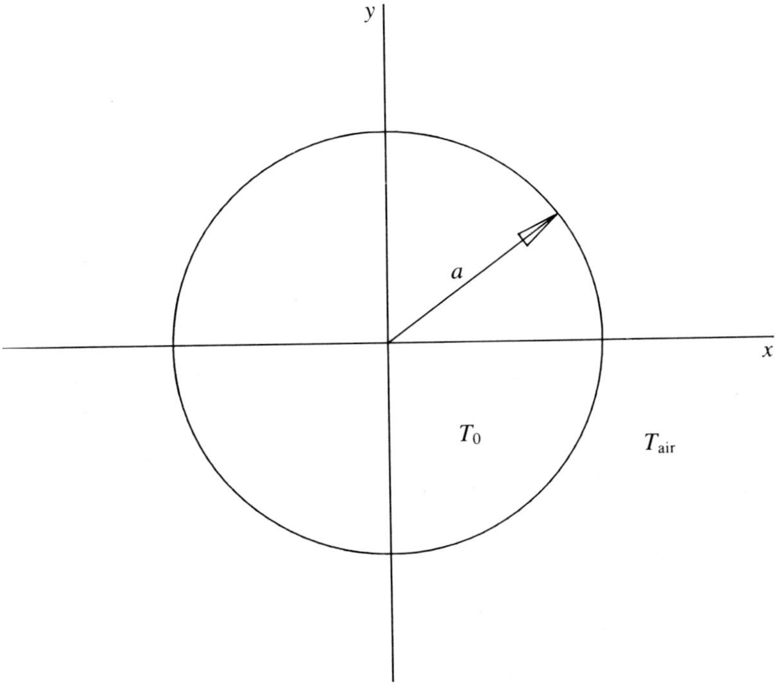

(*b*)

or

$$T_0 - T_{\text{air}} = \frac{Q_h}{2\pi a h} \tag{7.27}$$

per unit length, where h is the surface heat transfer coefficient.

The assumption is made that the energy is transferred radially. Therefore the area of concern is $2\pi a(1)$.

Example 7.7 A 2 cm diameter, 10 cm long connecting rod for an internal combustion engine is operating at a temperature of 150 °C. The (surrounding) engine oil is at 60 °C. The convection heat transfer coefficient of the oil is 10^3 W/m^2 °C. Calculate the heat transferred from the connecting rod to the oil. Also calculate the heat transfer density at surface of the rod.

From equation (7.26) the heat transferred is

$$\begin{aligned} Q_h &= hA\,\Delta T \\ &= 10^3 2\pi(10^{-2})(10^{-1})(150° - 60°) \\ &= 565.49 \text{ W} \end{aligned}$$

The heat flux density is

$$\begin{aligned} D_h &= \frac{Q_h}{A} \\ &= \frac{565.49}{2\pi(10^{-3})} \\ &= 90 \cdot 10^3 \text{ W/m}^2 \end{aligned}$$

Another common situation is one in which hot (or cold) water is transported through a pipe. There is a thermal transfer through the pipe and at the surface of the pipe. From Figure 7.10 we see that the transfer occurs at the inner surface, through the pipe and at the outer surface. Model this with three series resistors, that is, $R_T = R_1 + R_2 + R_3$. R_1 and R_3 represent convection heat and R_2 represents conductive heat (through the pipe). We write

$$T_\omega - T_1 = \frac{Q_h}{2\pi a h_1} \tag{7.28}$$

$$T_1 - T_2 = \frac{Q_h}{2\pi k} \ln \frac{2a}{a} \tag{7.29}$$

and

$$T_2 - T_{air} = \frac{Q_h}{2\pi(2a)h_2} \tag{7.30}$$

$$T_\omega - T_{air} = \frac{Q_h}{2\pi}\left[\frac{1}{ah_1} + \frac{\ln 2}{k} + \frac{1}{2ah_2}\right] \tag{7.31}$$

Let's attempt an example.

Example 7.8 Calculate the total heat transfer from the pipe carrying hot water ($T = 90\,°C$) to the ambient air at 5 °C. $h_1 = 300\ W/m^2\,°C$,

Figure 7.10. A pipe conveying fluid immersed in fluid (*a*) section of pipe, (*b*) model.

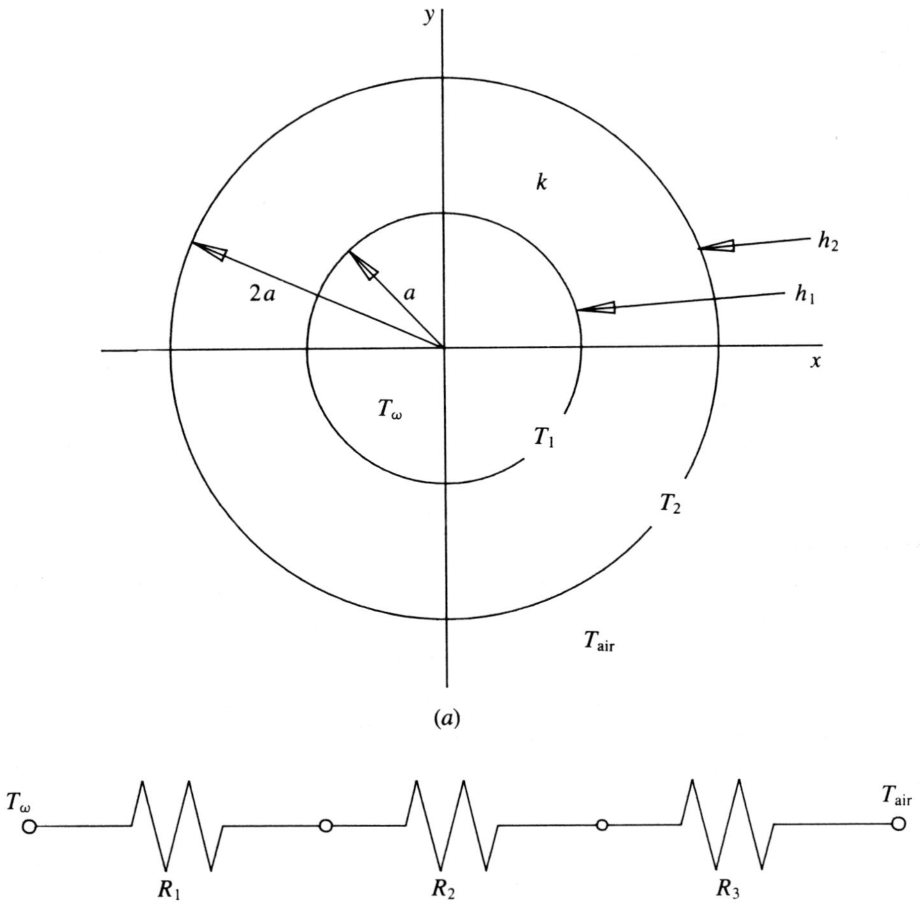

(*a*)

(*b*)

$T_{\text{air}} = 5\,°\text{C}$, $h_2 = 30\ \text{W/m}^2\,°\text{C}$, $a = 10^{-2}\ \text{m}$, $k = 300\ \text{W/m}\,°\text{C}$. From equation (7.31)

$$T_\omega - T_{\text{air}} = \frac{Q_h}{2\pi}\left[\frac{1}{10^{-2}(300)} + \frac{\ln 2}{300} + \frac{1}{2\cdot 10^{-2}(30)}\right]$$

$$Q_h = \frac{2\pi 85}{\left[\frac{1}{10^{-2}(300)} + \frac{\ln 2}{300} + \frac{1}{2\cdot 10^{-2}(30)}\right]}\ \text{W}$$

D7.8 A pipe of inner radius $r = 2$ cm and outer radius $r = 3$ cm contains hot water at a constant temperature of 95 °C. The surface coefficient for the hot water is 300 W/m² °C. The pipe is aluminum; $k = 204$ W/m °C. The forced air outside has a surface coefficient of 100 W/m² °C. Calculate the heat transfer, if the air temperature is 29 °C.
Answer. 829.4 W.

D7.9 Calculate the heat flux density at (a) $r = 2$ cm and (b) $r = 3$ cm for the pipe in **D7.8**.
Answer. (a) 6.6 kW/m² (b) 4.4 kW/m².

D7.10 From the results of **D7.8** calculate the thermal resistance of the pipe.
Answer. $8\cdot 10^{-2}$ °C/W.

7.8 Plane source

A third popular configuration is a planar source. Consider a wall separating a warm region from a cooler region as shown in Figure 7.11. Let's calculate the heat loss for three cases. First, consider the wall to be at a constant temperature T_2; this will be convection. Second, we assume a temperature gradient across the wall; this will be conduction. For the third case we will combine the first and second cases – convection and conduction. Note that the heat flow is in the x-direction. First consider the wall at a constant temperature T_2. The heat transfer to the T_3 region is by convection. We write

$$Q_h = hA(T_2 - T_3) \tag{7.32}$$

where h is the convection heat transfer coefficient and A is the cross-sectional area of the wall normal to the direction of heat flow.

Next consider the temperature at $x = x_1$ to be T_1 and the temperature at $x = x_2$ to be T_2. Therefore there is a gradient across the wall; there is heat loss in the wall through conduction. We write

$$Q_h = kA\frac{\Delta T}{\Delta x}$$

or

$$Q_h = kA\frac{T_1 - T_2}{x_1 - x_2} \tag{7.33}$$

Finally for the third case consider the two cases together. The heat transfer is of the two types – convection at the wall's surface and conduction through the wall. Manipulating and combining equations (7.32) and (7.33) we obtain

$$T_1 - T_2 + T_2 - T_3 = \frac{Q_h}{kA}(x_1 - x_2) + \frac{Q_h}{hA}$$

$$T_1 - T_3 = \frac{Q_h}{A}\left[\frac{x_1 - x_2}{k} + \frac{1}{h}\right]$$

The rate of heat transfer is

$$Q_h = (T_1 - T_3)A\left[\frac{x_1 - x_2}{k} + \frac{1}{h}\right]^{-1}$$

Figure 7.11. A planar wall separating regions of different temperatures.

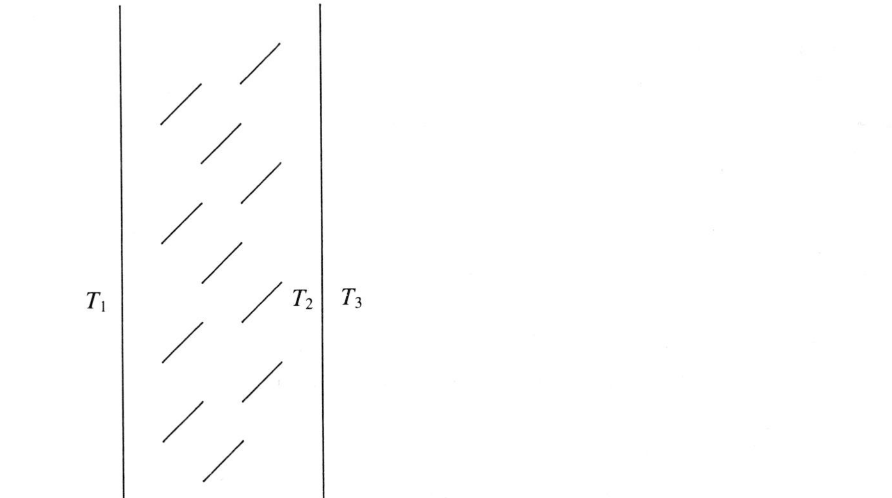

D7.11 Calculate the rate of heat transfer between a 0.5 m × 0.5 m door of an oven baking bread at a temperature of 450 °F and the kitchen air at a temperature of 80 °F. The door is 4 cm thick – 3.8 cm fiberglass $k = 0.30$; 0.2 cm steel $k = 50$ W/mK. The steel is on the kitchen side of the door. The inner surface temperature of the door is 450 °F. The outer surface coefficient is 20 W/m²K.
Answer. 290.7 W.

D7.12 Determine the temperature at the boundary between the fiberglass and the steel of **D7.11.**
Answer. 357.9 K.

7.9 Radiation

The third method of heat transfer is through radiation. It is a high frequency occurrence.

As with conduction and convection, heat transfer by radiation increases with temperature. It is not however a linear relationship. Radiative transfer varies with the fourth power of temperature. It is governed by the Stefan–Boltzmann law. The law for black body radiation is

$$D_h = \sigma T^4 \tag{7.34}$$

where σ is the Stefan–Boltzmann constant ($\sigma = 5.67 \cdot 10^{-8}$ W/m²K). A black body is one which absorbs all the energy it receives. It, in turn, radiates all the energy received when its temperature is higher than its environment. Both reception and radiation are governed by equation (7.34). A black body is an ideal model which is a convenience for calculations. All bodies absorb, reflect, and transmit radiated energy. We write

$$\alpha + \rho_r + \tau = 1 \tag{7.35}$$

where α is absorptivity, ρ_r is reflectivity and τ is transmissivity. (Note that the subscript 'r' is added to ρ so that it will not be confused with volume density which is used throughout the rest of the text.)

We can visualize equation (7.35) by noting Figure 7.12. We see that the received (incident) radiative heat is divided into that reflected by the surface of the body, that absorbed by the body and that which passes through the body.

It is the absorbed heat energy which can be radiated to other bodies when different temperature conditions exist. The term emissivity, e, is defined. In this text we assume the emitted energy is the same as the absorbed energy. None of the absorbed energy is retained. In other words

$$\alpha = e$$

As stated above, radiation is a high frequency occurrence. In addition it should be noted that emissivity is frequency and temperature dependent. Neither dependency will be considered in this text. Consider several examples.

Example 7.9 Calculate the rate of heat density radiated from a black body at 5000 K. From equation (7.34)

$$\begin{aligned} D_h &= \sigma T^4 \\ &= 5.67 \cdot 10^{-8}(5000)^4 \\ &= 3544 \cdot 10^4 \text{ W/m}^2 \end{aligned}$$

Figure 7.12. A body at temperature T_0 receiving radiative heat.

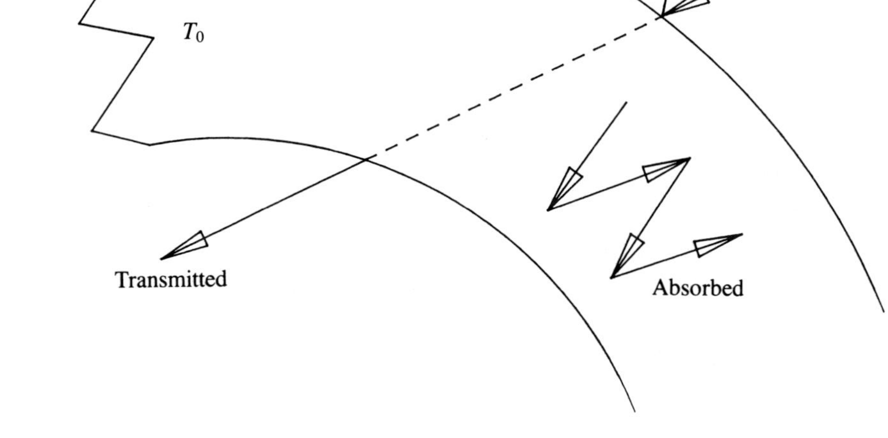

We multiply density by area to obtain power, or energy rate. That is

$$Q_h = D_h A$$

Example 7.10 Assume the sun is a black body radiating at 5760 K ($r_{sun} \approx 695 \cdot 10^6$ m). Calculate the rate of heat energy radiating from its surface.

$$\begin{aligned} Q_h &= \sigma A T^4 \\ &= 5.67 \cdot 10^{-8}(4\pi)(695 \cdot 10^6)^2(5760)^4 \\ &= 3.79 \cdot 10^{26}\ \text{W} \end{aligned}$$

Example 7.11 If the sun (a spherical source) in example 7.10 is radiating at 5760 K, calculate the received power on a 2 m × 2 m flat plate aboard a satellite 10^{12} m from the center of the sun.

From example 7.10 the sun's power is $3.79 \cdot 10^{26}$ W. Since it is a spherical source we can divide the power by the area $4\pi r^2$ to obtain the density. Also we can invoke Gauss' theorem and consider the sun as a point source of strength $3.79 \cdot 10^{26}$ W. The (imaginary) Gaussian surface intersects the plane of the flat plate. The area of this (spherical) surface is $4\pi(10^{12})^2$.

The power density at the flat plate

$$\begin{aligned} D_h &= \frac{Q_h}{4\pi r^2} \\ &= \frac{3.79 \cdot 10^{26}}{4\pi(10^{12})^2} \\ &= 0.302 \cdot 10^2\ \text{W/m}^2 \end{aligned}$$

The power on the 2 m × 2 m flat plate is

$$\begin{aligned} Q_h &= D_h(\text{Area}) \\ &= 1.208 \cdot 10^2\ \text{W} \end{aligned}$$

Note that this is an approximation because the Gaussian surface is curved while the flat plate is not. However, at these distances, 10^{12} m, the approximation is acceptable for our purposes.

D7.13 The sun is radiating at 5760 K. Calculate the received power on seven flat 3 m × 3 m collectors aboard a satellite 10^{13} m from the center of the sun.

Answer. 19 W.

Now let's consider other than radiating black bodies. Equation (7.34) must be modified or

$$D_h = C_1 \sigma T^4 \tag{7.36}$$

where C_1 is some constant $0 \leq C_1 \leq 1$

$$C_1 = 0 \Rightarrow \text{a perfect reflector}$$

$$C_1 = 1 \Rightarrow \text{a black body}$$

In addition when two bodies are at different temperatures equation (7.36) becomes

$$D_h = C_1 \sigma (T_1^4 - T_2^4) \tag{7.37}$$

That is, the heat transfer is dependent on the *difference* in temperatures.

Now let's consider a spherical source in more detail.

7.10 Spherical source

As we have seen in the above exercises we can use Gauss' theorem in radiative problems. We repeat it here for convenience

$$\Phi \equiv Q_h = \oint \mathbf{D}_h \cdot d\mathbf{A}$$

from which we obtain

$$\mathbf{D}_h = \frac{Q_h}{4\pi r^2} \hat{\mathbf{a}}_r \tag{7.38}$$

in spherical coordinates. Equation (7.38) is a most simple relationship; the source, Q_h, is a radiating sphere. What if Q_h ($\equiv \Phi$) is a receiving sphere? This complicates matters, for now we need to deal with areas and shape factors. For example consider Figure 7.13. Point A will receive more energy than any other point on the sphere. In addition the shaded half of the receiving sphere will receive no energy. Consider an example.

Example 7.12 Two concentric spheres are at temperatures T_1 and T_2 ($T_1 > T_2$) as shown in Figure 7.14. Calculate the total heat absorbed by the outside sphere. The inner sphere is radiating and the outer sphere is receiving. The inner sphere is a black body at temperature $T_1 = 600$ K; the outer sphere is made of aluminum at temperature $T_2 = 400$ K and has an absorptivity $\alpha = 0.08$. Assume $\tau = 0$ for the aluminum. Since

Figure 7.13. A sphere receiving radiative heat.

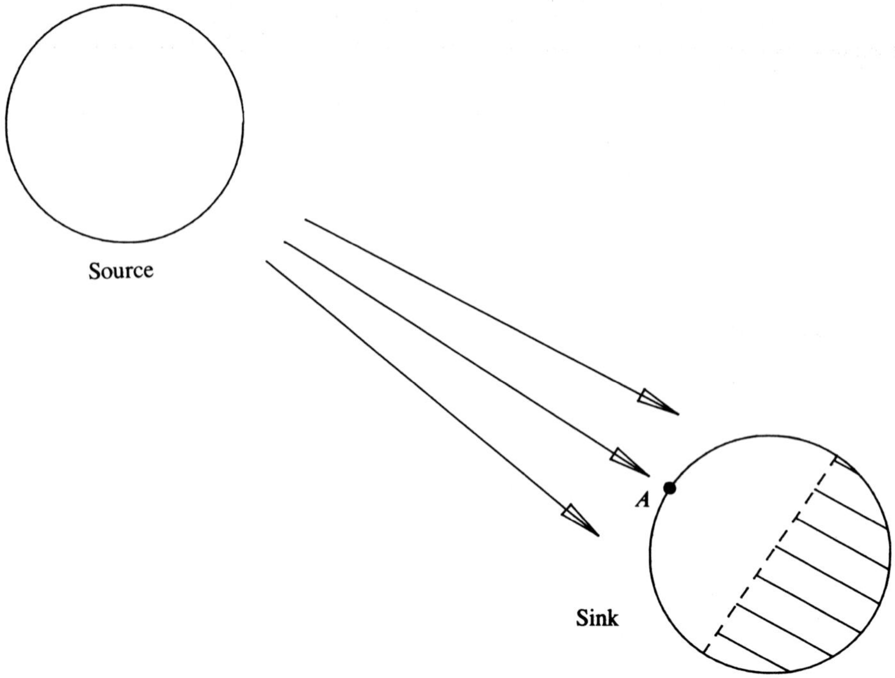

Figure 7.14. Two concentric spheres at different temperatures.

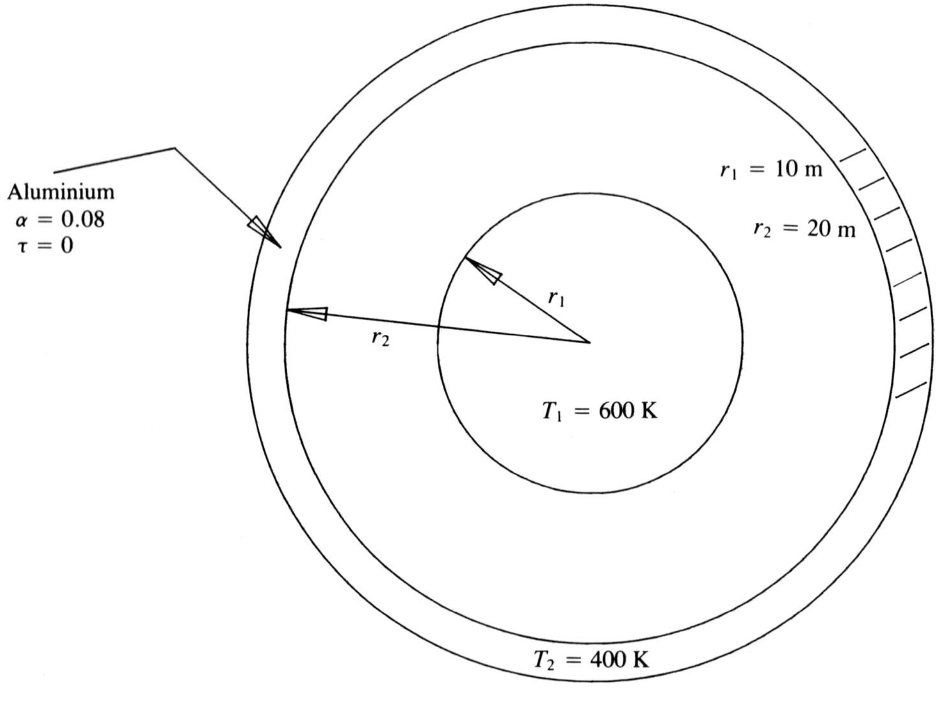

$\alpha + \rho_r + \tau = 1$, $\rho_r = 0.92$ and 92% of the energy is reflected toward the source by the aluminum. None is transmitted; eight percent is absorbed. From equation (7.34)

$$\begin{aligned} D_h &= \sigma T^4 \\ &= 5.67 \cdot 10^{-8}(600)^4 \\ &= 7348.3\ \mathrm{W/m^2} \end{aligned}$$

For the inner sphere

$$\begin{aligned} Q_h &= D_h 4\pi r_1^{\,2} \\ &= 7348.3(4\pi)(10)^2 \\ &= 9.23 \cdot 10^6\ \mathrm{W} \end{aligned}$$

For the outer sphere using equation (7.37)

$$\begin{aligned} D_h &= C_1\sigma(T_1^{\,4} - T_2^{\,4}) \\ &= 0.08(5.67 \cdot 10^{-8})(600^4 - 400^4) \\ &= 471.7\ \mathrm{W/m^2} \end{aligned}$$

The total energy absorbed by the outer sphere is

$$\begin{aligned} Q_h &= D_h 4\pi r_2^{\,2} \\ &= 471.7 \cdot 4\pi(20)^2 \\ &= 2.37 \cdot 10^6\ \mathrm{W} \end{aligned}$$

D7.14 Repeat example 7.12 if the outside sphere is made of wood where $\alpha = 0.86$.
Answer. 25.5 MW.

7.11 Cylindrical source

If the radiating source is cylindrical we can use Gauss' law again, assuming the cylinder is long compared to its radius, say 10:1. With this assumption fringing can be neglected.

$$\Phi \equiv Q_h = \oint \mathbf{D}_h \cdot d\mathbf{A}$$

and

$$\mathbf{D}_h = \frac{Q_h}{2\pi r(1)}\,\hat{\mathbf{a}}_r \tag{7.39}$$

For a cylinder of radius a

$$D_{\mathrm{h}} = \frac{Q_{\mathrm{h}}}{2\pi a}$$

Using equation (7.34),

$$D_{\mathrm{h}} = \sigma T^4$$

we obtain

$$\sigma T^4 = \frac{Q_{\mathrm{h}}}{2\pi a}$$

or

$$Q_{\mathrm{h}} = \sigma(2\pi a)T^4$$

for the power of the source per meter.

Example 7.13 Calculate the total power for a black body at 500 K. The body is 10 meters in length and has a radius of 0.5 m. From the above discussion

$$\begin{aligned} Q_{\mathrm{h}} &= \sigma(2\pi a)LT^4 \\ &= 5.67 \cdot 10^{-8}[2\pi(0.5)]10(500)^4 \\ &= 111.33\ \mathrm{kW} \end{aligned}$$

Example 7.14 Calculate the radiative power density at a radial distance of 150 meters for the black body of example 7.13. From the results of example 7.13 and equation (7.39)

$$\begin{aligned} \mathbf{D}_{\mathrm{h}} &= \frac{Q_{\mathrm{h}}}{2\pi r}\hat{\mathbf{a}}_r \\ &= 118.1\hat{\mathbf{a}}_r \end{aligned}$$

Example 7.15 Given a long, cylindrical heating cylinder mounted parallel to a concentric reflector as shown in Figure 7.15, calculate the power density at the reflector and the power per meter reflected by the perfect reflector. The radiator (source) is 750 W.

From equation (7.39) the power density at the reflector is

$$\begin{aligned} D_{\mathrm{h}} &= \frac{Q_{\mathrm{h}}}{2\pi r(1)} \\ &= \frac{750}{2\pi(0.01)} \end{aligned}$$

The power at the reflector is

$$Q_h = \int_0^1 \int_0^{120^\circ} \frac{750}{2\pi(0.01)} \hat{\mathbf{a}}_r \cdot r \,\mathrm{d}\phi \,\mathrm{d}z\hat{\mathbf{a}}_r$$

$$= 250 \text{ W}$$

Note that since $\rho_r = 1$ in example 7.15, $\tau = 0$ and $\alpha = 0$. No energy is transmitted or absorbed by the reflector.

D7.15 Calculate the power density at the reflector in example 7.15 if the source is 100 cm away from the reflector. Also calculate the angle of the reflector such that 100 watts is reflected.
Answer. 1193.7 W/m^2, 48°.

7.12 Planar source

The third and final configuration we study in this chapter is the planar. Examples of planar sources are hot plates, cooling fins and enclosures. We show some examples.

Figure 7.15. A cylindrical heater – see example 7.15.

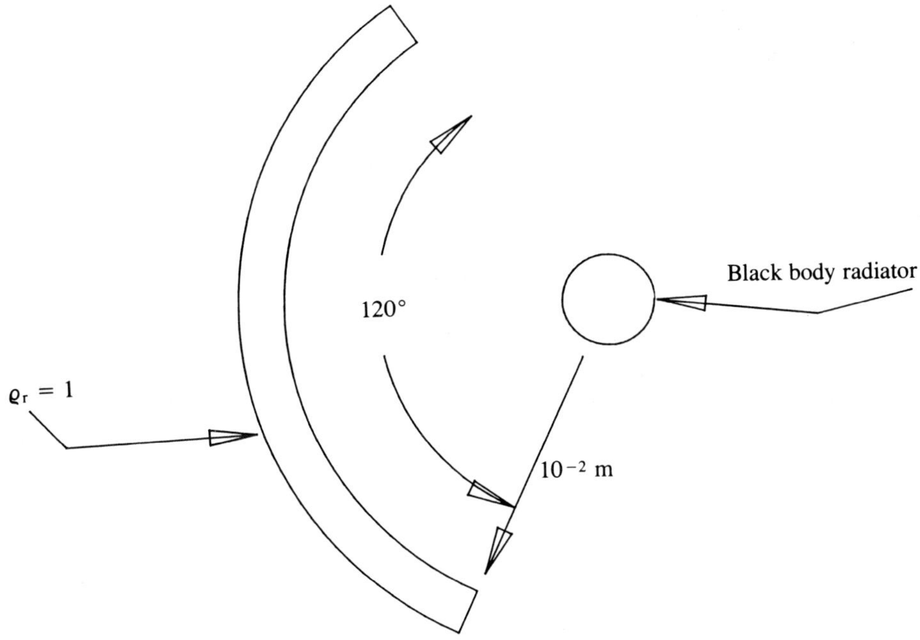

Example 7.16 Two flat plates, both one meter square, are parallel as shown in Figure 7.16. The top plate is a black body at temperature T_1; the bottom is at temperature T_2. Neglect fringing. If the bottom plate is porcelain with $\alpha = 0.8$ calculate the total heat absorbed by the bottom plate. Assume $\tau = 0$. You may recall that the flux from an infinite plate is *not* dependent on distance from the plate. Since we are neglecting fringing we may assume both plates are infinite.

From equation (7.34)

$$D_{\mathrm{h}} = \sigma T^4$$
$$= 5.67 \cdot 10^{-8}(1000)^4$$
$$= 5.67 \cdot 10^4 \text{ W/m}^2$$

and

$$Q_{\mathrm{h}} = D_{\mathrm{h}}(1)$$

where $A = 1$.

The heat flux density at the bottom plate is from equation (7.37)

$$D_{\mathrm{h}} = C_1\sigma(T_1^4 - T_2^4)$$

The total heat absorbed

$$Q_{\mathrm{h}} = D_{\mathrm{h}}(1)$$
$$= (0.8)5.67 \cdot 10^{-8}[1000^4 - 500^4] = 42525 \text{ W}$$

Figure 7.16. Two parallel plates – see example 7.16.

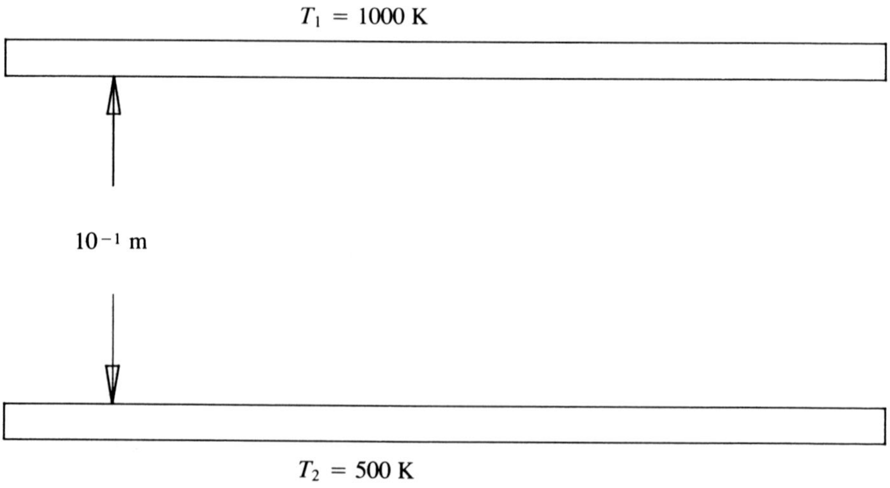

Example 7.17 Calculate the total heat absorbed by the bottom plate in example 7.16 if the bottom plate is tilted by 15° as shown in Figure 7.17. The receiving area for the bottom porcelain plate decreases by the cosine as the plate is tilted, so that

$$Q_h = Q_{h_0}(\cos 15°)$$

$$= 41076 \text{ W}$$

where $Q_{h_0} = 42525$ W represents the total heat when the plates are parallel.

If the top plate is tilted from its original position – in addition to the bottom plate – it is necessary to multiply the source heat by the cosine as well. This is illustrated in Figure 7.18. In other words

$$Q_h = Q_{h_0}(\cos \alpha)(\cos \beta)$$

D7.16 Calculate the total heat (*a*) absorbed by and (*b*) transmitted through the roof shown in Figure 7.19. Assume $\alpha = 0.9$ and $\tau = 0.01$ for the roof shingles. Assume the sun's temperature to be 5800 K.
Answer. (*a*) $8.8 \cdot 10^9$ W, (*b*) $9.78 \cdot 10^7$ W.

Much more could be written on radiation and heat transfer but let this chapter serve as a brief introduction to thermal fields.

Figure 7.17. See example 7.17.

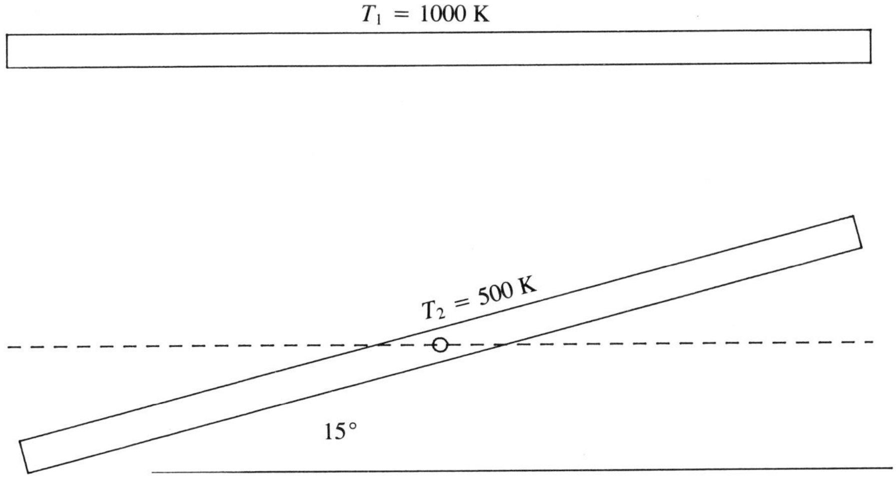

Figure 7.18. Two plates tilted at angles α and β.

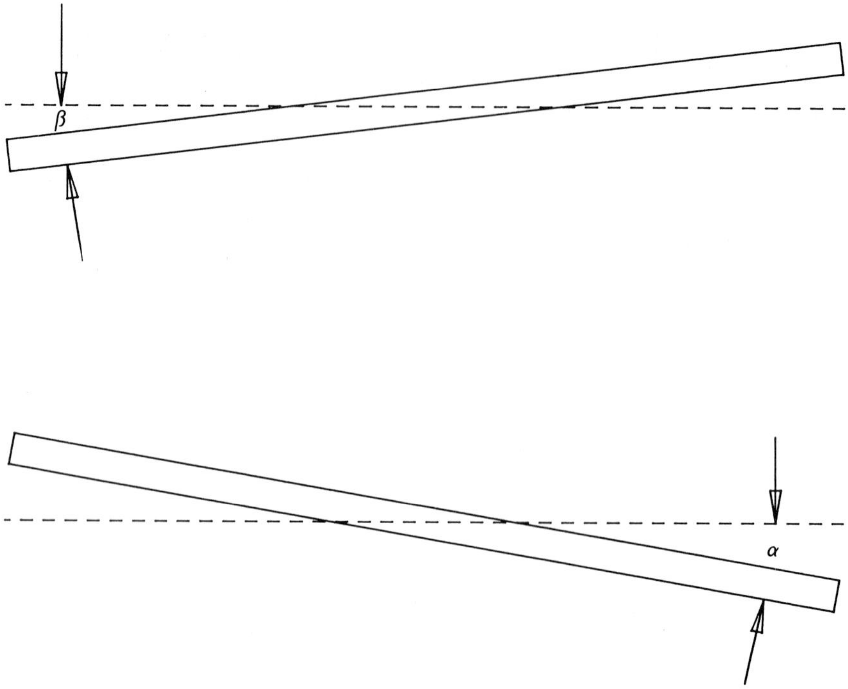

Figure 7.19. See **D7.16.**

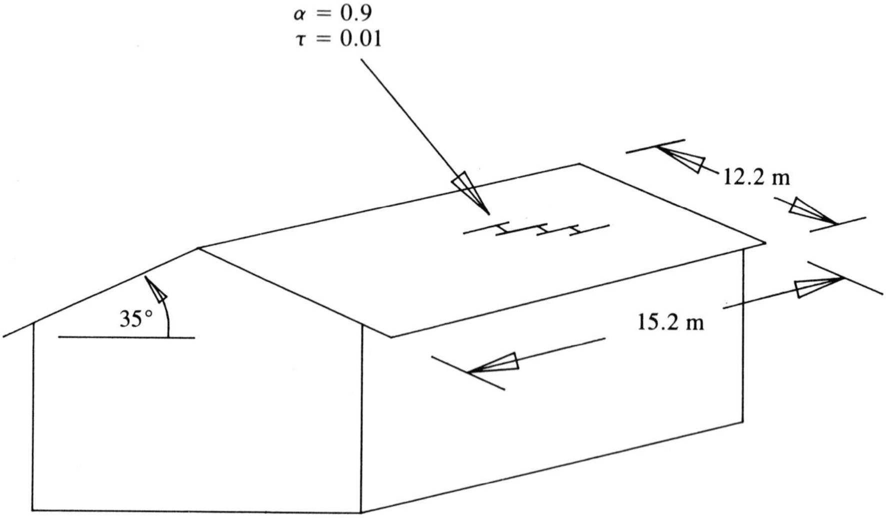

Problems

1. A flat aluminum plate ($k = 180$ W/m °C) is conducting heat along its longest dimension. It is $\frac{1}{2}$ cm × 5 cm × 0.3 m. If it is conducting 750 watts of heat at 275 °C determine the temperature of the cooler end.

2. A flat copper plate $\frac{1}{8}$ cm × 2 cm × 0.2 m in dimension is conducting heat along the 0.2 m dimension. If the heat density is 55 W/m^2 calculate the difference in temperature from end to end.

3. A tiny cylindrical rod of tungsten ($k = 160$ W/m °C) 4 cm long is used for the filament of a light bulb. Calculate the diameter of the filament if 40 W is generated by a temperature difference of 60 °C.

4. A 100 W spherical, incandescent light bulb is 5 cm in radius. If the glass ($k = 0.78$ W/m °C) is $1.7 \cdot 10^{-3}$ m thick calculate the thermal field intensity across the glass.

5. Calculate the heat potential necessary to provide a heat density of 15 W/m^2 on the surface of a copper sphere of radius $r = 3$ cm. The sphere is 0.5 cm thick; the source is within.

6. If $\mathbf{D} = \sin\theta\cos\phi/r\hat{\mathbf{a}}_r + \cos\theta\sin\phi\hat{\mathbf{a}}_\phi$ in spherical coordinates determine the heat (volume) density by taking the divergence of $\mathbf{D}$.

7. An igloo (hemisphere) is made of brick ($k = 0.69$ W/m °C) of thickness 4 cm. Calculate heat conducted radially if the temperature difference is 50 °C. The igloo is 3 m in height.

8. Design a Quonset hut (semi-cylindrical building) such that the heat density just on the outside does not exceed 20 W/m^2. The hut must be at least 2.5 m high in the center and 10 m in length. The minimum external temperature is 5 °C. Design against cold.

9. Continue the design of problem 8 by considering the constraints of economics and safety. Assume there is one order of magnitude difference between cardboard, pine wood and brick. That is, brick is 10 times more expensive than wood which is 10 times more expensive than cardboard. Also brick is 10 times ‘stronger’ than wood and so forth.

10. From Figure 7.20 design the size of the clamps such that the temperature at the handles is at most 45 °C. The handles are made of high temperature steel, $k = 50$ W/m °C. The heated piece is 140 °C.

11. Calculate the heat density of the handles if the temperature difference is 95 °C, the inside radius is 7 cm and the outside radius is 9 cm. Assume the handles are made of tungsten, $k = 160$ W/m °C.

12. Model the handles of Figure 7.20 as parallel resistors and derive a formula for the resistance of either side.

13. A thin window separates quiet air, $h = 5\ \mathrm{W/m^2\,°C}$ outside at 37 °C and forced air, $h = 300\ \mathrm{W/m^2\,°C}$, inside at 24 °C. Determine the heat density transferred and the surface temperatures.

14. A cylindrical pipe is carrying boiling water, $h = 3 \cdot 10^4\ \mathrm{W/m^2\,°C}$. The ambient temperature surrounding the pipe is 30 °C. The pipe's inside diameter is 3 cm and the outside diameter is 3.5 cm. Calculate the heat transfer rate neglecting conduction.

15. Repeat problem 14 including conduction heat transfer where the pipe is made of copper.

Figure 7.20. See problems 10–12.

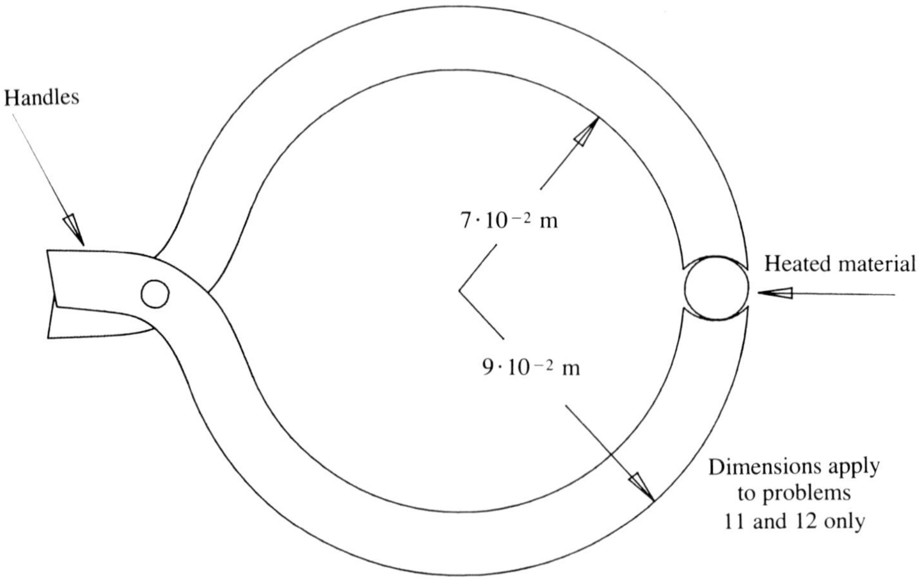

16. A cooling fin is radiating in two directions as shown in Figure 7.21. The fin is 3 cm × 10 cm and its temperature is 75 °C. Calculate the heat density in either direction and the total heat from the fin.

17. Calculate the heat received by radiation by the black body fin shown in Figure 7.21. $\alpha = 1$.

18. Calculate the heat received by the fin in Figure 7.21 if $\alpha = 0.5$ and $\tau = 0.5$. Also calculate the heat transmitted through the fin.

19. A cylindrical, black body $T = 0$, radius 5 m, is receiving radiation from a plane source at $T = 165$ °C. Calculate the heat transferred to the cylinder.

Figure 7.21. See problems 16–18.

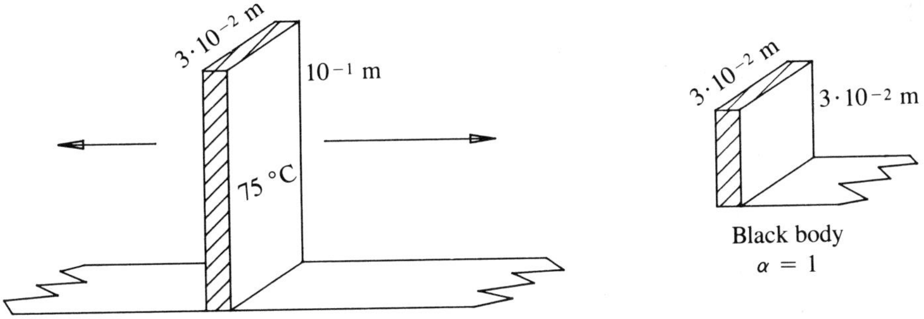

Figure 7.22. See problem 20.

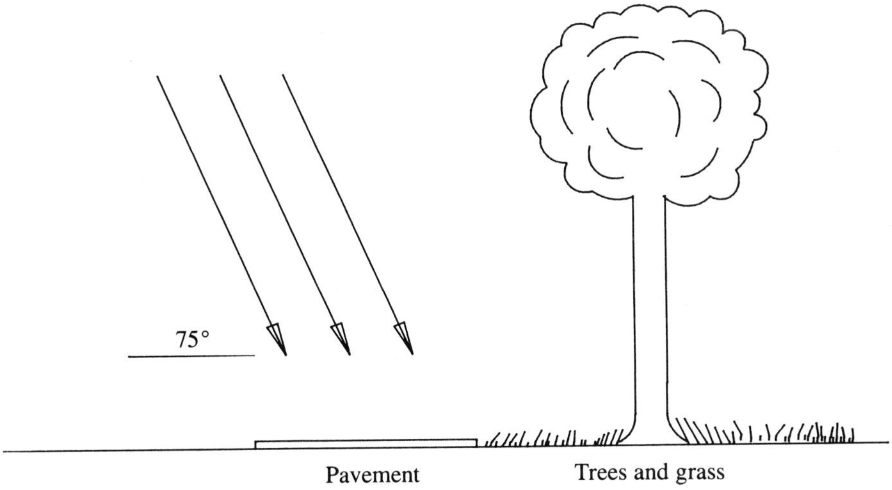

20. The sun is 75° above the horizon as shown in Figure 7.22. Compare the heat density re-radiated by the black pavement – $\alpha = 0.95$, $\tau = 0.05$ – with the heat re-radiated by the trees and grass – $\alpha = 0.5$, $\tau = 0.45$. The sun's temperature is 5760 K. Assume the environmental temperature to be 325 K. Assume emissivity equals absorptivity.

8

Other fields

8.1 Introduction

Other fields which have analogous quantities to those we have been studying are electrical current, flow through porous media, diffusion and acoustics. The fields due to electric currents, flow through porous media and diffusion are considered static fields, but acoustics is a time-varying field. Let us first examine fields due to electric currents.

8.2 Electrical current

Electrical current is defined as movement of charge per time, that is

$$i(t) \triangleq \frac{\mathrm{d}Q}{\mathrm{d}t}$$

Example 8.1 Calculate the current passing a point if $325 \cdot 10^3$ electrons travel past that point in 7 ms. The charge on an electron is $1.6 \cdot 10^{-19}$ C

$$\begin{aligned} i &\approx \frac{\Delta Q}{\Delta t} \\ &= \frac{325 \cdot 10^3 \cdot 1.6 \cdot 10^{-19}}{7 \cdot 10^{-3}} \\ &= 74.29 \cdot 10^{-13}\ \mathrm{A} \end{aligned}$$

If we consider the current as the flux, we write from Ohm's law

$$I = \frac{V}{R} \tag{8.1}$$

where $I(\equiv \Phi)$ is considered to be the flux, V is the potential or pressure and R is the resistance.

8.3 Spherical source

Consider a point source from which current is emanating. Visualize the point source as one electrode (the anode); the second electrode (the cathode) is concentric to the source. A resistive material lies between the two electrodes. Place a Gaussian surface around the resistor – inside the cathode – as shown in Figure 8.1. From Gauss' law

$$\Phi \equiv I = \oint \mathbf{J} \cdot \mathrm{d}\mathbf{A}$$

where d**A** is the cross-sectional area through which the flux is flowing. In this section the density is current density, $\mathbf{J}[A/\mathrm{m}^2]$. From Figure 8.1, we write

$$\Phi = \int_0^{2\pi} \int_0^{\pi} J_r \hat{\mathbf{a}}_r \cdot r^2 \sin\theta \, \mathrm{d}\theta \, \mathrm{d}\phi \hat{\mathbf{a}}_r$$

and

$$\mathbf{J} = \frac{\Phi}{4\pi r^2} \hat{\mathbf{a}}_r$$

From Ohm's law

$$\mathbf{J} = \sigma \mathbf{E}$$

Figure 8.1. A spherical current source.

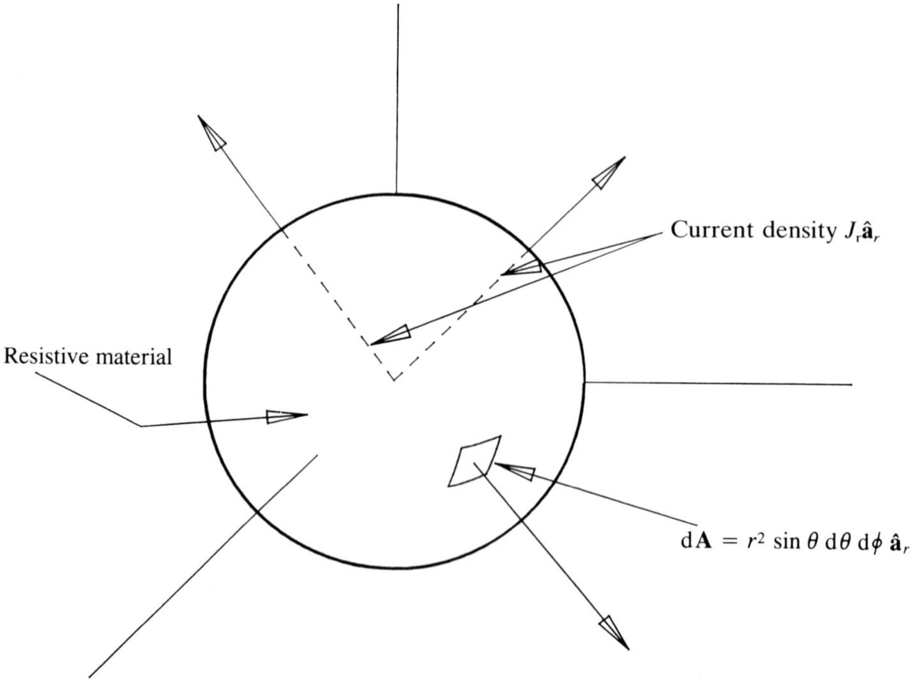

we obtain

$$\mathbf{E} = \frac{\Phi}{4\pi\sigma r^2}\,\hat{\mathbf{a}}_r \tag{8.2}$$

Example 8.2 Calculate the current density, **J**, at $r = 10^{-2}$ m for a spherical semiconductor material – radius 10^{-2} m. The current of 2 A is flowing radially.

$$\mathbf{J} = \frac{\Phi}{4\pi r^2}\,\hat{\mathbf{a}}_r = \frac{2\hat{\mathbf{a}}_r}{4\pi(10^{-2})^2} = 1592\hat{\mathbf{a}}_r \ \mathrm{A/m^2}$$

D8.1 A metal ball bearing radius 10^{-2} m, is conducting current radially; $I = 2$ A. The bearing is steel, $\sigma = 3 \cdot 10^4$ S. Calculate (*a*) **E** at $r = 0.5 \cdot 10^{-2}$ m (*b*) **J** at $r = 0.25 \cdot 10^{-2}$ m.
Answer. (*a*) $2.12 \cdot 10^{-1}\hat{\mathbf{a}}_r$ V/m (*b*) $2.55 \cdot 10^4\hat{\mathbf{a}}_r$ A/m^2.

8.4 Gradient

From Chapter 1 we write

$$\mathbf{E} = -\nabla V_i \tag{8.3}$$

Using equation (8.2) in equation (8.3) we obtain

$$\nabla V_i = -\frac{\Phi}{4\pi\sigma r^2}\,\hat{\mathbf{a}}_r \tag{8.4}$$

From Chapter 2 and using spherical coordinates

$$\nabla V_i = \frac{\partial V_i}{\partial r}\,\hat{\mathbf{a}}_r + \frac{1}{r}\frac{\partial V_i}{\partial \theta}\,\hat{\mathbf{a}}_\theta + \frac{1}{r\sin\theta}\frac{\partial V_i}{\partial \phi}\,\hat{\mathbf{a}}_\phi$$

We note from equation (8.4) that ∇V_i has an $\hat{\mathbf{a}}_r$ component only, thus

$$\frac{\partial V_i}{\partial r} = -\frac{\Phi}{4\pi\sigma r^2}$$

Integrating once we get

$$V_i = \frac{\Phi}{4\pi\sigma r} + \text{constant} \tag{8.5}$$

The equipotential surfaces are concentric spheres with the potential increasing as the radius decreases.

For two concentric spheres of radii r_1 and r_2 and voltages V_1 and V_2, respectively, we write

$$V_1 - V_2 = \frac{\Phi}{4\pi\sigma}\left(\frac{1}{r_1} - \frac{1}{r_2}\right) = \frac{\Phi(r_2 - r_1)}{4\pi\sigma r_1 r_2}$$

Since Φ is the current we obtain

$$R = \frac{r_2 - r_1}{4\pi\sigma r_1 r_2} \tag{8.6}$$

Recall the equation for dc resistance is

$$R = \frac{l}{\sigma A} \tag{8.7}$$

Comparing equations (8.6) and (8.7) we see that the length corresponds to $r_2 - r_1$ and the denominator in equation (8.6) resembles an area times a conductivity. Additionally if $r_1 \approx r_2$

$$R \approx \frac{\Delta r}{4\pi\sigma r^2} \tag{8.8}$$

Example 8.3 Calculate the resistance between two concentric metallic spheres where $\Delta r = r_2 - r_1$ of radii $r_1 = 2 \cdot 10^{-2}$ m and $r_2 = 2.1 \cdot 10^{-2}$ m separated by carbon, $\sigma = 10^{-5}$ s, using equations (8.6) and (8.8). From equation (8.8)

$$R = \frac{(2.1 - 2)10^{-2}}{4\pi\sigma(2 \cdot 10^{-2})^2} \approx 2 \cdot 10^4\ \Omega$$

and from equation (8.6)

$$R = \frac{10^{-3}}{4\pi\sigma(2.1)(2)10^{-4}} = 1.89 \cdot 10^4\ \Omega$$

D8.2 Determine the percentage error in making the approximation in example 8.3.
Answer. 1.06%

8.5 Cylindrical source

Consider a cylindrical source aligned with the z-axis as shown in Figure 8.2. We will consider two cases. First where the flow is radial and second where the flow is longitudinal.

Radial current flow

From Gauss' law the current flux per unit length is

$$\Phi \equiv I = \oint \mathbf{J} \cdot \mathrm{d}\mathbf{A}$$

$$= \int_0^1 \int_0^{2\pi} J_r \hat{\mathbf{a}}_r \cdot r \, \mathrm{d}\phi \, \mathrm{d}z \hat{\mathbf{a}}_r$$

or

$$\mathbf{J} = \frac{\Phi}{2\pi r} \hat{\mathbf{a}}_r \tag{8.9}$$

Again from Ohm's law

$$\mathbf{J} = \sigma \mathbf{E}$$

and

$$\mathbf{E} = \frac{\Phi}{2\pi\sigma r} \hat{\mathbf{a}}_r \tag{8.10}$$

Figure 8.2. A cylindrical current source.

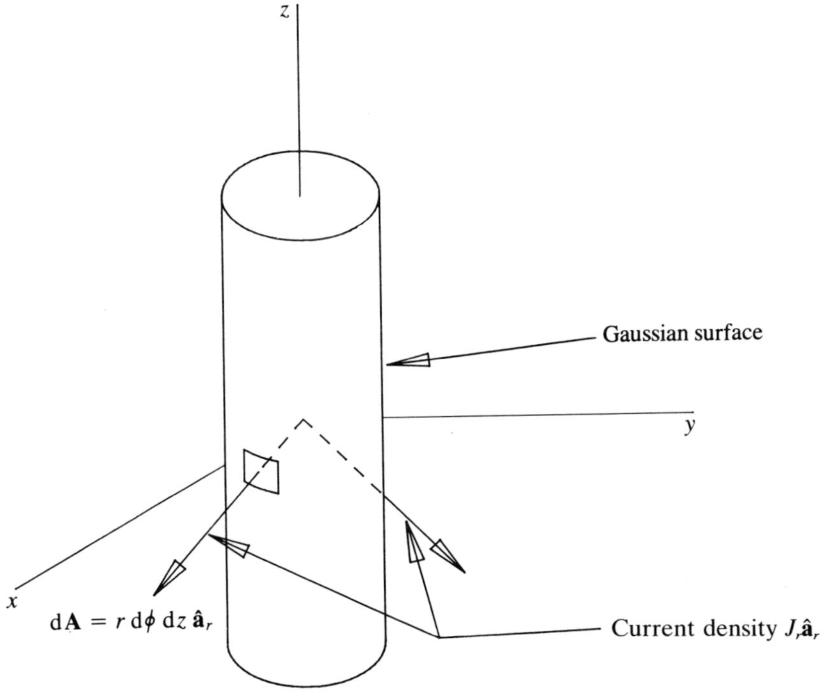

Also

$$\mathbf{E} = -\nabla V_i$$

Since **E** is in the radial direction, we use only that component of ∇V_i or

$$\nabla V_i = \frac{\partial V}{\partial r}\,\hat{\mathbf{a}}_r$$

and

$$\frac{\partial V_i}{\partial r} = -\frac{\Phi}{2\pi\sigma r}$$

Integrating both sides we obtain

$$V_i = -\frac{\Phi}{2\pi\sigma}\ln r + \text{constant}$$

For two coaxial cylinders of radii r_1 and r_2 and voltages V_1 and V_2, respectively, we obtain

$$V_1 - V_2 = \frac{\Phi}{2\pi\sigma}\ln\frac{r_2}{r_1} \tag{8.11}$$

Manipulating equation (8.11) we obtain

$$R = \frac{V_i}{\Phi} = \frac{\ln\dfrac{r_2}{r_1}}{2\pi\sigma}\ \Omega\ \text{m} \tag{8.12}$$

For $r_1 \approx r_2$ equation (8.12) can be written

$$R \approx \frac{2\left(\dfrac{r_2}{r_1} - 1\right)}{2\pi\sigma\left(\dfrac{r_2}{r_1} + 1\right)} = \frac{r_2 - r_1}{\pi\sigma(r_2 + r_1)}$$

Note that the units of resistance are ohms meters. To find the total resistance we must divide by the length of the coax.

Example 8.4 Given a coaxial cylindrical resistor with inner radius $r_1 = 10^{-2}$ m and outer radius $r_2 = 2 \cdot 10^{-2}$ m, the voltage between the radii varies such that $V = 10 \ln r + 46.052$ V. Determine the electric field, the current density and current as a function of distance. Also determine the total resistance for a 10 m length. The conductivity of the resistor material is 10^{-11} S. From the gradient we obtain

$$\mathbf{E} = -\nabla V_{\mathrm{i}} = -\frac{10}{r}\hat{\mathbf{a}}_r$$

and

$$\mathbf{J} = \sigma\mathbf{E} = -\frac{10}{r}\cdot 10^{-11}\,\hat{\mathbf{a}}_r$$

$$I = \iint \mathbf{J}\cdot \mathrm{d}\mathbf{A} = \int_0^1 \int_0^{2\pi} \frac{10^{-10}}{r}\, r\,\mathrm{d}\phi\,\mathrm{d}z$$

The resistance in ohms meters is

$$R = \frac{V_1 - V_2}{I} = \frac{10(\ln 10^{-2}) + 46.05 - [10(\ln 10^{-3}) + 46.05]}{10^{-10}\,2\pi} = 3.66\cdot 10^{10}\ \Omega\ \mathrm{m}$$

Finally

$$R_{\mathrm{T}} = \frac{R}{10} = 3.66\cdot 10^{9}\ \Omega$$

Longitudinal current flow

From Gauss' law the current flow per unit length is

$$\Phi \equiv I = \int_0^{2\pi}\int_0^{r} J_z\hat{\mathbf{a}}_z \cdot r\,\mathrm{d}r\,\mathrm{d}\phi\hat{\mathbf{a}}_z$$

we obtain

$$J_z = \frac{\Phi}{\frac{r^2}{2}2\pi}$$

From Ohm's law

$$E_z = \frac{\Phi}{\sigma\pi r^2}$$

From the gradient

$$\mathbf{E} = -\nabla V_{\mathrm{i}} = \frac{\Phi}{\sigma\pi r^2}\,\hat{\mathbf{a}}_z = -\frac{\partial V_{\mathrm{i}}}{\partial z}\,\hat{\mathbf{a}}_z$$

from which

$$|V_{\mathrm{i}}| = \frac{z\Phi}{\sigma\pi r^2}$$

As before, since $\Phi \equiv I$ the resistance of the cylinder is

$$R = \frac{z}{\sigma\pi r^2}$$

where $l \leftrightarrow z$ and $A \leftrightarrow \pi r^2$.

Example 8.5 Calculate the resistance of a cylindrical barium titanate resistor where $l = 3$ cm, $r = 0.5$ cm, and for barium titanate $\sigma \approx 10^{-11}$ S. From above

$$R = \frac{z}{\sigma\pi r^2} = \frac{3\cdot 10^{-2}}{10^{-11}\cdot \pi(0.5\cdot 10^{-2})^2} = 38.2\cdot 10^{12}\ \Omega$$

D8.3 A coaxial cable is manufactured using two layers of dielectric. The inner conductor radius is 10^{-2} m and the outer conductor radius is $2\cdot 10^{-2}$ m. The inner layer of dielectric extends from 10^{-2} m to $1.5\cdot 10^{-2}$ m. The conductivity is $\sigma_1 = 10^{-12}$ S. The outer dielectric's conductivity is $\sigma_2 = 10^{-13}$ S. Calculate the loss resistance of a 100 m length of coax.
Answer. $5.23\cdot 10^9\ \Omega$.

D8.4 A metal shaft made of copper is conducting a current of 1.5 A in the z-direction. The shaft is rectangular in cross-section, 0.5 cm × 0.3 cm, and is 5 cm in length. Calculate the resistance of the shaft and the voltage drop along it.
Answer. $5.75\cdot 10^{-5}\ \Omega$; $86.25\cdot 10^{-6}$ V.

8.6 Boundary conditions

The boundary conditions for electric current are similar to those for electric fields. Recall from Chapter 3 for two different media at the boundary

$$E_{\mathrm{t}_1} = E_{\mathrm{t}_2}$$

and

$$D_{n_1} = D_{n_2}$$

Let's first examine the *tangential* boundary conditions. Since

$$\mathbf{J} = \sigma\mathbf{E}$$

or

$$\mathbf{E} = \frac{\mathbf{J}}{\sigma}$$

we write

$$E_{t_1} = \frac{J_{t_1}}{\sigma_1}$$

Therefore the tangential boundary conditions are

$$E_{t_1} = E_{t_2} \tag{8.13}$$

and

$$\frac{J_{t_1}}{\sigma_1} = \frac{J_{t_2}}{\sigma_2} \tag{8.14}$$

Example 8.5 Determine **E** and **J** in both regions for the two media shown in Figure 8.3. From equation (8.13)

$$E_{t_1} = 3 = E_{t_2}$$

so that

$$J_{t_1} = \sigma_1 E_{t_1} = 10^2(3)$$

Figure 8.3. See example 8.5.

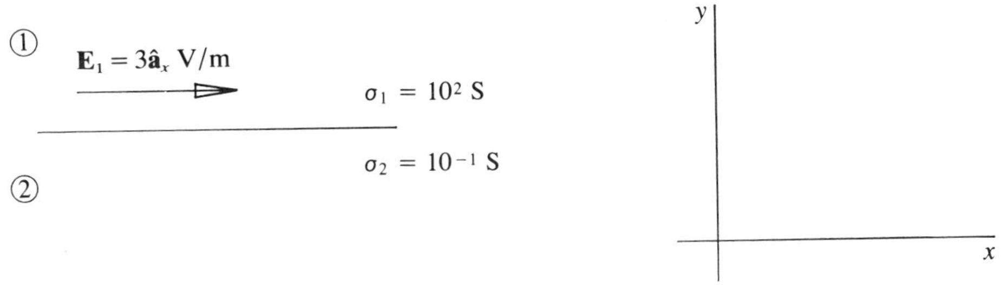

From equation (8.14)

$$\frac{J_{t_1}}{\sigma_1} = \frac{J_{t_2}}{\sigma_2}$$

$$\frac{10^2(3)}{10^2} = \frac{J_{t_2}}{10^{-1}}$$

or

$$J_{t_2} = 0.3 \text{ A/m}^2$$

Finally $\mathbf{E}_1 = 3\hat{\mathbf{a}}_x$, $\mathbf{E}_2 = 3\hat{\mathbf{a}}_x$ V/m, $\mathbf{J}_1 = 300\hat{\mathbf{a}}_x$ and $\mathbf{J}_2 = 0.3\hat{\mathbf{a}}_x$ A/m^2.

For the *normal* boundary conditions consider Figure 8.4. From Gauss' law we have

$$\Phi = \oint \mathbf{J} \cdot \mathrm{d}\mathbf{A} = I$$

We assume no current sources or sinks within the Gaussian surface so

Figure 8.4. Configuration for normal boundary conditions.

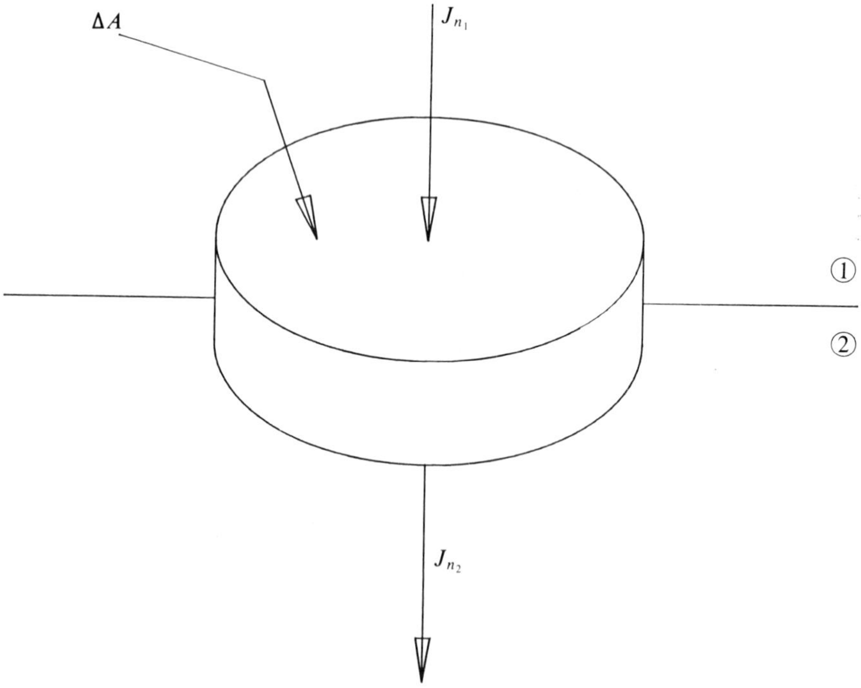

that

$$\oint \mathbf{J} \cdot \mathrm{d}\mathbf{A} = 0$$

For the three surfaces we have

$$\oint \mathbf{J} \cdot \mathrm{d}\mathbf{A} = \int_{\text{top}} + \int_{\text{side}} + \int_{\text{bottom}} = 0$$

Since **J** and d**A** are normal for the side of the Gaussian 'can' in Figure 8.4 then

$$\int_{\text{side}} = 0$$

In the limit as $\Delta A \rightarrow 0$

$$\mathbf{J} \cdot \Delta\mathbf{A} = J_{\mathrm{n}_1}\, \Delta A_{\text{top}} - J_{\mathrm{n}_2}\, \Delta A_{\text{bottom}} = 0$$

or

$$J_{\mathrm{n}_1} = J_{\mathrm{n}_2} \tag{8.15}$$

and

$$\sigma_1 E_{\mathrm{n}_1} = \sigma_2 E_{\mathrm{n}_2} \tag{8.16}$$

The normal electric fields are discontinuous.

Example 8.6 Determine **E** and **J** for both regions for the two media shown in Figure 8.5. From Ohm's law

$$J_{\mathrm{n}_1} = \sigma_1 E_{\mathrm{n}_1}$$

Figure 8.5. See example 8.6.

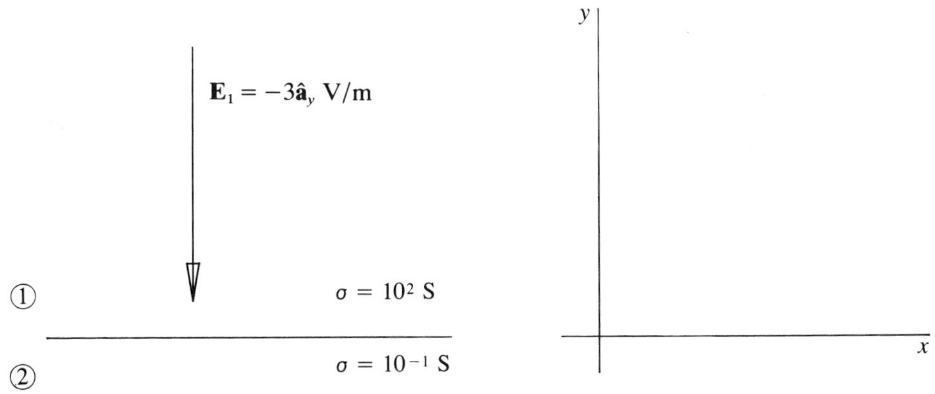

so that

$$\mathbf{J}_1 = 10^2(-3)\hat{\mathbf{a}}_y$$

From equation (8.15)

$$\mathbf{J}_2 = -300\hat{\mathbf{a}}_y \text{ A/m}^2$$

so that

$$\mathbf{E}_2 = \frac{\mathbf{J}_2}{\sigma_2}$$

$$= \frac{-300}{10^{-1}}\hat{\mathbf{a}}_y$$

$$= -3000\hat{\mathbf{a}}_y \text{ V/m}$$

D8.5 Determine the electric field, **E**, and the current density, **J**, for both regions where $\mathbf{E} = 2.5\hat{\mathbf{a}}_x + 5\hat{\mathbf{a}}_y - 2\hat{\mathbf{a}}_z$ in region 1. The boundary is the xz plane. For region 1 $\sigma = 2 \cdot 10^{-2}$ S, $y > 0$, and for region 2 $\sigma = 40$ S, $y < 0$.
Answer. $2.5\hat{\mathbf{a}}_x + 2.5 \cdot 10^{-3}\hat{\mathbf{a}}_y - 2\hat{\mathbf{a}}_z$ V/m; $\mathbf{J}_1 = [5\hat{\mathbf{a}}_x + 10\hat{\mathbf{a}}_y - 4\hat{\mathbf{a}}_z]10^{-2}$ A/m^2, $\mathbf{J}_2 = 100\hat{\mathbf{a}}_x + 0.1\hat{\mathbf{a}}_y - 80\hat{\mathbf{a}}_z$ A/m^2.

8.7 Flow through permeable media

The flow of fluid through a permeable medium has two characteristics with which we are familiar. These are streamlines and equipotential lines. In general the velocity of flow is slow enough that there is no turbulence. There is laminar flow. The fluid follows the streamlines or flow lines through the permeable medium. In addition there is a pressure head or energy head (with reference to some point) which causes the flow. We designate the lines where the pressure or energy is constant as equipotential. Typically the velocity is small so that the difference in velocities from point to point is negligible. The fluid flows through the interstices of the material. The more porous the material the higher the velocity of fluid flow. Porosity is the volume of interstitial spaces per volume of material. Naturally, the flow is not an uninterrupted flow but analogous to the movement of an electron through a resistance. The higher the resistance the more impeded the electron's path. In electrical conduction we consider average velocity because of the many collisions and various particle velocities. Likewise, in fluid flow through porous media we consider average velocity.

Therefore the flow lines are approximations to the flow; they represent the 'big picture'.

When a difference of pressure (or energy) exists across a material, any fluid in the material will move to the lower pressure. Consider a two-dimensional situation as shown in Figure 8.6. The higher pressure, $p_h(=p_l + \Delta p)$, causes the fluid to flow in the x-direction.

The relationship between velocity and pressure or potential is written

$$\mathbf{u} = -k\,\nabla p = -k\,\nabla V_f \tag{8.17}$$

where k is the permeability of the medium in m^4/Ns, p is pressure and V_f is potential.

If the material is tilted, as shown in Figure 8.7, so that the flow is not horizontal, then there will be a vertical component of flow and we

Figure 8.6. Fluid flow through a permeable medium.

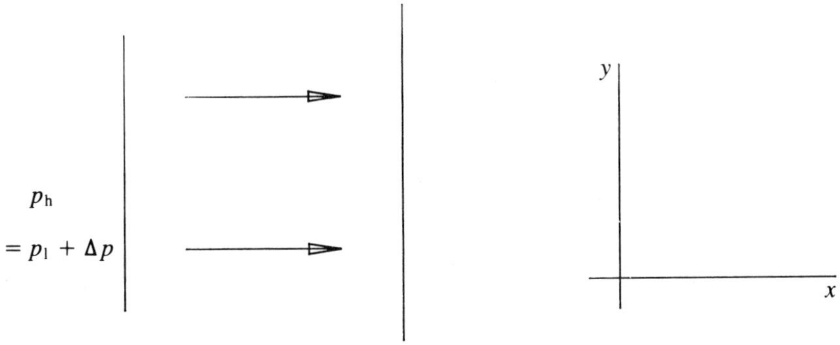

Figure 8.7. Fluid flow through a permeable medium including gravity.

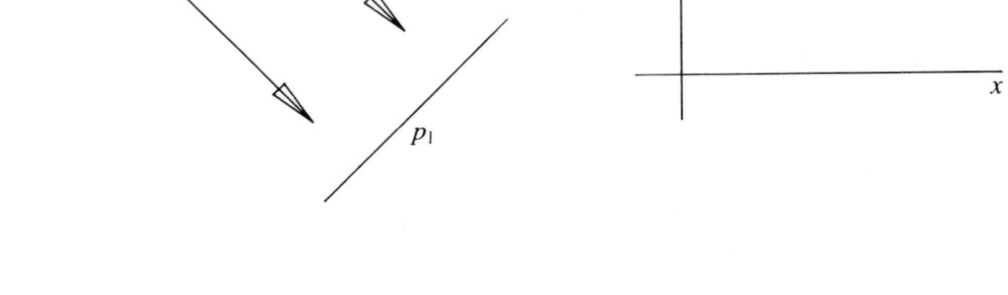

consider gravity. Thus, we write

$$\mathbf{u} = -k(\nabla p + \rho g \, \nabla h) \tag{8.18}$$

where h is the height above a datum and ∇h is the unit vertical vector. This is an experimental law, known as Darcy's law, named after a French scientist of the nineteenth century. If we neglect the gravitational effect, we can write equation (8.18) in integral form, that is

$$V_{f_1} - V_{f_2} = p_h - p_l = \frac{-1}{k} \int \mathbf{u} \cdot \mathrm{d}\mathbf{l} \tag{8.19}$$

Once again we note that the equipotential lines are orthogonal to the flow. Figure 8.8 shows a portion of material with the flow and equipotentials. The flow is in the x-direction while the equipotentials are in the y-direction.

D8.6 From Figure 8.8 assume pressure varies in the x-direction, $p(x) = -Ax + B$, in a sandy soil where $k = 3 \cdot 10^{-5}\,\mathrm{m^4/Ns}$. Calculate the velocity of fluid flow if $p(0) = 0$ and $p(1) = +0.3\,\mathrm{N/m^2}$.
Answer. $9 \cdot 10^{-6}\,\mathrm{m/s}$.

Figure 8.8. The orthogonality of streamlines and equipotentials.

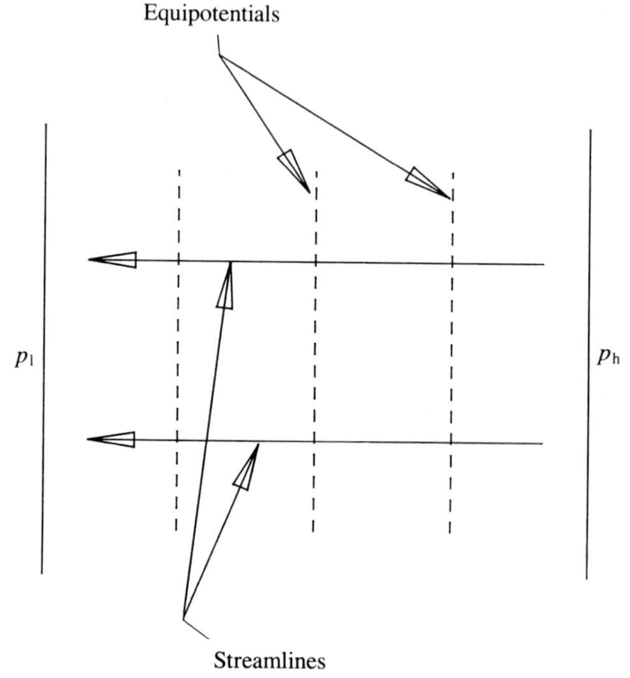

Consider a differential volume of fluid flowing in a permeable medium as shown in Figure 8.9. Once again from Gauss' law

$$Q_{\mathrm{f}} \equiv \Phi = \oint \mathbf{u} \cdot \mathrm{d}\mathbf{A} \tag{8.20}$$

If there are no sources or sinks in the volume then

$$\oint \mathbf{u} \cdot \mathrm{d}\mathbf{A} = 0$$

This also assumes a homogeneous medium. From equation (8.20) and Gauss' divergence theorem

$$\oint \mathbf{u} \cdot \mathrm{d}\mathbf{A} = \int_{\mathrm{Vol}} \nabla \cdot \mathbf{u} \, \mathrm{d}v \tag{8.21}$$

If the medium is not homogeneous we have

$$\int \rho\mathbf{u} \cdot \mathrm{d}\mathbf{A} = \int_{\mathrm{Vol}} \nabla \cdot \rho\mathbf{u} \, \mathrm{d}v \tag{8.22}$$

where ρ is volume density in $\mathrm{kg/m^3}$.

Recall the continuity equation. In fluid flow we have

$$\nabla \cdot \rho\mathbf{u} = -\frac{\partial \rho}{\partial t} \tag{8.23}$$

Figure 8.9. A differential volume of fluid flow.

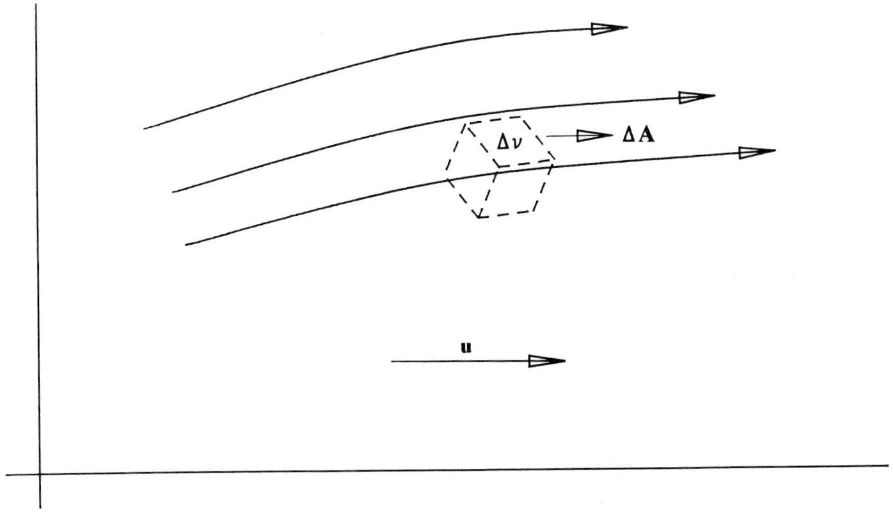

We assume conservation of matter, so that the flux of fluid flowing from within the area must be equal to the time rate of decrease of mass within the volume enclosed by the area. In integral form this becomes

$$\int \rho\mathbf{u} \cdot \mathrm{d}\mathbf{A} = -\frac{\mathrm{d}}{\mathrm{d}t}\int \rho \,\mathrm{d}v \tag{8.24}$$

where $\rho\mathbf{u}$ is a flow rate density.

Define a flow rate density $\mathbf{F}$ as

$$\mathbf{F} = \rho\mathbf{u} \tag{8.25}$$

Equation (8.25) is analogous to $\mathbf{D} = \epsilon\mathbf{E}$, $\mathbf{B} = \mu\mathbf{H}$, and so forth. Equation (8.24) is the continuity equation in integral form. For steady flow the density is constant with time; the right side of equation (8.23) is zero and

$$\oint \rho\mathbf{u} \cdot \mathrm{d}\mathbf{A} = 0$$

For steady flow equation (8.23) becomes

$$\nabla \cdot \rho\mathbf{u} = 0 \tag{8.26}$$

In rectangular coordinates this is written – if ρ is constant with position

$$\rho\left(\frac{\partial u_x}{\partial x} + \frac{\partial u_y}{\partial y} + \frac{\partial u_z}{\partial z}\right) = 0 \tag{8.27}$$

ρ cannot be zero, therefore

$$\frac{\partial u_x}{\partial x} + \frac{\partial u_y}{\partial y} + \frac{\partial u_z}{\partial z} = 0 \tag{8.28}$$

From equation (8.17)

$$u_x\hat{\mathbf{a}}_x + u_y\hat{\mathbf{a}}_y + u_z\hat{\mathbf{a}}_z = -k\left[\frac{\partial V_\mathrm{f}}{\partial x}\hat{\mathbf{a}}_x + \frac{\partial V_\mathrm{f}}{\partial y}\hat{\mathbf{a}}_y + \frac{\partial V_\mathrm{f}}{\partial z}\hat{\mathbf{a}}_z\right] \tag{8.29}$$

Substituting equation (8.29) into equation (8.28) we obtain

$$\frac{\partial^2 V_\mathrm{f}}{\partial x^2} + \frac{\partial^2 V_\mathrm{f}}{\partial y^2} + \frac{\partial^2 V_\mathrm{f}}{\partial z^2} = 0 \tag{8.30}$$

or

$$\nabla^2 V_\mathrm{f} = 0$$

which is Laplace's equation. In cylindrical coordinates this is

$$\frac{1}{r}\frac{\partial}{\partial r}\left(r\frac{\partial V_{\mathrm{f}}}{\partial r}\right)+\frac{1}{r^2}\frac{\partial^2 V_{\mathrm{f}}}{\partial \phi^2}+\frac{\partial V_{\mathrm{f}}}{\partial z^2}=0 \tag{8.31}$$

In spherical coordinates this is

$$\frac{1}{r^2}\frac{\partial}{\partial r}\left(r^2\frac{\partial V_{\mathrm{f}}}{\partial r}\right)+\frac{1}{r^2\sin\theta}\frac{\partial}{\partial\theta}\left(\sin\theta\frac{\partial V_{\mathrm{f}}}{\partial\theta}\right)+\frac{1}{r^2\sin^2\theta}\frac{\partial^2 V_{\mathrm{f}}}{\partial\phi^2}=0 \tag{8.32}$$

Example 8.7 The pressure in a sandy soil medium varies in three dimensions as $p(x, y, z)=xyz$. If $k=0.5$ determine the velocity.

From equation (8.29)

$$u_x=-k\frac{\partial p}{\partial x}$$
$$=-0.5yz$$

$$u_y=-k\frac{\partial p}{\partial y}$$
$$=-0.5xz$$

and

$$u_z=-k\frac{\partial p}{\partial z}$$
$$=-0.5xy$$

In other words

$$\mathbf{u}=-0.5(yz\hat{\mathbf{a}}_x+xz\hat{\mathbf{a}}_y+xy\hat{\mathbf{a}}_z)$$

D8.7 Plot several equipotentials and several velocity vectors from example 8.7.

Now let's consider three configurations – spherical, cylindrical, and planar.

8.8 Spherical source

For a spherical source (or sink) we begin with equation (8.22)

$$\int \rho\mathbf{u}\cdot \mathrm{d}\mathbf{A}=\int_{\mathrm{Vol}}\nabla\cdot\rho\mathbf{u}\,\mathrm{d}v \tag{8.22}$$

From the left side of equation (8.24)

$$Q \equiv \Phi = \oint \mathbf{F} \cdot \mathrm{d}\mathbf{A}$$

$$= \int_0^{2\pi} \int_0^{\pi} F_r \hat{\mathbf{a}}_r \cdot r^2 \sin\theta \, \mathrm{d}\theta \, \mathrm{d}\phi \hat{\mathbf{a}}_r$$

and the flow is radial. The density is

$$\mathbf{F} = \frac{\Phi}{4\pi r^2} \hat{\mathbf{a}}_r \tag{8.33}$$

In addition

$$p_\mathrm{h} - p_\mathrm{l} = \frac{-1}{k} \int \mathbf{u} \cdot \mathrm{d}\mathbf{l} \tag{8.19}$$

$$= \frac{-1}{k} \int \frac{\Phi}{4\pi\rho r^2} \hat{\mathbf{a}}_r \cdot \mathrm{d}r\hat{\mathbf{a}}_r$$

$$= \frac{\Phi}{4\pi k\rho r} + \text{constant} \tag{8.34}$$

Example 8.8 Calculate the density, $\mathbf{F}$, and the velocity of fluid, $\mathbf{u}$, at $r = 6.5$ m for a point source of strength $\Phi = 3.5$ kg/s. The density of water is $\rho = 10^3$ kg/m^3.

From equation (8.33)

$$\mathbf{F} = \frac{3.5}{4\pi(6.5)^2} \hat{\mathbf{a}}_r$$

$$= 0.0066\hat{\mathbf{a}}_r \text{ kg/m}^2\text{s}$$

and from equation (8.25)

$$\mathbf{u} = \frac{\mathbf{F}}{\rho}$$

$$= 6.6 \cdot 10^{-6} \hat{\mathbf{a}}_r \text{ m/s}$$

Example 8.9 Calculate the pressure difference necessary to achieve the velocity of fluid flow in example 8.8 in gravel ($k = 10^{-1}$ m^4/Ns).

From equation (8.19)

$$p_{\mathrm{h}} - p_{\mathrm{l}} = \frac{-1}{k} \int_{\infty}^{6.5} \frac{3.5}{\rho 4\pi r^2} \, \mathrm{d}r$$

$$= 4.3 \cdot 10^{-4} \, \mathrm{N/m^2} \qquad (8.19)$$

D8.8 Calculate the volumetric flow rate at $x = 2$ in the x-direction for the velocity, $\mathbf{u} = xyz\hat{\mathbf{a}}_x$, of fluid through 1 m^2.
Answer. $\frac{1}{2}$ $\mathrm{m^3/s}$.

8.9 Cylindrical source

For a cylindrical source (or sink) we again begin with equation (8.22). From the left side of equation (8.22) for radial flow

$$Q = \oint \mathbf{F} \cdot \mathrm{d}\mathbf{A}$$

$$= \int_0^1 \int_0^{2\pi} F_r \hat{\mathbf{a}}_r \cdot r \, \mathrm{d}\phi \, \mathrm{d}z \hat{\mathbf{a}}_r$$

and

$$\mathbf{F} = \frac{\Phi}{2\pi r(1)} \hat{\mathbf{a}}_r \qquad (8.35)$$

Also

$$\mathbf{u} = \frac{\Phi}{2\pi\rho r} \hat{\mathbf{a}}_r$$

We can determine the pressure (potential) difference from equation (8.19)

$$p_{\mathrm{h}} - p_{\mathrm{l}} = \frac{-1}{k} \int \mathbf{u} \cdot \mathrm{d}\mathbf{l}$$

$$= \frac{-1}{k} \frac{\Phi}{2\pi\rho} \ln r + \text{constant} \qquad (8.36)$$

It should be clear that the loci of equations (8.35) and (8.36) are orthogonal. Equation (8.35) describes streamlines in the $\hat{\mathbf{a}}_r$ direction while equation (8.36) describes equipotential (constant pressure) lines which are concentric about the z-axis.

Example 8.10 A long, deep water well of diameter 10^{-1} m is modeled by a cylindrical source of strength $\Phi = 2\pi\rho$. The well is dug in a sand, silt, clay mixture where $k = 10^{-4}\,m^4/Ns$. If $p_h = 10^5\,N/m^2$ at $r = 0.05$ m determine p_l at $r = 0.5$ m.

From equation (8.36)

$$\Delta p = \frac{-1}{k}\frac{\Phi}{2\pi\rho}\ln r + \text{constant}$$

From the boundary conditions

$$p = 0 = \frac{-1}{k}\ln(1) + \text{constant}$$

$$\text{constant} = 0$$

therefore

$$\Delta p = \frac{-1}{10^{-4}}\ln(0.5)$$

$$= 3.01 \cdot 10^3\ N/m^2 \quad \text{and} \quad p_l = 10^5 - 3 \cdot 10^3$$

D8.9 Repeat example 8.10 if the soil changes at $r = 0.2$ m to $k = 10^{-2}$.
Answer. 7093 N/m^2.

8.10 Planar source (or sink)

When the flow is uniform in one direction we consider a planar source in Cartesian coordinates (see Figure 8.10). From Gauss' law

$$\Phi = \oint \mathbf{F} \cdot d\mathbf{A}$$

Consider the flow through the surface at $x = x_0$ and

$$\Phi = \int_0^1 \int_0^1 F_x\hat{\mathbf{a}}_x \cdot dy\, dz\hat{\mathbf{a}}_x$$

$$\Phi = F_x$$

In addition

$$u_x = \frac{F_x}{\rho}$$

In other words the streamlines are in the x-direction. For the equipotentials we write

$$\Delta p = \frac{-1}{k}\int \mathbf{u} \cdot d\mathbf{l}$$

where

$$\mathbf{dl} = dx\hat{\mathbf{a}}_x$$

thus

$$\Delta p = \frac{-1}{k} u_x \, \Delta x$$

Example 8.11 For a pressure difference of 13.5 N/m^2 over a distance of 8 m in the x-direction calculate the flow velocity in the x-direction in a clay soil where $k = 10^{-10}$ m^4/Ns.

From above

$$u_x = -k \frac{\Delta p}{\Delta x}$$

$$= -10^{-10} \cdot \frac{13.5}{8}$$

$$|u_x| = 1.69 \cdot 10^{-10} \text{ m/s}$$

As expected the velocity through clay soil is typically slow.

Figure 8.10. A planar source.

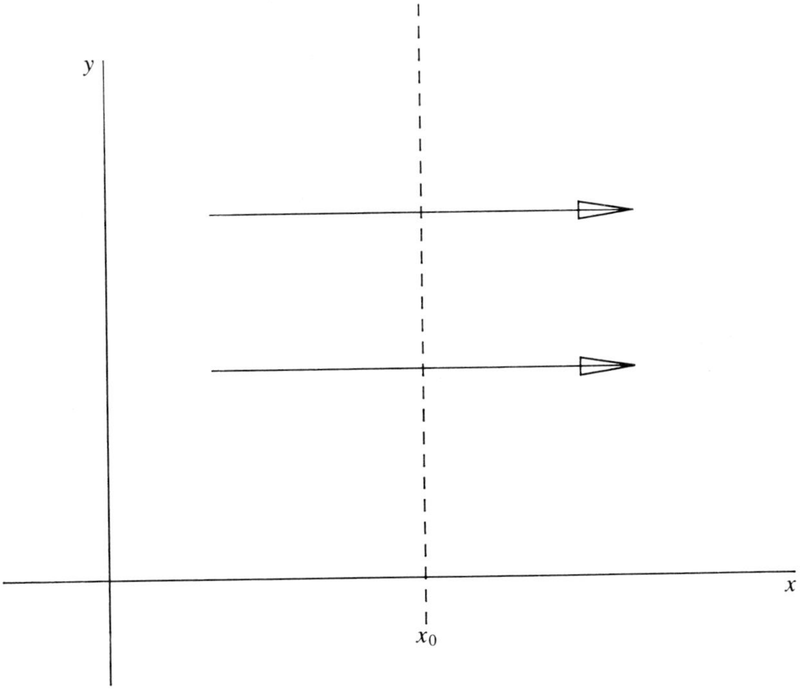

Example 8.12 A city water filtration system consists of a large pool filled with a sand–gravel mixture which is 2.5 meters deep. $k \approx 8$. The water level is 2.5 meters above the mixture. Calculate the pressure head and velocity if the bottom of the pool is the reference. The pressure head is

$$\begin{aligned} p &= \rho g h \\ &= 10^3 \cdot 9.81 \cdot 5 \\ &= 49.05 \cdot 10^3 \text{ N/m}^2 \end{aligned}$$

From above

$$\Delta p = \frac{-1}{k} u_y \, \Delta y$$

$$\begin{aligned} u_y &= (8) \frac{49 \cdot 10^3}{2.5} \\ &= 156.8 \cdot 10^3 \text{ m/s} \end{aligned}$$

D8.10 If a swimming pool 3 m deep is to be built over a clay ground how deep must the clay be to limit the fluid velocity to 0.001 m/s?
Answer. $3 \cdot 10^{-3}$ m.

8.11 Diffusion

Another field which is analogous to the flows we have been studying is the diffusive field. The mixing of fluids is an example. Another is the diffusion of electrons in a semiconductor material. In diode construction two different semiconductor materials are combined. The semiconductors are formed by adding dopants to both materials. Currents are induced by applying various voltages across the junction of the two materials. Picture a semiconductor material as shown in Figure 8.11. A typical semiconductor diode joins a material which has a surplus of electrons – n in Figure 8.11 – to a material which has a surplus of holes – p in Figure 8.11. Diffusion will occur at the junction of the n and p regions.

Diffusion processes are governed by the diffusion equation. The higher concentration in one region causes a concentration gradient which causes the flow to a region of lower concentration. Thus the concentration is analogous to a pressure or a potential which is the 'force' behind a flow. We call this potential V_d.

Consider a concentration of particles along the x-axis as shown in Figure 8.12. N_1 is the particle concentration per volume at x_1 and N_2 is the particle concentration per volume at x_2. Assume $N_1 > N_2$ so that flow is in the $+x$-direction. The difference in concentrations is written

$$N_1 - N_2 = -d\frac{\Delta N}{\Delta x} \tag{8.37}$$

where d is the distance $x_2 - x_1$. The negative sign arises because $N_1 > N_2$ and the gradient, $\Delta N/\Delta x$ is negative. We now define a particle flux density, $\mathbf{D}_\mathrm{d}$, in particles/m$^2 \cdot$ s. Thus we can write

$$\mathbf{D}_\mathrm{d} = k(N_1 - N_2)\hat{\mathbf{a}}_x \tag{8.38}$$

where k has dimensions m/s and is independent of N_1, N_2. Combining equations (8.37) and (8.38) we obtain in the limit

$$\mathbf{D}_\mathrm{d} = -kd\frac{\partial N}{\partial x}\hat{\mathbf{a}}_x \tag{8.39}$$

If we assume the particles diffuse the same in any direction we can write

$$\mathbf{D}_\mathrm{d} = -k_\mathrm{d}\nabla N \tag{8.40}$$

Figure 8.11. A semiconductor material.

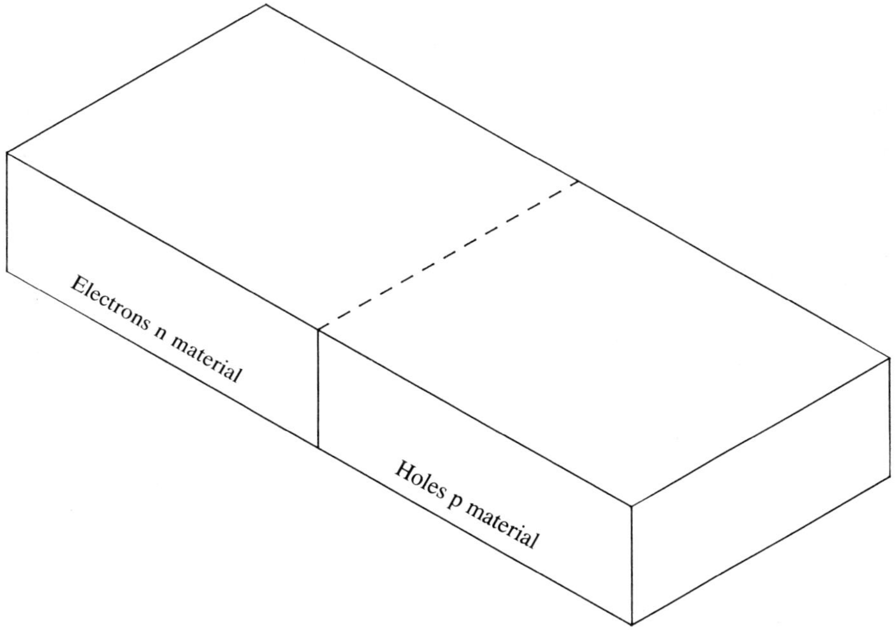

where $k_d = kd$ is called the diffusion coefficient in m^2/s. Equation (8.40) is called Fick's law.

Now recall the continuity equation from Chapter 6

$$\nabla \cdot \mathbf{J} = -\frac{\partial \rho}{\partial t}$$

In analogous fashion we write

$$\nabla \cdot \mathbf{D}_d = -\frac{\partial N}{\partial t} \tag{8.41}$$

Substituting equation (8.40) into equation (8.41) we obtain

$$\nabla \cdot (-k_d \nabla N) = -\frac{\partial N}{\partial t}$$

or

$$k_d \nabla^2 N = \frac{\partial N}{\partial t} \tag{8.42}$$

Figure 8.12. A one-dimensional semiconductor.

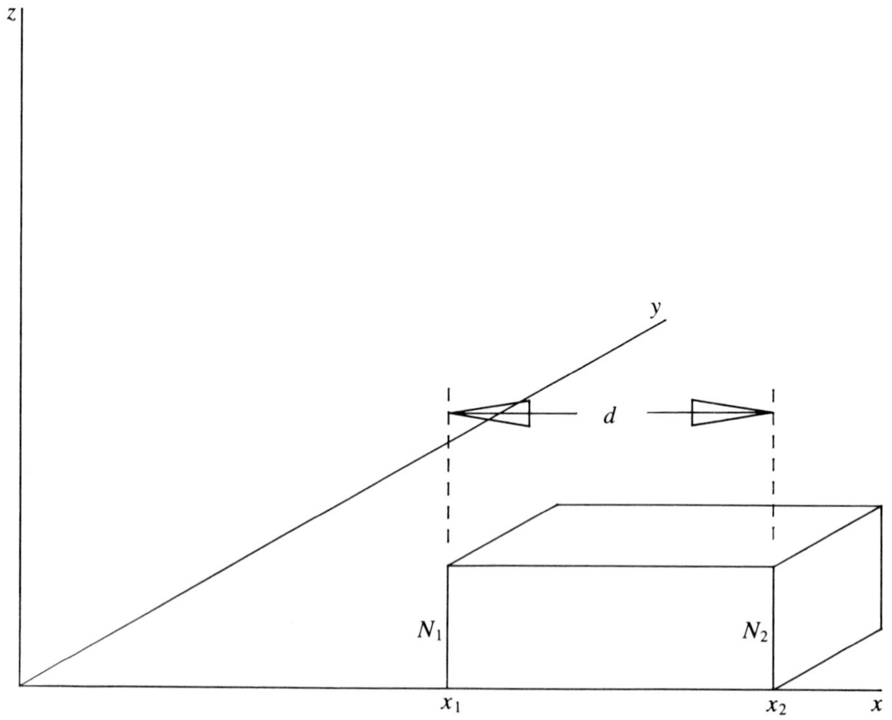

Equation (8.42) is one form of the diffusion equation. If there exist sources and sinks in the region, equation (8.43) is written

$$k_d \nabla^2 N = \frac{\partial N}{\partial t} + \Phi_{\text{source}} + \Phi_{\text{sink}} \tag{8.43}$$

Equation (8.43) is also called the diffusion equation. If the concentration is not a function of time $\partial N/\partial t = 0$ then

$$\nabla^2 N = \frac{\Phi_{\text{source}} + \Phi_{\text{sink}}}{k_d} \tag{8.44}$$

which is Poisson's equation. Finally, if there are no sources or sinks in the region

$$\nabla^2 N = 0 \tag{8.45}$$

which is Laplace's equation.

We may also illustrate the versatility of Gauss' law by applying it to the diffusion field. Recall $\Phi = \oint \mathbf{D} \cdot d\mathbf{A}$ which now becomes

$$Q_d \equiv \Phi = \oint \mathbf{D}_d \cdot d\mathbf{A} \tag{8.46}$$

where Q_d is the particle concentration rate in particles per second. These are the particles crossing the Gaussian surface every second.

Example 8.13 A long cylindrical bar of soap is positioned in the center of a tub of water. Assume the tub is large enough so that the boundaries do not appreciably affect the diffusion of soap throughout the water. This is illustrated in Figure 8.13. Determine the soap particle flux density at $r = a$ for a bar of radius $r = a/2$ m.

From equation (8.46)

$$Q_d = \iint D_d \hat{\mathbf{a}}_r \cdot \left(\frac{a}{2}\right) d\phi \, dz \hat{\mathbf{a}}_r$$

$\mathbf{D}_d$ is radial since for this process the density varies only with radius and

$$\mathbf{D}_d = \frac{Q_d}{2\pi\left(\frac{a}{2}\right)(1)} \hat{\mathbf{a}}_r \tag{8.47}$$

But what about Q_d in equation (8.47)? Similar to previous calculations we write

$$Q_d = \iiint \rho_d \, dv \tag{8.48}$$

where ρ_d is particle concentration per volume per second (part./m^{3}s). For this example

$$Q_d = \int_0^1 \int_0^{2\pi} \int_0^{a/2} \rho_d \, r \, dr \, d\phi \, dz$$

$$= \rho_d \frac{\left(\frac{a}{2}\right)^2}{2} (2\pi)(1) = \rho_d \frac{\pi a^2}{4} \tag{8.49}$$

where ρ_d is assumed uniform. This is the total (source) particle concentration per second within the Gaussian surface. The soap bar may be replaced by a line source of value $Q_d = \rho_d \pi a^2/4$ particles per second per meter.

Combining equations (8.47) and (8.49) we obtain

$$\mathbf{D}_d = \frac{Q_d}{2\pi a} \hat{\mathbf{a}}_r = \frac{\rho_d a}{8} \hat{\mathbf{a}}_r \tag{8.50}$$

Figure 8.13. A long cylindrical bar of soap in water.

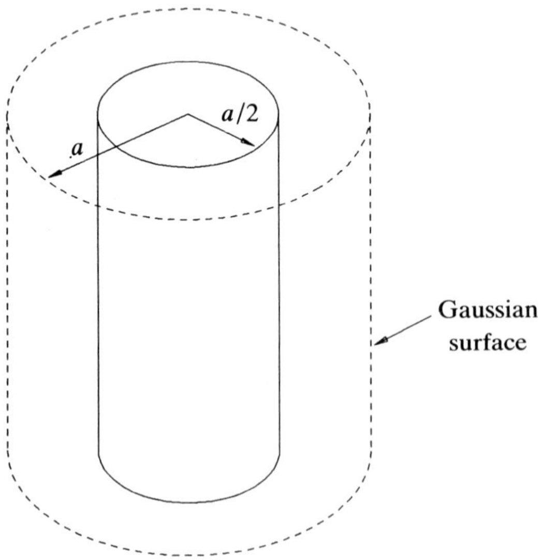

D8.11 The electron particle density in a p-type silicon material varies as $N(x) = 10^{23}e^{-x}$; $k_d = 0.08\ m^2/s$. Calculate the particle flux density, $\mathbf{D}_d$, as a function of x.
Answer. $8 \cdot 10^{21}\ e^{-x}$ part./m^2.

D8.12 The hole particle density in an n-type germanium semiconductor material varies as $N(x) = 3x^{-2}$; $k_d = 0.04\ m^2/s$. Calculate the particle flux density at $x = 3.6 \cdot 10^{-2}$.
Answer. 5144 part./m^2.

8.12 Acoustics

Acoustics is an engineering field which utilizes compressible fluids to propagate sound waves. The medium typically is air and the sound waves are the result of a source or sources which transmit high and low pressure acoustic waves. This modulation carries acoustic information; the medium is assumed to be a homogeneous fluid.

The behavior of sound waves is governed by the wave equation

$$\nabla^2 p = \frac{1}{c_s^2}\frac{\partial^2 p}{\partial t^2} \tag{8.51}$$

where p is pressure in N/m^2 and c_s is the velocity of sound in m/s. Transforming equation (8.51) to phasor domain we obtain

$$\nabla^2 p = -\frac{\omega^2}{c_s^2} p \tag{8.52}$$

where p can be complex. Let's assume the pressure varies only in the x-direction. Equation (8.52) becomes

$$\frac{d^2 p}{dx^2} + \left(\frac{\omega}{c_s}\right)^2 p = 0 \tag{8.53}$$

A solution to equation (8.53) is

$$p(x) = Ae^{-j(\omega x/c_s)} \tag{8.54}$$

$$= A\left(\cos\frac{\omega x}{c_s} - j\sin\frac{\omega x}{c_s}\right) \tag{8.55}$$

Now for sinusoidal variation, $d\mathbf{u}/dt$ transforms to $j\omega\mathbf{u}$. In other words the pressure varies sinusoidally in the x direction; so that $\mathbf{F}/\text{Vol.} = \text{m/vol}\ (\text{d}\mathbf{u}/\text{d}t) = \rho_0\, \text{d}\mathbf{u}/\text{d}t$, where $\mathbf{F}$ is a force vector and $\mathbf{u}$ is the local velocity vector.

D8.13 Prove that equation (8.54) is a solution to equation (8.53).

8.13 Velocity

Consider a volume of air moving in a frictionless medium where the sound pressure changes at a space rate of

$$\nabla p = \frac{\partial p}{\partial x}\hat{\mathbf{a}}_x + \frac{\partial p}{\partial y}\hat{\mathbf{a}}_y + \frac{\partial p}{\partial z}\hat{\mathbf{a}}_z$$

and p is the pressure at a point. This is modeled in Figure 8.14. Assume the pressure is higher to the right. The force accelerating the volume to the left is

$$\mathbf{F} = -\left(\frac{\partial p}{\partial x}\Delta x\right)\Delta y\ \Delta z\hat{\mathbf{a}}_x - \left(\frac{\partial p}{\partial y}\Delta y\right)\Delta x\ \Delta z\hat{\mathbf{a}}_y - \left(\frac{\partial p}{\partial z}\Delta z\right)\Delta x\ \Delta y\hat{\mathbf{a}}_z$$

Figure 8.14. A differential volume of air.

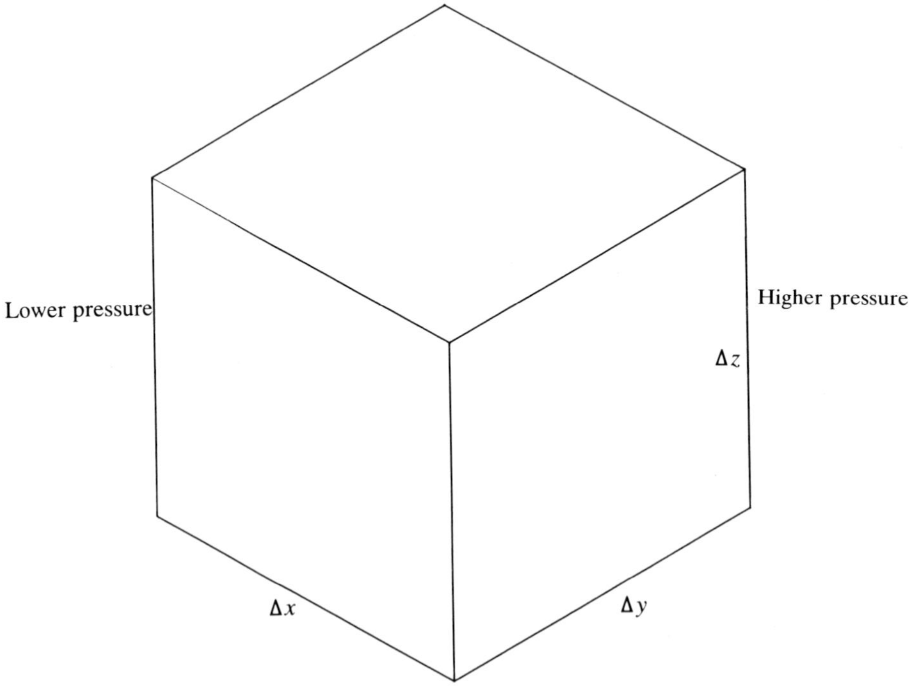

Dividing both sides by volume gives $\mathbf{F}/\text{Vol} = -\nabla p$. Also $\mathbf{F} = m\, d\mathbf{u}/\mathrm{d}t$. Thus the local velocity of a sound wave is related to the pressure by

$$\mathbf{u} = -\frac{1}{\mathrm{j}\omega\rho_0}\nabla p \tag{8.56}$$

where ρ_0 is the density of the medium in kg/m^3. Compare equation (8.56) with $\mathbf{E} = -\nabla V$. Integrating equation (8.56) we get

$$p_2 - p_1 = -\mathrm{j}\omega\rho_0 \int \mathbf{u}\cdot \mathrm{d}\mathbf{l} \tag{8.57}$$

D8.14 A sound wave propagating in the x-direction in air ($c_s = 350\ \text{m/s}$) at 1000 Hz has a velocity, $\mathbf{u}$. Calculate magnitude of the pressure difference across Δx if $\rho_0 = 1.3\ \text{kg/m}^3$ and $\mathbf{u} = 200\mathrm{e}^{-\mathrm{j}(\omega x/c_s)}$.
Answer. $91\ \text{kN/m}^2$.

Consider equation (8.54)

$$p(x) = A\mathrm{e}^{-\mathrm{j}(\omega x/c_s)} \tag{8.54}$$

which we write in phasor notation

$$p(x) = A\angle -\frac{\omega x}{c_s} \tag{8.58}$$

We can obtain the velocity from equation (8.56); that is,

$$\mathbf{u} = \frac{-1}{\mathrm{j}\omega\rho_0}\nabla p \tag{8.56}$$

Now

$$\nabla p = \frac{\partial}{\partial x} A\mathrm{e}^{-\mathrm{j}(\omega x/c_s)}\hat{\mathbf{a}}_x$$

$$= -\mathrm{j}\frac{\omega}{c_s} A\mathrm{e}^{-\mathrm{j}(\omega x/c_s)}\hat{\mathbf{a}}_x$$

or

$$\mathbf{u} = \left(\frac{-1}{\mathrm{j}\omega\rho_0}\right)\left(-\mathrm{j}\frac{\omega}{c_s}\right)A\mathrm{e}^{-\mathrm{j}(\omega x/c_s)}\hat{\mathbf{a}}_x$$

$$= \frac{A}{\rho_0 c_s}\mathrm{e}^{-\mathrm{j}(\omega x/c_s)}\hat{\mathbf{a}}_x \tag{8.59}$$

Notice the velocity vectors (streamlines) are in the x-direction. (By convention $e^{-j(\omega x/c_s)}$ shows propagation in the positive x-direction while $e^{j(\omega x/c_s)}$ shows propagation in the negative x-direction.) The lines of constant pressure (equipotentials) are shown as dashed lines in Figure 8.15. Figure 8.15 actually shows plane waves traveling to the right and to the left; as it proceeds the wave front compresses and rarifies, at a frequency of ω.

8.14 Spherical coordinates

In spherical coordinates equation (8.51) becomes

$$\frac{1}{r^2}\frac{\partial}{\partial r}\left(r^2\frac{\partial p}{\partial r}\right)+\frac{1}{r^2\sin\theta}\frac{\partial}{\partial\theta}\left(\sin\theta\frac{\partial p}{\partial\theta}\right)+\frac{1}{r^2\sin\theta}\frac{\partial^2 p}{\partial\phi^2}=\frac{1}{c_s^{\,2}}\frac{\partial^2 p}{\partial t^2} \qquad (8.60)$$

For waves propagating radially only the first term on the left side of

Figure 8.15. Planar waves propagating in the $\pm x$-directions.

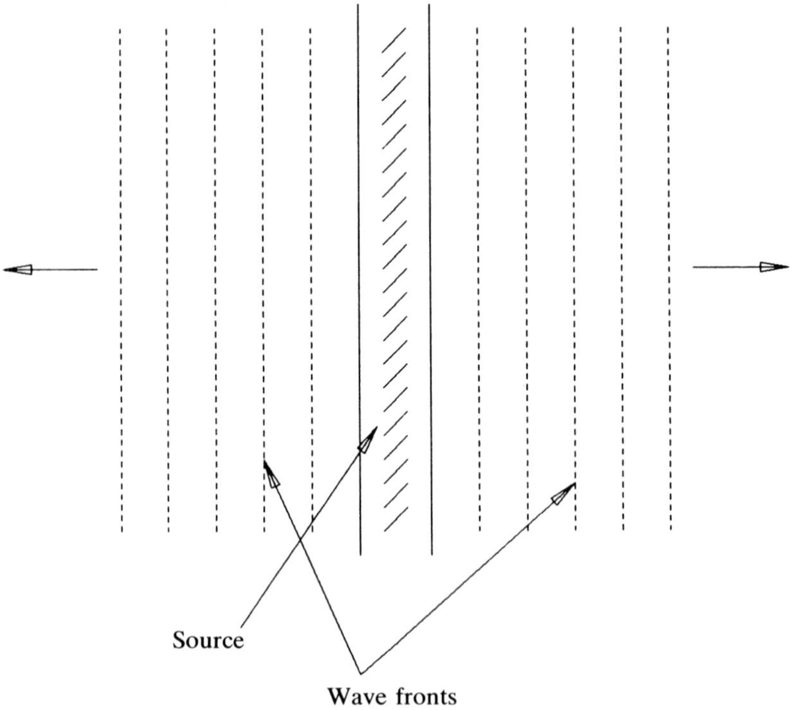

equation (8.60) is non-zero; that is,

$$\frac{1}{r^2}\frac{\partial}{\partial r}\left(r^2\frac{\partial p}{\partial r}\right)=\frac{1}{c_s^2}\frac{\partial^2 p}{\partial t^2} \tag{8.61}$$

In phasor notation equation (8.61) becomes

$$\frac{1}{r^2}\frac{\partial}{\partial r}\left(r^2\frac{\partial p}{\partial r}\right)+\left(\frac{\omega}{c_s}\right)^2 p=0 \tag{8.62}$$

A solution to equation (8.62) is

$$p(r)=\frac{A}{r}\mathrm{e}^{-\mathrm{j}(\omega r/c_s)} \tag{8.63}$$

D8.15 Prove that equation (8.63) is a solution to equation (8.62).

Now let's determine the velocity from equation (8.63). Since the pressure varies radially we write

$$\begin{aligned}\nabla p &= \frac{\partial p}{\partial r}\hat{\mathbf{a}}_r \\ &= \frac{\partial}{\partial r}\left[\frac{A}{r}\mathrm{e}^{-\mathrm{j}(\omega r/c_s)}\right]\hat{\mathbf{a}}_r \\ &= -\frac{A\mathrm{e}^{-\mathrm{j}(\omega r/c_s)}}{r}\left[\frac{1}{r}+\mathrm{j}\frac{\omega}{c_s}\right]\hat{\mathbf{a}}_r\end{aligned}$$

Finally

$$\mathbf{u}=\frac{A\mathrm{e}^{-\mathrm{j}(\omega r/c_s)}}{\omega\rho_0 r}\left[\frac{1}{\mathrm{j}r}+\frac{\omega}{c_s}\right]\hat{\mathbf{a}}_r \tag{8.64}$$

D8.16 Calculate the magnitude of the velocity of an acoustic wave in water at a distance of $r =$ (*a*) 1 m, (*b*) 0.1 m, (*c*) 0.01 m, from a point source if $\rho_0 = 10^3$ kg/m^3 and the frequency is 5 Hz. The velocity of sound is 300 m/s.
Answer. (*a*) $0.32 \cdot 10^{-4}A$ (*b*) $0.32 \cdot 10^{-2}A$ (*c*) $0.32A$ m/s.

It should be clear from equations (8.63) and (8.64) that the constant pressure lines (equipotentials) and velocity vectors (streamlines) are orthogonal. This is shown in Figure 8.16.

Example 8.14 In underwater acoustics research, explosives are used to determine propagation paths for underwater communication. An explosive is lowered into the water from a surface craft and detonated. The acoustic wavefronts propagate radially in all directions. Listening devices from other boats, planes or submarines receive the results of the explosion. Determine an expression for the pressure as a function of distance.

A spherical impulse function models the explosive charge. The charge is a spherical source at the origin at $t = 0$ for radial propagation. The wave equation is, from equation (8.61),

$$\frac{1}{r^2}\frac{\partial}{\partial r}\left(r^2\frac{\partial p}{\partial r}\right) = \frac{1}{c_s^{\,2}}\frac{\partial^2 p}{\partial t^2} \qquad t > 0 \tag{8.61}$$

This can also be written

$$\frac{\partial^2 p}{\partial r^2} + \frac{2}{r}\frac{\partial p}{\partial r} = \frac{1}{c_s^{\,2}}\frac{\partial^2 p}{\partial t^2} \qquad t > 0 \tag{8.65}$$

Figure 8.16. A spherical source showing orthogonality of constant pressure lines and streamlines.

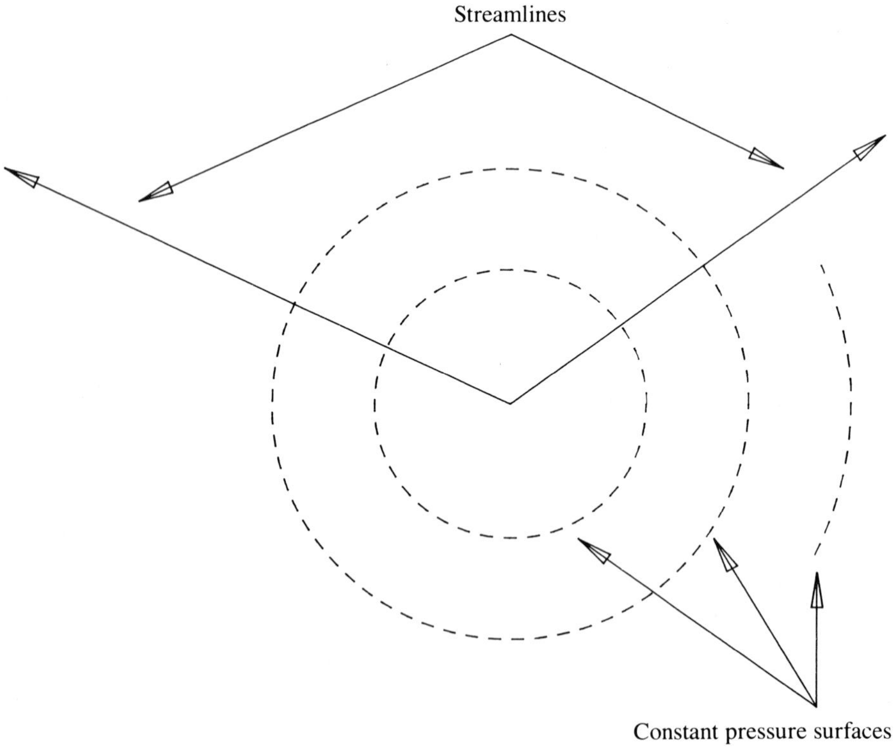

A solution to equation (8.65) is

$$p(r, t) = \frac{1}{r} e^{-j\omega[t-(r/c_s)]} \qquad t > 0 \tag{8.66}$$

Examining equation (8.66) we see that the pressure is time and space dependent. For a specific time and distance, the pressure has a value $p(r_1, t_1)$. We can find another time and distance $p(r_2, t_2)$ where the pressure has the same value. Therefore

$$t - \frac{r}{c_s} = \text{constant}$$

and

$$dt - \frac{dr}{c_s} = 0$$

or

$$\frac{dr}{dt} = c_s$$

Hence the pressure wave is a traveling wave.

If there were a boundary, reflections would occur and would be indicated by a $(t + r/c_s)$ term. The solution would have the form $e^{-j\omega[(t-r/c_s)]} + e^{-j\omega[(t+r/c_s)]}$. The first term indicates propagation *away* from the source and the second term indicates propagation *toward* the source.

Also note that there is no attenuation term, $e^{-\alpha r}$. We have neglected the attenuation of water. Actually the attenuation of water increases with frequency.

Example 8.15 As a final example consider an acoustic speaker modeled in Figure 8.17. Note that energy propagates in both directions – to the right (the intended direction) and to the left (the unintended direction). The edge or the rim of the speaker is fastened so that only the diaphragm moves the fluid. The diaphragm moves in both directions. In most speaker systems the wave to the left is attenuated by an enclosure of some sort. Without attenuation the wave to the left will cancel some of the waves to the right. Model the acoustic speaker with an electric dipole and determine an equation for the pressure as a function of distance. This is modeled by a dipole as shown in Figure 8.18. The dipole consists of a spherical point source and point sink

Figure 8.17. A model of an acoustic speaker.

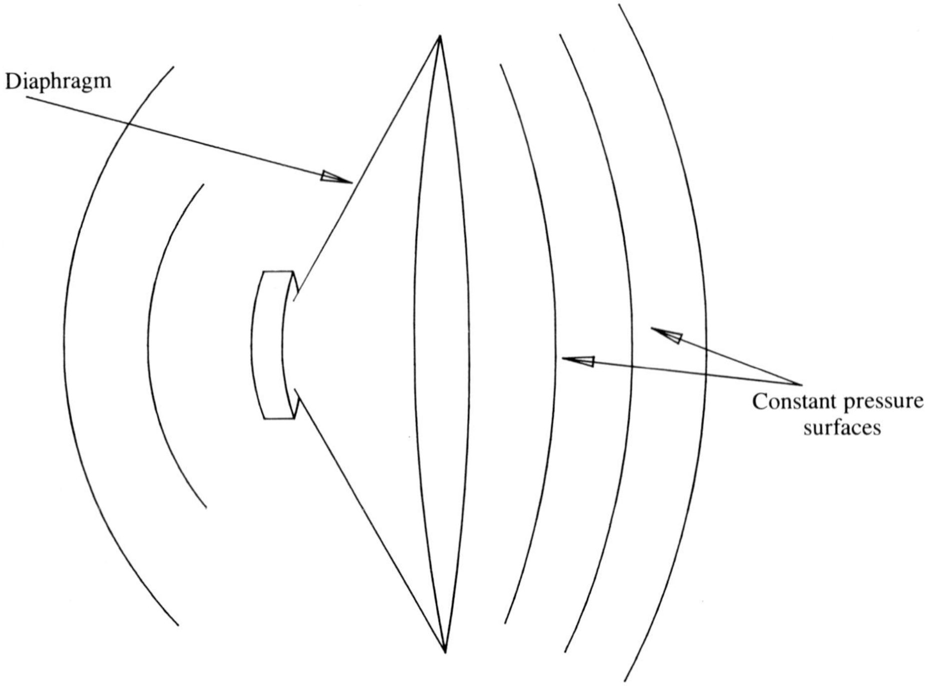

Figure 8.18. An electric dipole modeling an acoustic speaker.

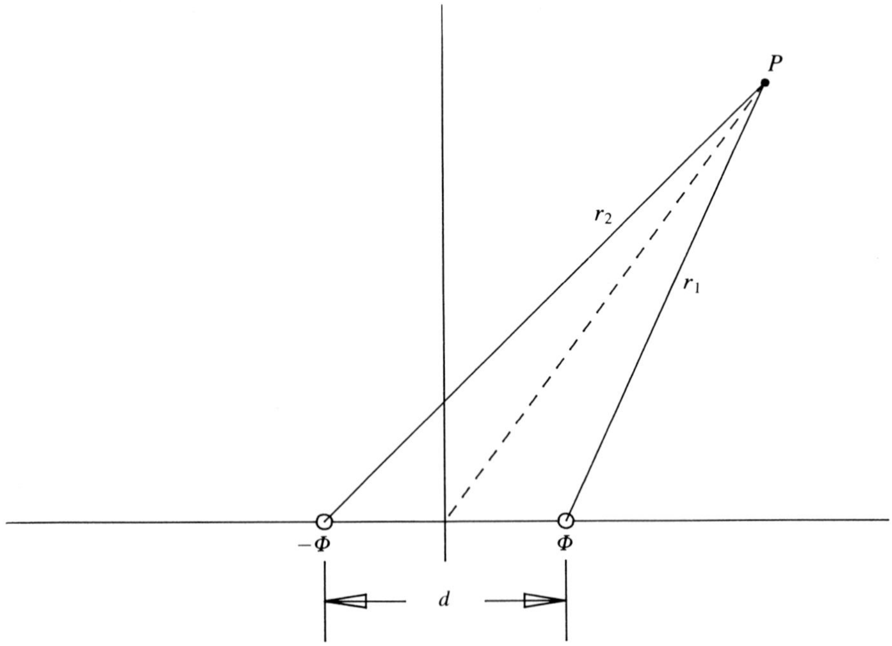

separated by a small distance d. The 'sources' propagate radially. Equation (8.63) above is a *phasor* solution to the phasor wave equation, equation (8.62), assuming sinusoidal variation.

$$p(r) = \frac{A}{r} e^{-j(\omega r/c_s)} \tag{8.63}$$

Using superposition we obtain

$$p = \frac{A}{r_1} e^{-j(\omega r_1/c_s)} - \frac{A}{r_2} e^{-j(\omega r_2/c_s)} \tag{8.67}$$

For $r \gg d$ equation (8.67) becomes

$$p \approx \frac{A}{r} [e^{-j(\omega r_1/c_s)} - e^{-j(\omega r_2/c_s)}] \tag{8.68}$$

For a more detailed analysis see Chapter 3 on electric dipoles. Figure 8.19 is a plot of the loci of constant pressure lines caused by the dipole. The surfaces are analogous to the pressure 'surfaces' emanating from an acoustic speaker. This concludes the discussion of four engineering fields – electric currents, fluid flow through a permeable medium, diffusion and acoustics.

The next chapter is a study of numerical and laboratory methods for solving boundary value problems.

D8.17 Derive an approximation to equation (8.68) in terms of d, r and ϕ.

Figure 8.19. Illustrating constant pressure loci for an acoustic speaker.

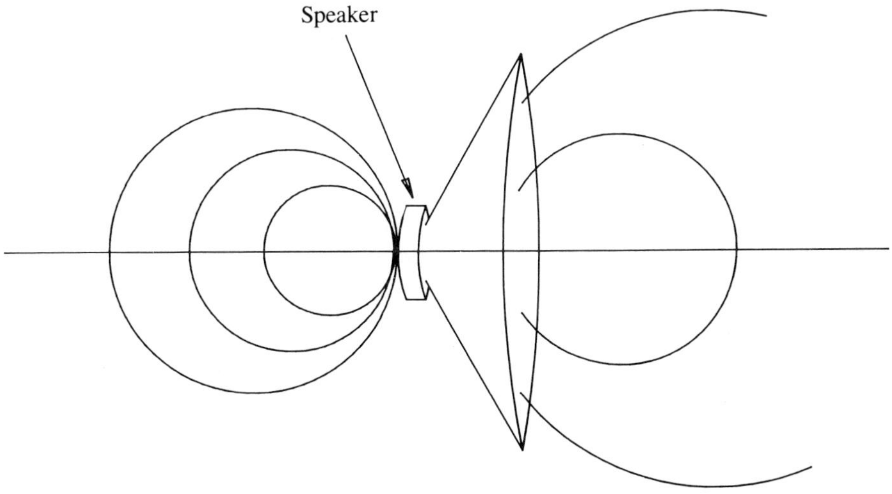

Problems

1. How much charge corresponds to a current of 35 mA flowing for 10 μs?

2. If the alternating current along a certain wire is $i(t) = 10 \sin \omega t$ mA calculate the charge moving along the wire as a function of time.

3. A spherical semiconductor material is conducting a current of 0.1 A radially. At what radius will the current density be 0.1 A/m^2?

4. Determine the electric field at (1, 1, 1) for the following potential fields: (*a*) $V = 3r^2 + 2 \sin \phi \cos \theta$ (spherical), (*b*) $V = 3r^2 + 2z \sin \phi$ (cylindrical), (*c*) $V = 3x^2 + 2y^2 + 3z$ (Cartesian).

5. Calculate the resistance of a carbon resistor which is 7 cm long and 1 cm in diameter. $\sigma = 5.8 \cdot 10^{-9}$ S.

6. If a 1.5 V battery is placed across the ends of the resistor of problem 5, calculate the current density midway between the ends.

7. The boundary between regions 1 and 2 is the xy plane. For $z > 0$ $\sigma = 10^{-1}$ S and for $z < 0$ $\sigma = 10^{-5}$ S. If $\mathbf{E}_1 = x\hat{\mathbf{a}}_x + y\hat{\mathbf{a}}_y + z\hat{\mathbf{a}}_z$ determine **E** and **J** everywhere.

8. In cylindrical coordinates the boundary between regions 1 and 2 is the $r = 2$ cylinder. For $r < 2$ m, $\sigma = 10^5$ S and for $r > 2$ m, $\sigma = 10^{-2}$ S. If $\mathbf{E}_1 = 2.5\hat{\mathbf{a}}_r + 3.5\hat{\mathbf{a}}_\phi - \hat{\mathbf{a}}_z$ V/m, $r < 2$, determine **E** and **J** everywhere.

9. In spherical coordinates the boundary between regions 1 and 2 is the $r = 1.5$ sphere. For $r < 1.5$ m, $\sigma = 10^3$ S and for $r > 1.5$ m, $\sigma = 10^{-4}$ S. If $\mathbf{E}_1 = 1.3\hat{\mathbf{a}}_r - 2.3\hat{\mathbf{a}}_\theta + 3.3\hat{\mathbf{a}}_\phi$ V/m for $r < 1.5$ m determine **E** and **J** everywhere.

10. In cylindrical coordinates the boundary between regions 1 and 2 is the $r = 0.5$ cylinder. For $r < 0.5$ m, $\sigma = 10^2$ S and for $r > 0.5$ m, $\sigma = 10^{-1}$ S. If $\mathbf{E}_1 = x\hat{\mathbf{a}}_x + y\hat{\mathbf{a}}_y + z\hat{\mathbf{a}}_z$ determine **E** and **J** everywhere. Leave answers in cylindrical coordinates.

11. In a cylindrical, vertical well the pressure varies according to $p = \sin \phi / r$. Determine the velocity field if the medium has an average permeability of 10^{-2} m^4/Ns.

12. Repeat problem 11 if the fluid in the well is water and gravitational effects are included.

13. The fluid velocity in a gravel well, $k = 10^2$, is $\mathbf{u} = (1/r)\hat{\mathbf{a}}_r + \cos\phi\hat{\mathbf{a}}_\phi + z\hat{\mathbf{a}}_z$ $z < 0$, $\mathbf{u} = 0$, $z \geq 0$. Determine an equation for the pressure field.

14. The density in a non-homogeneous fluid medium varies in the xy plane; $\rho = 3x + 2y$. The velocity of the fluid also varies non-uniformly; $\mathbf{u} = x\hat{\mathbf{a}}_x + y\hat{\mathbf{a}}_y$. Determine $\nabla \cdot \rho\mathbf{u} = \mathbf{u} \cdot \nabla\rho + \rho\nabla \cdot \mathbf{u}$.

15. In cylindrical coordinates $\rho = r\sin\phi$ and $\mathbf{u} = (\sin\phi/r)\hat{\mathbf{a}}_r + \cos\phi\hat{\mathbf{a}}_\phi$. Using Gauss' law calculate the flux per unit length of a cylinder about the z-axis. The radius of the cylinder is 0.5 m.

16. Two deep water wells in gravel can be modeled as point sources. One is at $(1, 1, 0)$ and has a strength of $\Phi_1 = 3.5$ kg/s. The other is at $(1, -1, 0)$ and has a strength of $\Phi_2 = 7$ kg/s. Calculate the density, $\mathbf{F}$, at the origin.

17. Three long cylindrical water wells of equal strength are parallel. They are at $(1, 0, 0)$ $(1, 1, 0)$ and $(0, 1, 0)$. Calculate the velocity (in vector form) at the origin.

18. The concentration of particles in a material varies as (*a*) $N = N_0 x$, (*b*) $N = N_0 x^2$, (*c*) $N = N_0 x^2 + N_1 y^2 + N_2 z^2$. Determine the density, $\mathbf{D}_\mathrm{d}$, where k_d is known.

19. In a given material $N = 10^{10}$ particles/m^3 at $x = 10^{-2}$ m, $N = 10^{11}$ particles/m^3 at $x = 2 \cdot 10^{-2}$ m, $N = 10^{12}$ particles/m^3 at $x = 3 \cdot 10^{-2}$ m, $N = 10^{13}$ particles/m^3 at $x = 4 \cdot 10^{-2}$ m. Determine an equation for $\mathbf{N}$ and the density, $\mathbf{D}_\mathrm{d}$, $k_\mathrm{d} = 10^{-2}$ m^2/s.

20. In cylindrical coordinates $\mathbf{D}_\mathrm{d} = 10\mathrm{e}^{-r\sin\phi}(\hat{\mathbf{a}}_r + \hat{\mathbf{a}}_\phi)$ part./m^{2}s. Determine the concentration N as a function of time.

21. From equation (8.68) derive an expression for the pressure in terms of the sine function. Sketch the resultant pattern as a function of r and ϕ.

22. Find the gradient of the expression found in problem 21.

9

Applications

In the previous chapters we have attempted to show that many of the engineering fields exhibit the same equations in explaining the behavior of the fields.

Most of us prefer and are accustomed to closed form problems. Most of us prefer analysis over synthesis (design). Most of us prefer exact answers to exact problems. However, many problems cannot be solved in closed form. In other words there are two types of problems and solutions – approximate problems with exact solutions and exact problems with approximate solutions.

We are interested in the second type in this chapter. Examples of approximate solutions are finite differences, finite elements, separation of variables, method of moments, and graphical techniques. In addition, there are laboratory solutions such as potential measurements on Teledeltos paper or in conductive solutions. These and other methods are discussed in Chapter 9. In all, there are 15 exercises included.

9.1 Laplace's equation – finite differences

Consider a bounded two-dimensional region in which the potential is required. For example consider a region bounded by four walls at different temperatures. Using the finite differences method we can calculate the temperature at any point within the boundaries. We begin with Laplace's equation

$$\nabla^2 T = 0 \tag{9.1}$$

In Cartesian coordinates, in two dimensions, Laplace's equation becomes

$$\frac{\partial^2 T}{\partial x^2} + \frac{\partial^2 T}{\partial y^2} = 0 \tag{9.2}$$

In the x-direction we write

$$\frac{\partial^2 T}{\partial x^2} \approx \frac{\Delta\left(\dfrac{\Delta T}{\Delta x}\right)}{\Delta x} \tag{9.3}$$

and in the y-direction we write

$$\frac{\partial^2 T}{\partial y^2} \approx \frac{\Delta\left(\frac{\Delta T}{\Delta y}\right)}{\Delta y}$$

Figure 9.1 shows a bounded Laplacian region in which we are to determine the temperatures. The region is divided into an arbitrary number of meshes. We are to find the temperatures at the intersections. Figure 9.2 shows an expanded view of several meshes. The origin is arbitrarily chosen at T_0. From Figure 9.2 in the x-direction we write

$$\left.\frac{\Delta T}{\Delta x}\right|_{x=-(d/2)} = \frac{T_A - T_0}{d} \tag{9.4}$$

and

$$\left.\frac{\Delta T}{\Delta x}\right|_{x=d/2} = \frac{T_0 - T_C}{d} \tag{9.5}$$

Figure 9.1. A bounded Laplacian region.

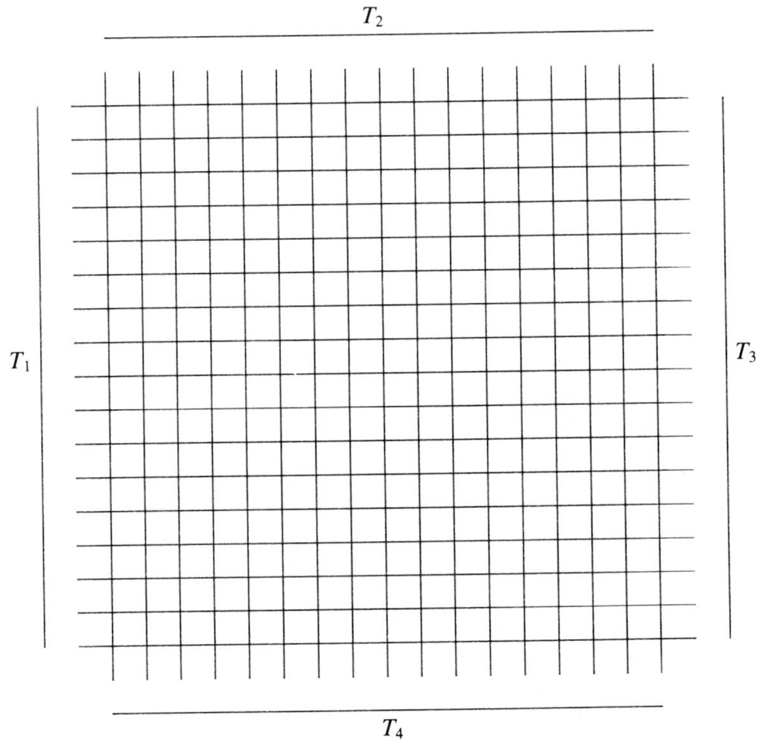

Substituting equations (9.4) and (9.5) into equation (9.3) we obtain

$$\frac{\Delta\left(\dfrac{\Delta T}{\Delta x}\right)}{\Delta x} = \frac{\left.\dfrac{\Delta T}{\Delta x}\right|_{-d/2} - \left.\dfrac{\Delta T}{\Delta x}\right|_{d/2}}{d} = \frac{\dfrac{T_A - T_0}{d} - \dfrac{T_0 - T_C}{d}}{d} \tag{9.6}$$

Likewise in the y-direction we obtain

$$\frac{\Delta\left(\dfrac{\Delta T}{\Delta y}\right)}{\Delta y} = \frac{\left.\dfrac{\Delta T}{\Delta y}\right|_{-d/2} - \left.\dfrac{\Delta T}{\Delta y}\right|_{d/2}}{d} = \frac{\dfrac{T_B - T_0}{d} - \dfrac{T_0 - T_D}{d}}{d} \tag{9.7}$$

Figure 9.2. An expanded view of several meshes.

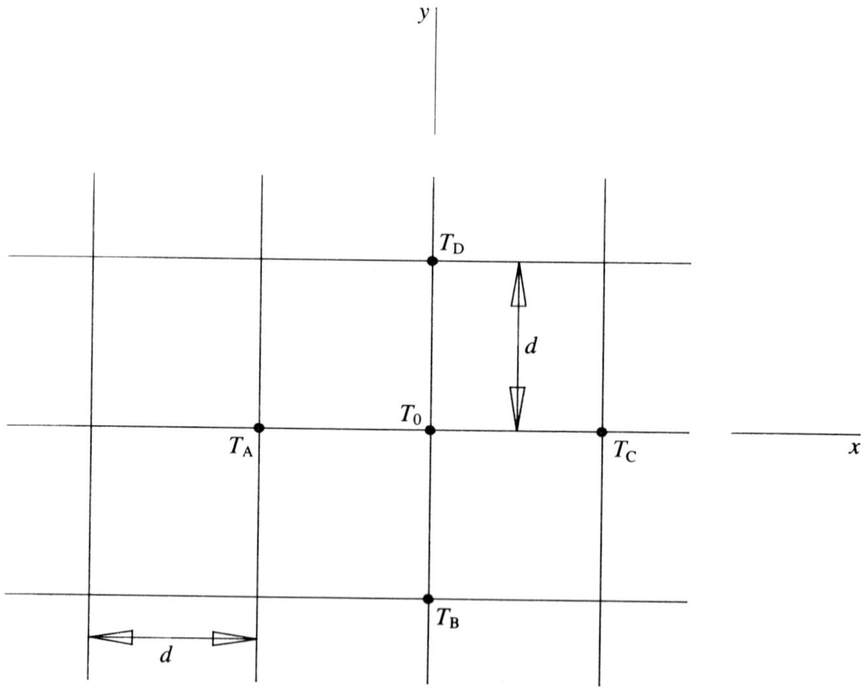

Combining equations (9.6) and (9.7) we obtain

$$T_A + T_B + T_C + T_D = 4T_0 \tag{9.8}$$

So for any point (0, 0) in our region we can calculate T_0 knowing T_A, T_B, T_C and T_D. The problem is we usually don't know T_A, T_B, T_C and T_D! There are several ways of finding T_0 in spite of that. We will use two of them.

1. Finite differences – matrix solution

In this method we first label each node in the region including those on the boundaries. Then we write an equation at each node similar to equation (9.8). The total number of equations will equal the number of unknowns. Any reasonable computer will calculate the temperatures at the nodes.

Example 9.1 Let's illustrate the matrix solution with an example. Consider Figure 9.3 where the boundary temperatures T_0, T_1 and T_2 are

Figure 9.3. A typical Laplacian region.

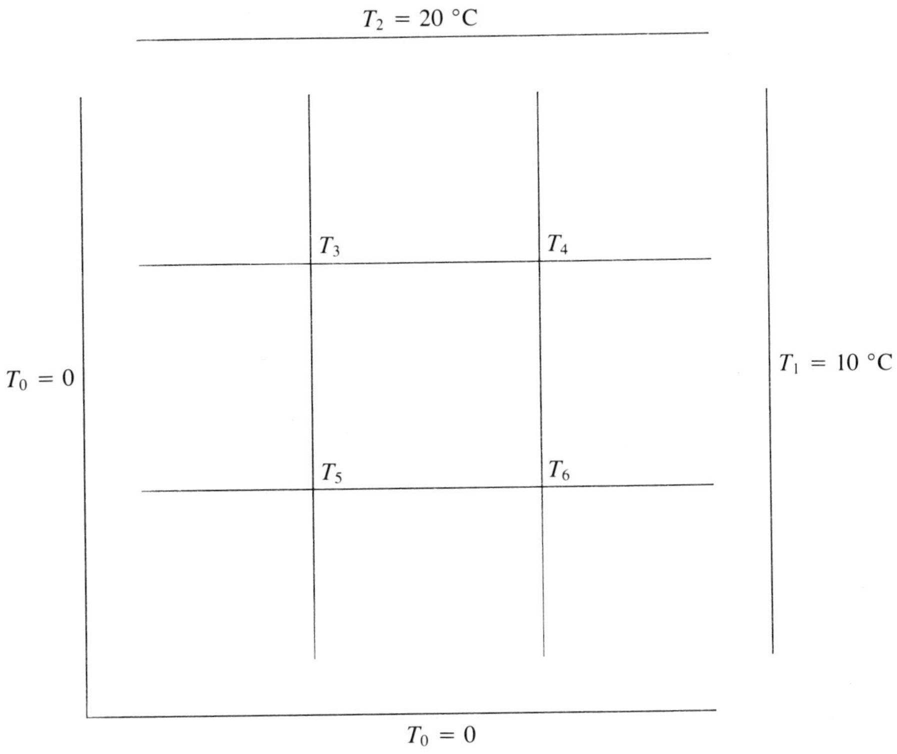

known. We want to find the temperature at any point within the boundaries, say points T_3, T_4, T_5 and T_6. Now use equation (9.8) at each point as follows: For T_3

$$T_0 + T_2 - 4T_3 + T_4 + T_5 = 0$$

for T_4

$$T_1 + T_2 + T_3 - 4T_4 + T_6 = 0$$

for T_5

$$2T_0 + T_3 - 4T_5 + T_6 = 0$$

and for T_6

$$T_0 + T_1 + T_4 + T_5 - 4T_6 = 0$$

Again this 4×4 is solved readily by computer.

2. Finite differences – iterative solution

In this method an iteration technique is used to find the temperatures within our region. Again, consider Figure 9.3. We begin by assuming any temperatures for T_3, T_4, T_5 and T_6. For convenience $T_3 = T_4 = T_5 = T_6 = 0$. Using equation (9.8) and starting at the T_3 position we write

$$\begin{matrix} (T_0) & & (T_2) & & (T_4) & & (T_5) & \\ 0 & + & 20 & + & 0 & + & 0 & = 4T_3 \end{matrix}$$

$$T_3 = 5$$

Using our new value for T_3 we write for the T_4 position

$$\begin{matrix} (T_3) & & (T_2) & & (T_1) & & (T_6) & \\ 5 & + & 20 & + & 10 & + & 0 & = 4T_4 \end{matrix}$$

$$T_4 = \frac{35}{4}$$

and for the T_5 position

$$\begin{matrix} (T_3) & & (T_6) & & (T_0) & & (T_0) & \\ 5 & + & 0 & + & 0 & + & 0 & = 4T_5 \end{matrix}$$

$$T_5 = \frac{5}{4}$$

Finally for T_6

$$\begin{matrix} (T_4) & & (T_1) & & (T_0) & & (T_5) \\ \dfrac{35}{4} & + & 10 & + & 0 & + & \dfrac{5}{4} & = 4T_6 \end{matrix}$$

$$T_6 = 5$$

Figure 9.4 shows our region with the first two values for each position. The iteration continues until the desired accuracy is achieved. It is important to maintain the same 'pattern' when iterating.

9.2 Laplace's equation – finite differences – relaxation method of solution

The relaxation method of solving Laplace's equation is related to the previous finite differences method. However, the relaxation method approaches the actual values more quickly. Another difference is that a *residue* remains after each iteration. Consider again Figure 9.3

Figure 9.4. Laplacian region showing the results of the first iteration.

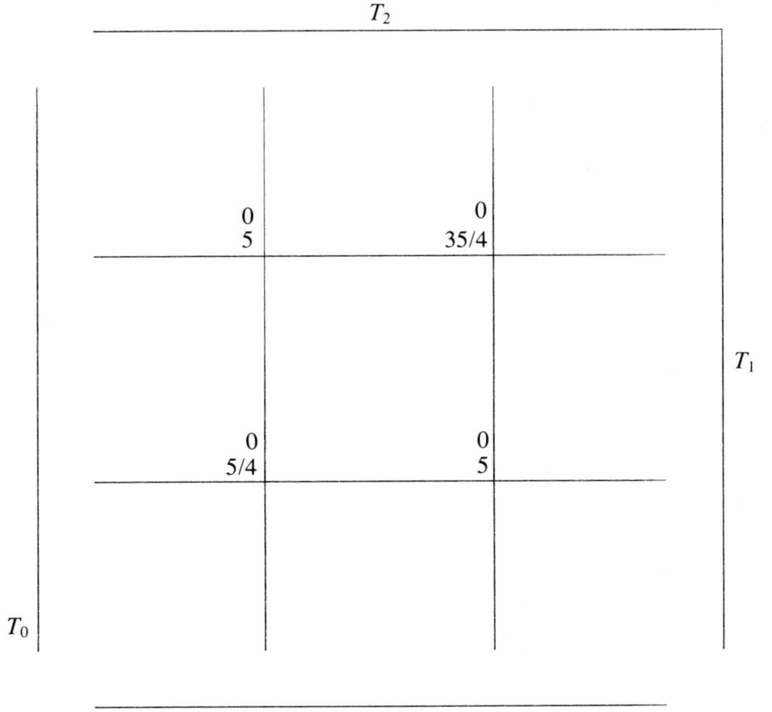

and equation (9.8). Equation (9.8) is rewritten

$$T_A + T_B + T_C + T_D - 4T_0 = 0 \tag{9.8}$$

We now replace the zero with an R, or residue,

$$T_A + T_B + T_C + T_D - 4T_0 = R \tag{9.9}$$

The objective of the relaxation method is to minimize the residue or preferably to reduce it to zero. There are several variations of the relaxation method. Only one will be shown here. This method consists of reducing the largest residue. After one iteration of the entire region we inspect the residues to see which is the largest. If there is more than one 'largest R' choose one arbitrarily and reduce the residue to zero. This new value will of course affect the surrounding grid points. It follows that the whole region will change; all grid points are recalculated. Each time the largest residue is attacked. It is important to repeat the point that there are other methods of relaxation. An illustration follows:

Example 9.2 We begin by guessing at grid temperatures. (We could guess zero but that would mean more computation time.) We place the grid values to the left of each point and the residues to the right. We show that $T_3 = 10$, $T_4 = 15$, $T_5 = 5$ and $T_6 = 7$. From equation (9.9), for point 3

$$0 + 20 + 15 + 5 - 4(10) = 0$$

for point 4

$$10 + 20 + 10 + 7 - 4(15) = -13$$

for point 5

$$0 + 10 + 7 + 0 - 4(5) = -3$$

and for point 6

$$5 + 15 + 10 + 0 - 4(7) = 2$$

Since point 4 has the largest (magnitude) residue we reduce there first. From equation (9.9) note that the temperature at 4 is multiplied by 4. Thus we reduce there by adding $-13/4$ and our new equation is

$$10 + 20 + 10 + 7 - 4(11.75) = 0$$

For point 3

$$0 + 20 + 5 + 11.75 - 4(10) = -3.25$$

Now reduce the temperature at point 3 by $3.25/4 \approx 0.8$

$$0 + 20 + 5 + 11.75 - 4(9.2) = -0.05$$

We proceed to points 4, 5 and 6. For point 4

$$9.2 + 20 + 10 + 7 - 4(11.75) = -0.8$$

For point 5

$$0 + 9.2 + 7 + 0 - 4(5) = -3.8$$

For point 6

$$5 + 11.75 + 10 + 0 - 4(7) = -1.25$$

Figure 9.5 shows the results to this point. How far we relax depends on accuracy desired.

We see that point 5 has the largest residue, −3.8, after this cycle of iteration. There are several ways to reduce the residue; that is, there are several methods of relaxation. For example one method is to

Figure 9.5. Example 9.2 using the relaxation method.

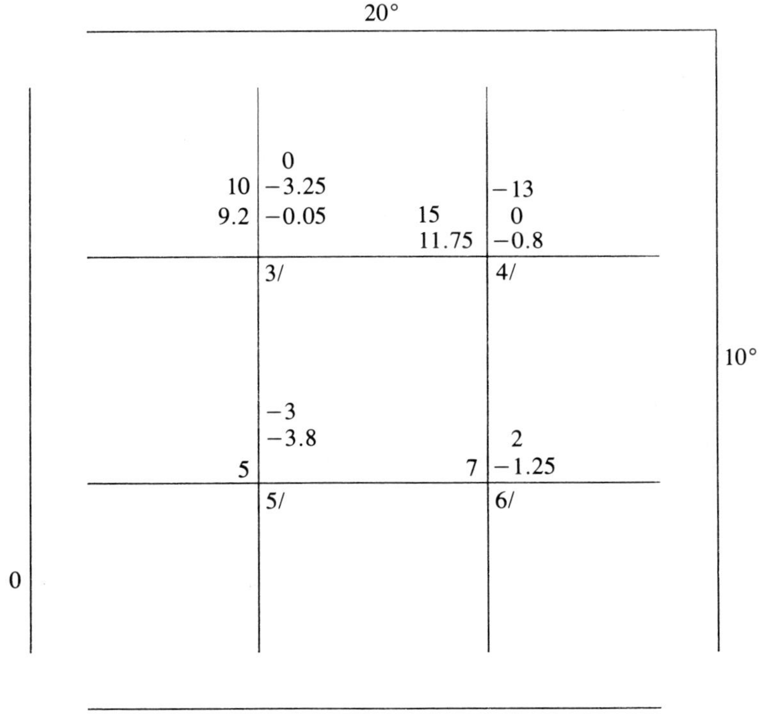

over-relax. By this method the residue is reduced beyond zero. Consider point 5 again. If we reduce the potential there by $3.8/4 = 0.95$ the residue will be zero.

However, if we over-relax and reduce the potential by say 1.4, then the residue is not zero but 1.8. That is at point 5

$$0 + 9.2 + 7 + 0 - 4(3.6) = 1.8$$

Another choice left to the student is whether to reduce all of the residues and then iterate the whole region or to reduce one residue and then iterate the region, or some combination of these.

Once the relaxation method is understood we can extend this to Poisson's equation. For electrostatics we have

$$\nabla^2 V = -\frac{\rho}{\epsilon} \tag{9.10}$$

In two dimensions

$$\frac{\partial^2 V}{\partial x^2} + \frac{\partial^2 V}{\partial y^2} = -\frac{\rho}{\epsilon} \tag{9.11}$$

Comparing equation (9.8) with equation (9.11) we write

$$\frac{\partial^2 V}{\partial x^2} + \frac{\partial^2 V}{\partial y^2} \approx \frac{V_{\mathrm{A}} + V_{\mathrm{B}} + V_{\mathrm{C}} + V_{\mathrm{D}} - 4V_0}{d^2}$$

or

$$\frac{V_{\mathrm{A}} + V_{\mathrm{B}} + V_{\mathrm{C}} + V_{\mathrm{D}} - 4V_0}{d^2} = -\frac{\rho}{\epsilon}$$

$$\frac{V_{\mathrm{A}} + V_{\mathrm{B}} + V_{\mathrm{C}} + V_{\mathrm{D}} - 4V_0}{d^2} + \frac{\rho}{\epsilon} = 0 \tag{9.12}$$

Now combining equation (9.12) with the relaxation form above we obtain

$$\frac{V_{\mathrm{A}} + V_{\mathrm{B}} + V_{\mathrm{C}} + V_{\mathrm{D}} - 4V_0}{d^2} + \frac{\rho}{\epsilon} = R$$

Generally ρ and ϵ are known at a point. In other words we are simply adding a constant ρ/ϵ to an equation for which we have a solution. So we have a solution to Poisson's equation using the relaxation technique. We'll return to Poisson's equation later. For now we go on to the separation of variables technique.

9.3 Separation of variables

Another method of solving a Laplacian region is the separation of variables method. Here we assume the potential along one axis is independent of the potential in the other two directions for an orthogonal coordinate system. Let's ignore the z-direction and work in two dimensions. Thus

$$\nabla^2 V = \frac{\partial^2 V}{\partial x^2} + \frac{\partial^2 V}{\partial y^2} = 0 \tag{9.13}$$

We assume

$$V = V_x(x)V_y(y) \tag{9.14}$$

Substituting equation (9.14) into (9.13) we have

$$V_y \frac{\partial^2 V_x}{\partial x^2} + V_x \frac{\partial^2 V_y}{\partial y^2} = 0$$

and

$$\frac{1}{V_x}\frac{\partial^2 V_x}{\partial x^2} = -\frac{1}{V_y}\frac{\partial^2 V_y}{\partial y^2} \tag{9.15}$$

Now the left-hand side of equation (9.15) is independent of y and the right-hand side is independent of x, therefore let

$$\frac{1}{V_x}\frac{\mathrm{d}^2 V_x}{\mathrm{d}x^2} = k \tag{9.16}$$

where k is a constant and

$$\frac{1}{V_y}\frac{\mathrm{d}^2 V_y}{\mathrm{d}y^2} = -k \tag{9.17}$$

From equation (9.16)

$$\frac{\mathrm{d}^2 V_x}{\mathrm{d}x^2} = kV_x \tag{9.18}$$

A solution to equation (9.17) is

$$V_y(y) = K_1 \cos \sqrt{k}\, y + K_2 \sin \sqrt{k}\, y \tag{9.19}$$

and a solution to equation (9.18) is

$$V_x(x) = K_3 \cosh \sqrt{k}\, x + K_4 \sinh \sqrt{k}\, x \tag{9.20}$$

Substituting equations (9.19) and (9.20) into equation (9.14) we obtain

$$V = (K_1 \cosh \sqrt{k}\,x + K_2 \sinh \sqrt{k}\,x)(K_3 \cos \sqrt{k}\,y + K_4 \sin \sqrt{k}\,y) \tag{9.21}$$

Equation (9.21) is a general solution to Laplace's equation, equation (9.13). A particular solution is found from the boundary conditions. Let's consider an example.

Example 9.3 From Figure 9.6 $V = 0$ at $x = 0$ which gives $K_1 = 0$ and $V = 0$ at $y = 0$ which gives $K_3 = 0$. Equation (9.21) becomes

$$V = K_2 K_4 \sinh \sqrt{k}\,x \sin \sqrt{k}\,y$$

or

$$V = K_5 \sinh \sqrt{k}\,x \sin \sqrt{k}\,y \tag{9.22}$$

where

$$K_5 = K_2 K_4$$

Figure 9.6. The Laplacian region of example 9.3.

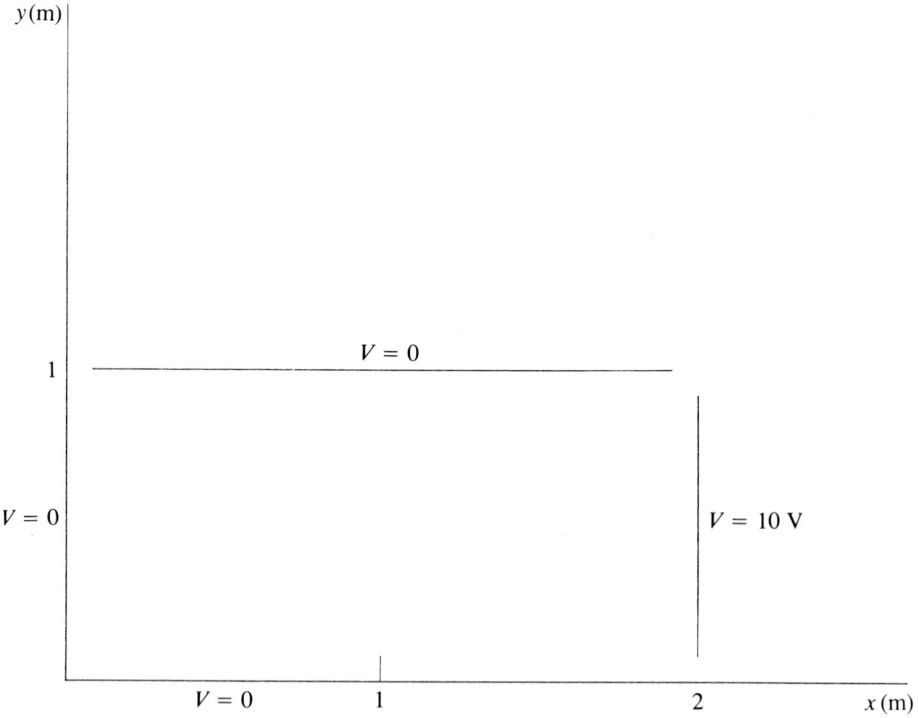

Next we note that $V = 0$ at $y = 1$ m

$$\sin \sqrt{k}\,(1) = 0$$

or

$$\sqrt{k} = m\pi \qquad m = 0, 1, 2, \ldots$$

Equation (9.22) becomes

$$V = K_5 \sinh m\pi x \sin m\pi y$$

At $x = 2$ $V = 10$ or $10 = K_5 \sinh m\pi 2 \sin m\pi y$. This equation cannot be satisfied for every y – unless we rely on a series solution, that is

$$V = \sum_{-\infty}^{\infty} K_m \sinh m\pi x \sin m\pi y \tag{9.23}$$

At $x = 2$, $V = 10$ which yields

$$10 = \sum_{-\infty}^{\infty} K_m \sinh m\pi 2 \sin m\pi y \tag{9.24}$$

Now let's resort to a bit of trickery. Imagine that a potential $V = -10$ V exists at $x = 2$ and $1 < y < 2$. In other words we have a periodic function. That is, $V = 10$ V for $x = 2$ and $0 < y < 1$; $V = -10$ V for $x = 2$ and $1 < y < 2$; $V = 10$ V for $x = 2$ and $2 < y < 3$ and so on. Recall that any periodic function $f(t)$ can be written as a trigonometric Fourier series or

$$f(t) = a_0 + \sum_{n=1}^{\infty} a_n \cos n\omega_0 t + \sum_{n=1}^{\infty} b_n \sin n\omega_0 t \tag{9.25}$$

Comparing equations (9.24) and (9.25)

$$f(t) \leftrightarrow V(x, y)$$

$$a_0 = a_n = 0$$

$$n\omega_0 t \leftrightarrow m\pi y$$

and

$$b_n \leftrightarrow K_m \sinh m\pi 2$$

It can be shown that for m odd

$$K_m = \frac{4(10)}{m\pi \sinh m\pi 2}$$

The final solution is

$$V(x, y) = \sum_{m \text{ odd}} \frac{40 \sinh m\pi x}{m\pi \sinh m\pi 2} \sin m\pi y \tag{9.26}$$

Remember this applies for the specific region shown in Figure 9.6. Equation (9.26) needs to be modified for other regions. For example, if in the previous example $V = 20$ V for $y = 1$ as shown in Figure 9.7, we would use superposition for our solution. That is, we would find separate solutions for Figure 9.7(a) and (b) and add the solutions. The sum would give us the solution for Figure 9.7(c).

9.4 Resistive analog

Another method for solving Laplace's or Poisson's equation is by the use of the resistive analog. The resistive analog is based on the equation used in the relaxation method which was presented previously. This is depicted in Figure 9.8(a). The justification follows: From Figure 9.8(b) we can write a node equation (assuming all currents

Figure 9.7. Illustrating a superposition solution.

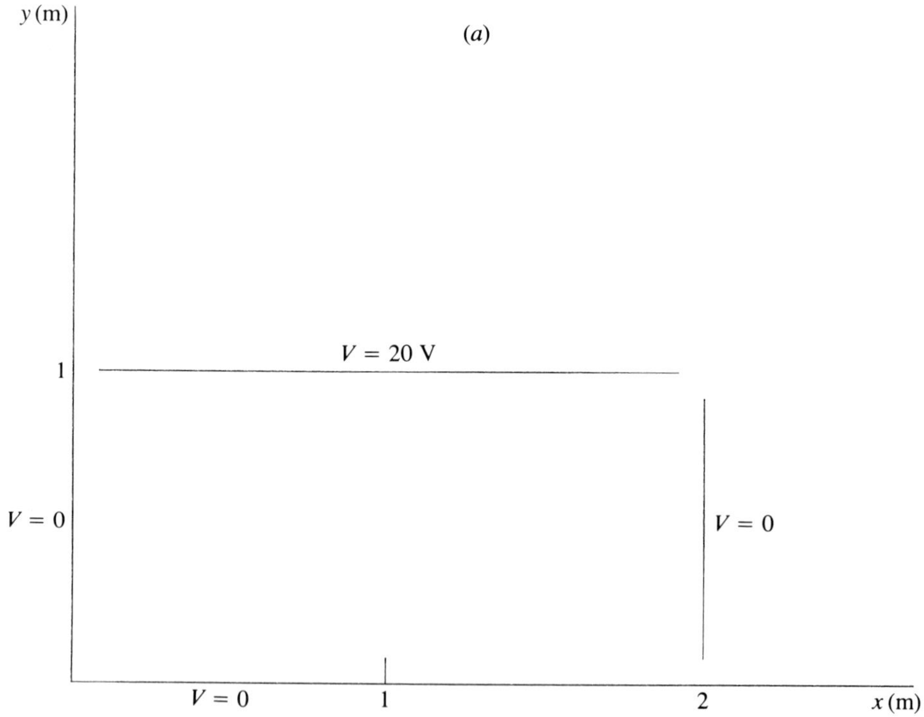

Figure 9.7. (*Continued*)

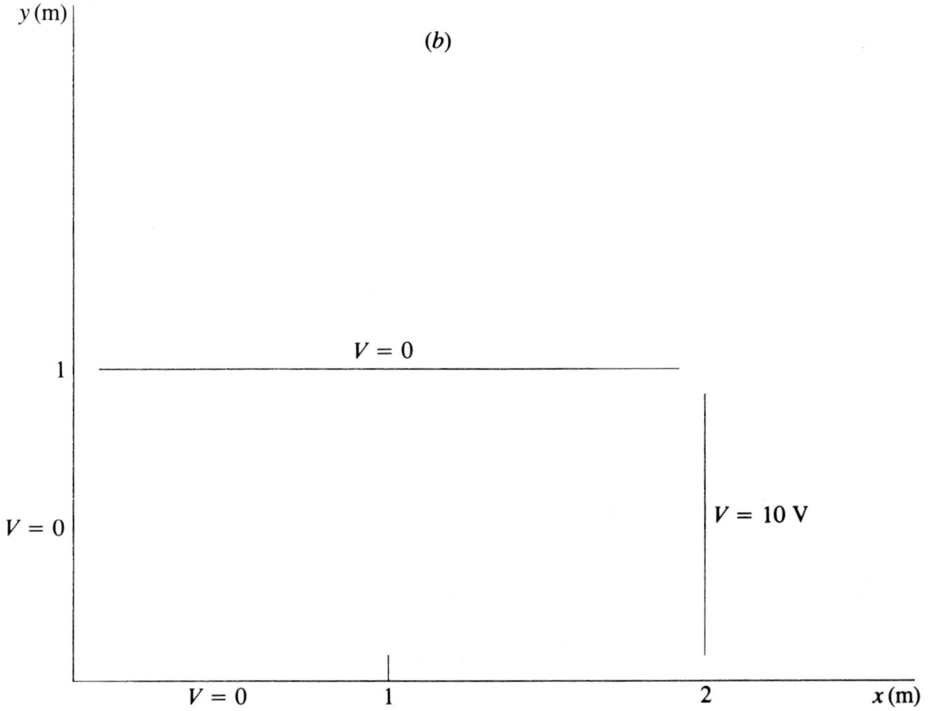

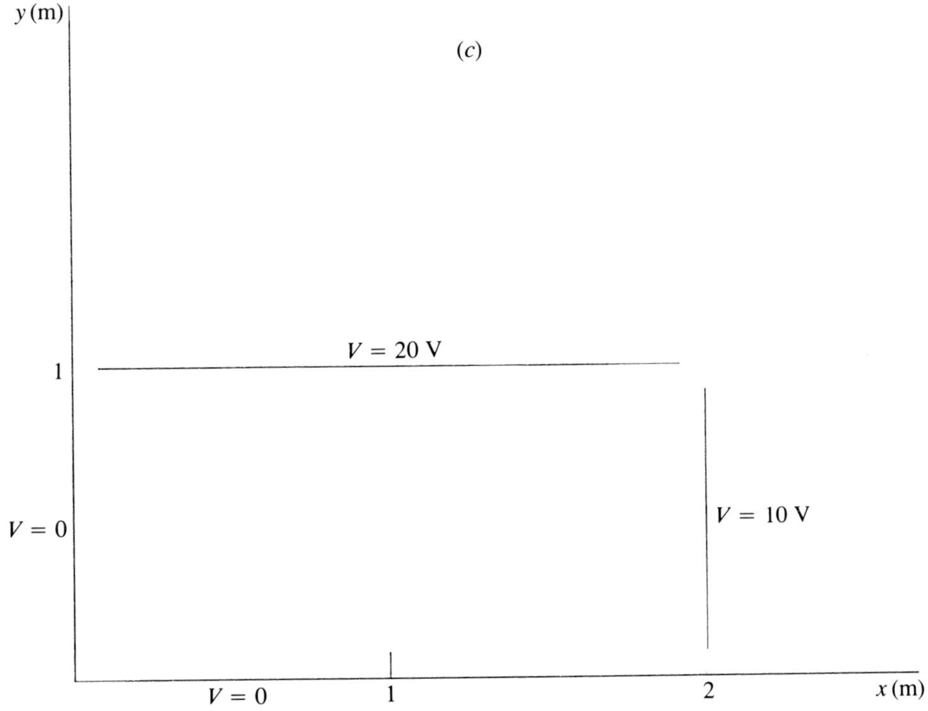

enter the node) that is

$$\frac{V_A - V_0}{R} + \frac{V_B - V_0}{R} + \frac{V_C - V_0}{R} + \frac{V_D - V_0}{R} + I = 0$$

or

$$V_A + V_B + V_C + V_D - 4V_0 = IR \tag{9.27}$$

Comparing equations (9.12) and (9.27) we see it is the same equation we used for the relaxation method.

In this method we divide the bounded region into foursided polygons. Consider a thermal conductor as shown in Figure 9.9. In Figure 9.9(*a*) the conductor is divided into two regions and in Figure 9.9(*b*) the conductor is divided into three regions. In Figure 9.9(*a*) the resistors are in series and $R_T = R_1 + R_2$. In Figure 9.9(b) there is a series–parallel combination and $R_T = R_1 + R_2R_3/(R_2 + R_3)$.

We calculate the resistance of each polygon. After the region is modeled by resistors in series and parallel we apply a temperature differential across the region.

Figure 9.8. A resistive mesh (*a*) Laplace region, (*b*) one mesh node.

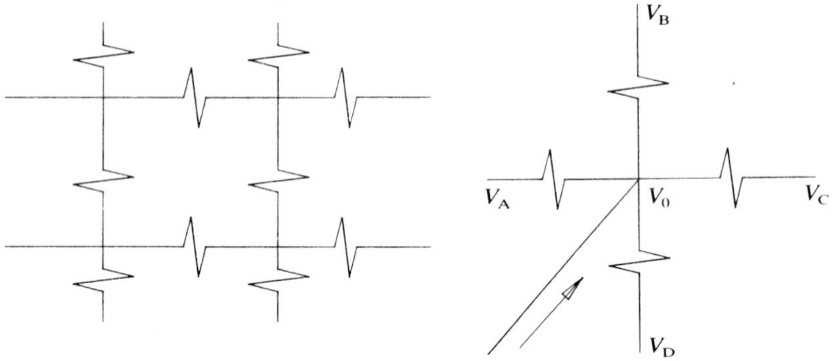

Figure 9.9. Two multi-material thermal conductors. (*a*) series $R_T = R_1 + R_2$, (*b*) series/parallel $R_T = R_1 + (R_2R_3)/(R_2 + R_3)$.

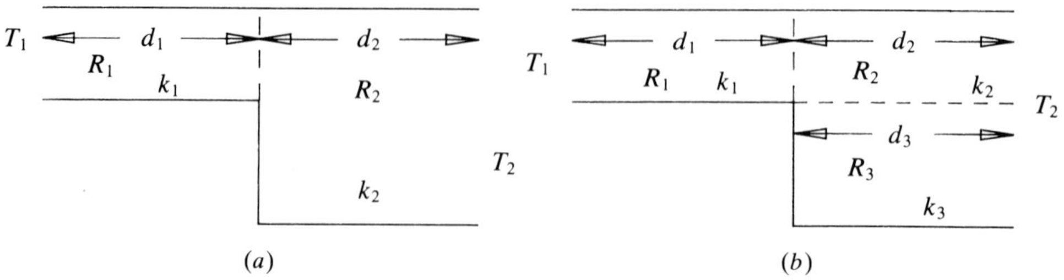

An equation of the form $Q = \Sigma T/R$ is written at each node. The resultant system of equations is solved to obtain the temperature throughout the region. We can either measure or calculate the temperatures at the connections. Let's illustrate with an example.

Example 9.4 Find the isotherms for the regions shown in Figure 9.9. Determine the resistances of the thermal conductors in Figure 9.9. Both figures have been divided into distinct rectangles. Figure 9.9(a) has two and Figure 9.9(b) has three. The greater the number of rectangles the greater the accuracy. Each rectangle will be modeled by a resistor connected as shown. For a thermal resistance $R = d/kA$. Compare this formula with that of an electrical resistance, that is, $R = d/\sigma A$. If we assume a unit depth, $R = d/kW$, where W is the width of the rectangle. This is shown in Figure 9.10.

Returning to Figure 9.9 we note that $R_T = R_1 + R_2$ for Figure 9.9(a), and $R_T = R_1 + R_2R_3/(R_2 + R_3)$ for Figure 9.9(b). Therefore, the total thermal resistance for Figure 9.9(a) is $R_T = d_1/k_1W_1 + d_2/k_2W_2$. The

Figure 9.10. One of the polygons in the region.

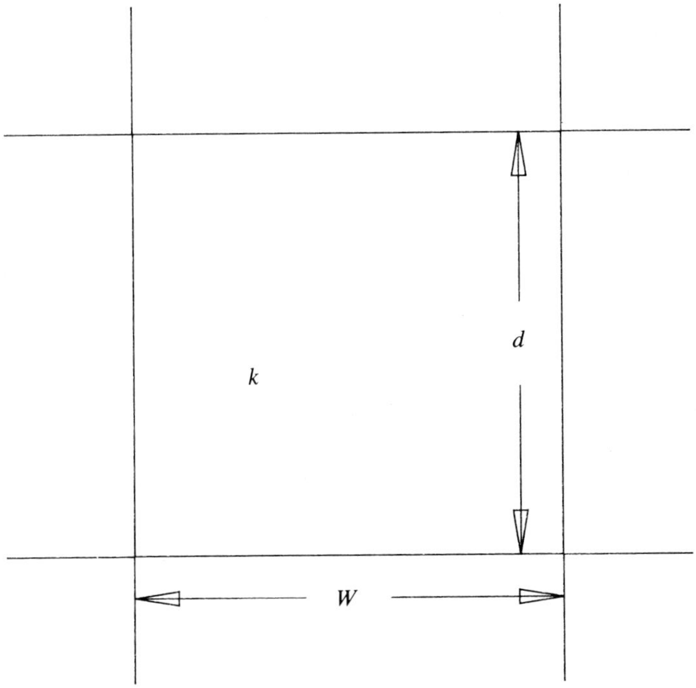

total resistance for Figure 9.9(b) is

$$R_{\mathrm{T}} = \frac{d_1}{k_1 W_1} + \frac{\dfrac{d_2}{k_2 W_2} \cdot \dfrac{d_3}{k_3 W_3}}{\dfrac{d_2}{k_2 W_2} + \dfrac{d_3}{k_3 W_3}}$$

Once the resistances are calculated we can solve the region using the system of equations mentioned above. One method for solving these is the Gauss–Seidel method. Another method is a circuit analysis program like PSPICE. Still a third method would involve laboratory measurements. This would involve several resistors, a dc source, a circuit board and a multimeter. Any of these methods would give the requisite data points. In any case it is a relatively simple task to plot the isotherms (equipotentials).

9.5 Potential and electric field measurements in a conductive solution

A convenient laboratory setup for solving a Laplacian region consists of measuring the potential between boundaries in a conductive solution (Figure 9.11). The solution is in a tub or pan. The boundaries are *in* the solution so that there are potentials at any point within the

Figure 9.11. Laboratory setup for potential and electric field measurements in a conductive solution.

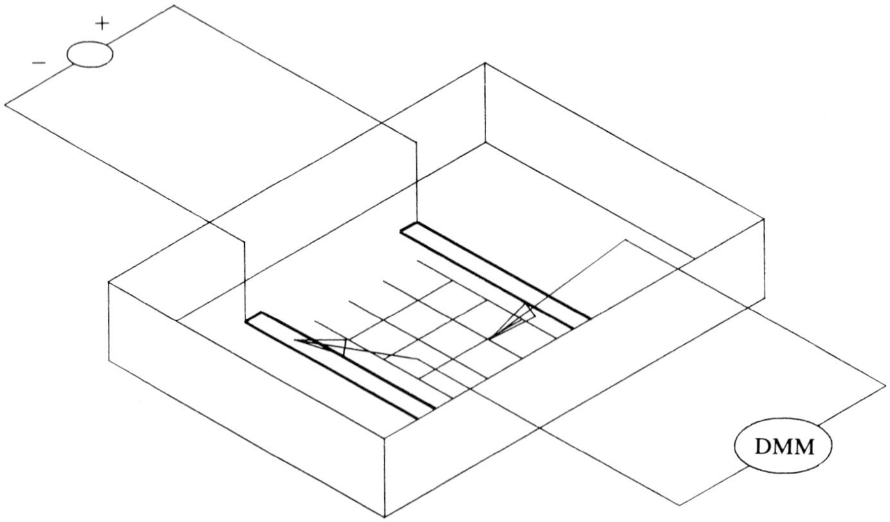

enclosure. We can measure the electric field and the potential at any point. For convenience a grid is laid out on the bottom of the pan before the solution is added. The potential measurements are made with one probe at one of the boundaries and the second probe at various points in the region. In addition, electric field measurements are made by applying each probe at any two points. Knowing the probe separation and the potential we can calculate the electric field; that is from

$$\mathbf{E} = -\nabla V$$

we get

$$E = \frac{V}{d} \tag{9.28}$$

where d is the probe separation in meters. It becomes apparent that if $E = 0$ in equation (9.28) then $V = 0$ or the probes are along an equipotential line. Thus by rotating the probes (clamping or taping the probes together is convenient) we observe anything from a maximum to a minimum electric field. A maximum means the probes are along a streamline or a flux line; a minimum means the probes are along an equipotential. Flux lines and potential lines are orthogonal. Therefore we can sketch flux lines and equipotential lines. We should be able to extend this method to any shape boundary and three or more potentials (temperatures). We can use a voltage divider to apply several boundary potentials. It is important that the pan or tank be level so that the depth of the solution remain constant. If the depth is not constant the conduction paths between boundaries will differ.

In some cases the pan is tilted intentionally to model other situations. For example, some antenna insulators are tapered, as shown in Figure 9.12(a). In this case there is rotational symmetry. The insulator can be

Figure 9.12. An antenna insulator and its model. (a) Antenna insulator, (b) insulator model.

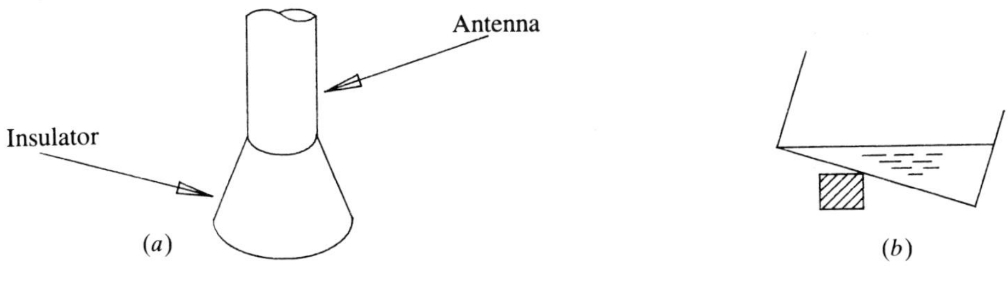

modeled by tilting the pan as shown in Figure 9.12(*b*). Another example of modeling by varying the depth of the conductive solution is one in which the dielectric permittivity of the insulator varies.

In the next section we continue with laboratory models.

9.6 Potential measurements on resistive paper

In this section we model a region on a resistive paper called Teledeltos paper. We model the boundaries by painting the boundaries on the Teledeltos paper with a conductive paint. With the two-dimensional region in place we measure with a digital multimeter the potential at various grid points on the paper. The equipotential lines may be sketched on the Teledeltos paper or on a companion paper. As in the previous section the boundaries may be placed anywhere on the paper. There will be fringing in both cases and whether this is tolerable depends on the problem. Indeed, these methods could be used to measure fringing. In addition these methods could be used to calculate the capacitance of a configuration. Let's derive an equation for calculating capacitance.

From Chapter 3

$$Q \equiv \Phi = \oint \mathbf{D} \cdot \mathrm{d}\mathbf{A} \tag{9.29}$$

and

$$\mathbf{E} = -\nabla V \tag{9.30}$$

Assuming linearity we write

$$Q = \epsilon E A \tag{9.31}$$

where A is the cross-sectional area through which the flux lines pass. Also

$$E = \frac{V}{d} \tag{9.32}$$

where d is the length along a flux line. Substituting equation (9.32) into equation (9.31)

$$Q = \epsilon \frac{V}{d} A$$

or

$$\frac{Q}{V} = \epsilon \frac{A}{d}$$

However

$$C = \frac{Q}{V}$$

thus

$$C = \frac{\epsilon A}{d} \tag{9.33}$$

9.7 Orthogonality of flux lines and equipotentials

In this section we examine the orthogonality of flux lines and equipotentials. Assume a two-dimensional Laplacian region.

Consider a function $F(c) = A + jB$ in the complex plane where $c = x + jy$. The function $F(c)$ can be written as the sum of a real part and an imaginary part, each a function of x and y; that is,

$$F(c) = A(x, y) + jB(x, y)$$

The complex function F is the sum of two real functions, A and B. If F is analytic, then the Cauchy–Riemann conditions hold. These are

$$\frac{\partial A}{\partial x} = \frac{\partial B}{\partial y} \tag{9.34}$$

$$\frac{\partial B}{\partial x} = -\frac{\partial A}{\partial y} \tag{9.35}$$

From equation (9.34)

$$\frac{\partial^2 A}{\partial x^2} = \frac{\partial^2 B}{\partial x\, \partial y} \tag{9.36}$$

and from equation (9.35)

$$\frac{\partial^2 B}{\partial x\, \partial y} = -\frac{\partial^2 A}{\partial y^2} \tag{9.37}$$

Substituting equation (9.37) into equation (9.36) we get

$$\frac{\partial^2 A}{\partial x^2} + \frac{\partial^2 A}{\partial y^2} = 0 \tag{9.38}$$

Similarly

$$\frac{\partial^2 B}{\partial x^2} + \frac{\partial^2 B}{\partial y^2} = 0 \tag{9.39}$$

Equations (9.38) and (9.39) indicate that both A and B satisfy Laplace's equation. Now let $A(x, y)$ be the potential function and $B(x, y)$ be the stream or flux function. If $A(x_0, y_0)$ and $B(x_0, y_0)$ obey the relationship

$$\left[\frac{-\partial A/\partial x}{\partial A/\partial y}\right]\left[\frac{-\partial B/\partial x}{\partial B/\partial y}\right] = -1 \tag{9.40}$$

at some point (x_0, y_0) they are considered orthogonal, and the two families are orthogonal. (Equation (9.40) is proved in most advanced mathematics books.) Let's try an example.

Example 9.5 Let $A(x, y) = (x^2 - y^2)$ and $B(x, y) = 2xy$. Show that $A(x, y)$ and $B(x, y)$ are orthogonal. Also show that both obey Laplace's equation.

First taking the partial derivatives, we obtain

$$\frac{\partial A}{\partial x} = 2x, \qquad \frac{\partial A}{\partial y} = -2y, \qquad \frac{\partial B}{\partial x} = 2y \quad \text{and} \quad \frac{\partial B}{\partial y} = 2x$$

Substituting these derivatives into equation (9.40) we obtain

$$\left[\frac{-2x}{-2y}\right]\left[\frac{-2y}{2x}\right] = -1$$

Thus A and B are orthogonal. In addition

$$\frac{\partial^2 A}{\partial x^2} = 2, \qquad \frac{\partial^2 A}{\partial y^2} = -2 \quad \text{and} \quad \frac{\partial^2 A}{\partial x^2} + \frac{\partial^2 A}{\partial y^2} = 0$$

and Laplace's equation is satisfied. Similarly $B(x, y)$ satisfies Laplace's equation. As an example consider the parallel plate capacitor in Figure 9.13.

Figure 9.13. A parallel plate capacitor in the (a) y-direction and (b) x-direction.

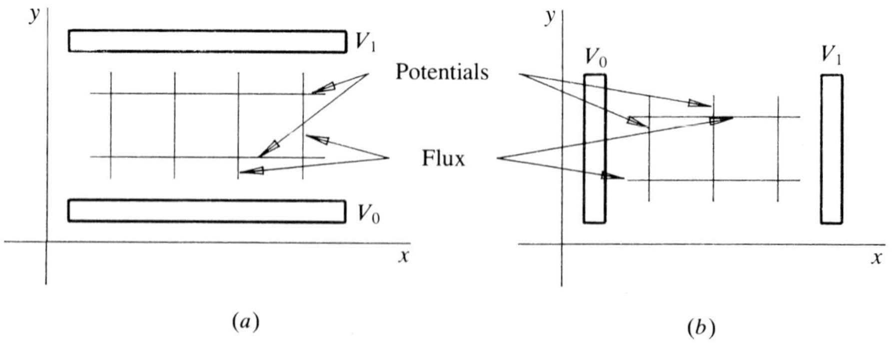

In Figure 9.13(*a*) the potential function is parallel to the x-axis and the flux is parallel to the y-axis. In Figure 9.13(*b*) the potential function is parallel to the y-axis and the flux function is parallel to the x-axis. Both functions satisfy Laplace's equation in both regions. In both figures the potential and flux functions are orthogonal everywhere.

In the next section we use this orthogonality property to calculate the capacitance and/or resistance of Laplacian regions.

9.8 Method of curvilinear squares

In section 9.7 we determined that the potential function and the flux function are orthogonal. In many cases it is not possible to write a neat analytical expression for a Laplacian region. For these cases other techniques are needed. In this section we illustrate the method of curvilinear squares, and we will use this fact to calculate capacitance and resistance in two-dimensional problems.

We know from the preceding chapters that the flux is the result of the charge. From Gauss' law the flux

$$Q \equiv \Phi = \oint \mathbf{D} \cdot \mathrm{d}\mathbf{A}$$

Assuming linearity we get

$$Q = \epsilon E A \tag{9.41}$$

Recall the area A normal to the Laplacian region. Let's assume a depth of one meter so that

$$A = \Delta L_{\mathrm{t}}(1)$$

where ΔL_{t} is perpendicular to the electric field (flux) lines or tangential to the boundary. Equation (9.41) becomes

$$\Delta Q = \epsilon E \, \Delta L_{\mathrm{t}}$$

or

$$E = \frac{1}{\epsilon} \frac{\Delta Q}{\Delta L_{\mathrm{t}}} \tag{9.42}$$

From Chapter 3

$$\mathbf{E} = -\nabla V$$

or

$$E = \frac{\Delta V}{\Delta L_{\mathrm{n}}} \tag{9.43}$$

where ΔL_n is *normal* to the boundary. From equations (9.42) and (9.43)

$$E = \frac{1}{\epsilon}\frac{\Delta Q}{\Delta L_t} = \frac{\Delta V}{\Delta L_n} \tag{9.44}$$

$$\frac{\Delta L_t}{\Delta L_n} = K = \frac{1}{\epsilon}\frac{\Delta Q}{\Delta V}$$

Let $K = 1$ and

$$\frac{1}{\epsilon}\frac{\Delta Q}{\Delta V} = 1 \tag{9.45}$$

Notice by choosing $K = 1$, $\Delta L_t = \Delta L_n$ and the region is divided into squares.

Now the total charge on a boundary is

$$Q = N_Q \, \Delta Q \tag{9.46}$$

and the total potential across the region is

$$V = N_V \, \Delta V \tag{9.47}$$

Combining equations (9.45), (9.46) and (9.47) we get

$$C = \frac{Q}{V} = \epsilon \frac{N_Q}{N_V} \tag{9.48}$$

Recall from section 3.20 equation (3.82)

$$RC = \epsilon/\sigma$$

or

$$\frac{C}{\epsilon} = \frac{1}{R\sigma} \tag{9.49}$$

Combining equations (9.48) and (9.49) we get

$$R = \frac{N_V}{\sigma N_Q} \tag{9.50}$$

In summary, by dividing the region into squares, we can, by counting the squares, calculate the capacitance or resistance of any Laplacian region. Consider an example.

Example 9.6 Calculate the capacitance and resistance of one section of a coaxial conductor shown in Figure 9.14(a). Recall $C = \Phi/V$ and

$R = V/\Phi$ or, from equation (9.48),

$$C = \epsilon \frac{N_Q}{N_V} \tag{9.48}$$

and from equation (9.50)

$$R = \frac{N_V}{\sigma N_Q} \tag{9.50}$$

From Figure 9.14(a) we note that $N_V = 3$ and $N_Q = 4$ so that $C = \frac{4}{3}\epsilon$ and $R = 3/(4\sigma)$ for a one meter depth.

There are instances when it is difficult to divide the entire Laplacian region into squares. When this happens, we need to divide the troublesome areas into smaller squares as shown in Figure 9.14(b). The smaller the areas become the more likely we are to obtain a square.

We next discuss the sketching of field maps.

Graphical field mapping

In order to calculate capacitance from equation (9.48) or resistance from equation (9.50) we must know N_V and N_Q. These are obtained from the field maps. The field map is sketched as follows:

1. Divide the region into an integer number of guessed equipotentials.
2. Sketch the equipotentials beginning where the fields are well known. Proceed to the more difficult areas, hugging convex corners and spreading the lines in the area of concave corners. Note Figure 9.15.

Figure 9.14. A section of a coaxial cable from example 9.6. (a) Section of coaxial conductor, (b) expanded view of section.

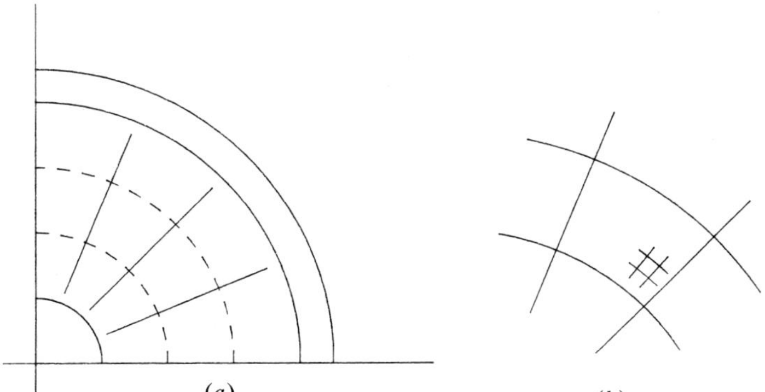

(a) (b)

3. Next sketch the streamlines, maintaining orthogonality of potential and streamlines throughout. These will form curvilinear squares.
4. In some places of low field intensity there will be figures with more than four sides. These can be subdivided to form squares. The streamlines and the potentials must be divided by the same factor as the difficult areas are subdivided.
5. Continue to erase and sketch until satisfactory results are obtained.

In the next section we use this orthogonality property to calculate the capacitance and/or resistance of Laplacian regions.

9.9 The magnetic vector potential

Up to this point we have only considered Laplace's equation, $\nabla^2 V = 0$. Consider Poisson's equation, $\nabla^2 V = -\rho/\epsilon$. We derive a similar equation for magnetics. Let's begin with Ampere's law

$$\nabla \times \left(\frac{\mathbf{B}}{\mu}\right) = \mathbf{J}$$

Figure 9.15. Field map sketching.

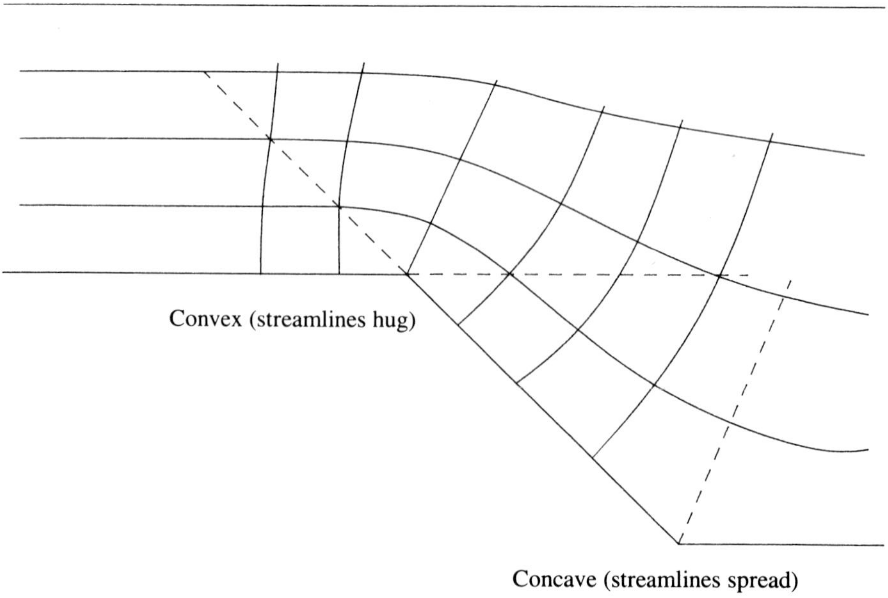

and the magnetic vector potential definition

$$\mathbf{B} = \nabla \times \mathbf{A}$$

that is

$$\nabla \times \left(\frac{1}{\mu} \nabla \times \mathbf{A}\right) = \mathbf{J} \tag{9.51}$$

We assume the permeability is a function of position in two dimensions, $\mu = \mu(x, y)$, and that $\mathbf{J} = J_z \hat{\mathbf{a}}_z$ and $\mathbf{A} = A_z \hat{\mathbf{a}}_z$

$$\nabla \times \mathbf{A} = \begin{vmatrix} \hat{\mathbf{a}}_x & \hat{\mathbf{a}}_y & \hat{\mathbf{a}}_z \\ \dfrac{\partial}{\partial x} & \dfrac{\partial}{\partial y} & 0 \\ 0 & 0 & A_z \end{vmatrix}$$

$$= \frac{\partial A_z}{\partial y} \hat{\mathbf{a}}_x - \frac{\partial A_z}{\partial x} \hat{\mathbf{a}}_y$$

and

$$\nabla \times \left(\frac{1}{\mu} \nabla \times \mathbf{A}\right) = \begin{vmatrix} \hat{\mathbf{a}}_x & \hat{\mathbf{a}}_y & \hat{\mathbf{a}}_z \\ \dfrac{\partial}{\partial x} & \dfrac{\partial}{\partial y} & 0 \\ \dfrac{1}{\mu}\dfrac{\partial A_z}{\partial y} & \dfrac{-1}{\mu}\dfrac{\partial A_z}{\partial x} & 0 \end{vmatrix}$$

$$= -\hat{\mathbf{a}}_z \left[\frac{\partial}{\partial x}\left(\frac{1}{\mu}\frac{\partial A_z}{\partial x}\right) + \frac{\partial}{\partial y}\left(\frac{1}{\mu}\frac{\partial A_z}{\partial y}\right)\right]$$

which gives

$$\frac{\partial}{\partial x}\left(\frac{1}{\mu}\frac{\partial A_z}{\partial x}\right) + \frac{\partial}{\partial y}\left(\frac{1}{\mu}\frac{\partial A_z}{\partial y}\right) = -J_z \tag{9.52}$$

A finite differences solution to equation (9.52) for the mesh of Figure 9.16 is*

$$A_x = \frac{J\dfrac{d^2}{2} + \displaystyle\sum_n \frac{A_n}{\mu_x + \mu_n}}{\displaystyle\sum_n \frac{1}{\mu_x + \mu_n}} \tag{9.53}$$

* From Neff (12), p. 275.

Equation (9.53) is a numerical solution to equation (9.52). From Figure 9.16, A is the desired potential; J is the current density in the region. The distance between nodes is d meters. The permeability is indicated by μ_x and μ_n, where μ_x is at the point of inquiry and μ_n is at the surrounding nodes. Let's try an example in one dimension.

Figure 9.16. A magnetic region illustrating the numerical solution from equation (9.53).

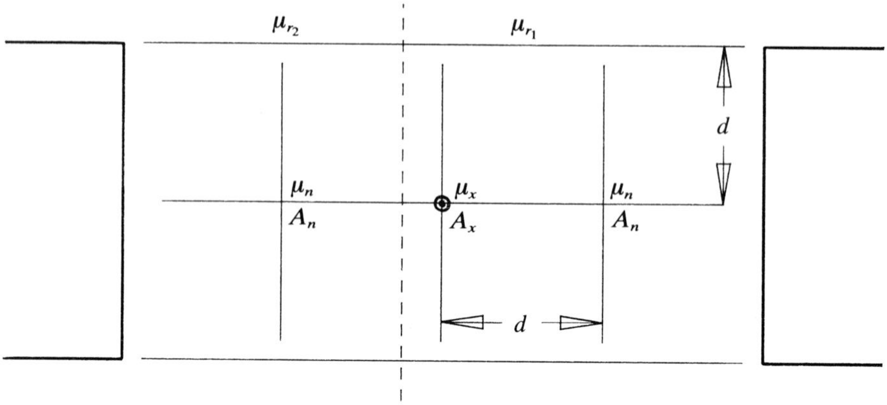

Figure 9.17. Magnetic region for example 9.7.

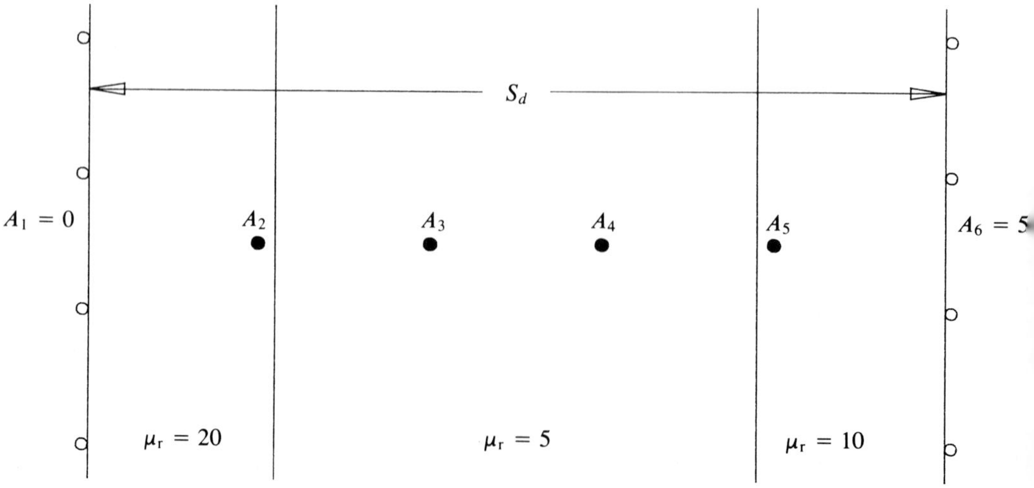

Example 9.7 Calculate the magnetic vector potential in the region shown in Figure 9.17. From equation (9.53) we write, at each node

$$A_2 = \frac{\dfrac{0}{\mu_1 + \mu_2} + \dfrac{A_3}{\mu_2 + \mu_3}}{\dfrac{1}{\mu_1 + \mu_2} + \dfrac{1}{\mu_2 + \mu_3}}$$

$$= \frac{\dfrac{A_3}{20 + 5}}{\dfrac{1}{20 + 20} + \dfrac{1}{20 + 5}}$$

$$A_3 = \frac{\dfrac{A_2}{20 + 5} + \dfrac{A_4}{5 + 5}}{\dfrac{1}{20 + 5} + \dfrac{1}{5 + 5}}$$

$$A_4 = \frac{\dfrac{A_3}{5 + 5} + \dfrac{A_5}{5 + 10}}{\dfrac{1}{5 + 5} + \dfrac{1}{5 + 10}}$$

$$A_5 = \frac{\dfrac{A_4}{5 + 10} + \dfrac{A_6}{10 + 10}}{\dfrac{1}{5 + 10} + \dfrac{1}{10 + 10}}$$

Thus we have calculated the magnetic vector potential in a region subjected to a magnetic field. We can extend this to two and three dimensions. In two dimensions the summations in equation (9.53) are from one to four; in three dimensions the summations are from one to six.

The magnetic vector potential is a convenience. With it we are able to calculate flux density **B** and field intensity **H**.

9.10 The torsion/elastic membrane/electric potential analogy

Prandtl in the early 1900s introduced an experimental method for solving two-dimensional torsion problems. The method was the membrane analogy. Torsion problems which are symmetric about the centroid of a shaft are solved in most mechanics texts. However, problems which are not symmetric are not solved so easily. The elastic membrane analogy is one way of solving the stress problem in a shaft where the membrane assumes the shape of the plane of shaft under torsion. The elastic membrane is secured around the edges and subjected to a uniform upward pressure; correspondingly there is no stress along the edges of the shaft.

Hoole (6) cites the analogy as well. Both torsion and membrane problems satisfy Poisson's equation. For the two-dimensional torsion problem we write

$$\nabla^2 V_\mathrm{T} = -2\theta G \tag{9.54}$$

where V_T is the stress function, θ is the angle of twist per length, and G is Young's modulus. For the elastic membrane problem we write

$$\nabla^2 V_\mathrm{E} = F/T \tag{9.55}$$

where V_E is the transverse deflection, F is the transverse load and T is the membrane tension.

We assume that the deflections in the membrane are small and that the slope of the membrane is proportional to the shearing stress and that twice the volume under the membrane equals the torque.

From Chapter 3 we have learned that electric potential satisfies Poisson's equation for a conservative field. In general we write

$$\nabla^2 V = -\frac{\rho}{\epsilon} \tag{9.56}$$

Since the vertical displacement of the elastic membrane satisfies in two dimensions

$$\frac{\partial^2 V_\mathrm{E}}{\partial x^2} + \frac{\partial^2 V_\mathrm{E}}{\partial y^2} = 0$$

and the potential satisfies

$$\frac{\partial^2 V}{\partial x^2} + \frac{\partial^2 V}{\partial y^2} = 0$$

we can say

$$V_\mathrm{E} \propto V$$

Let's try an example.

Example 9.8 Since this analogy is applicable to two-dimensional problems we further simplify by switching to cylindrical coordinates. Assume variation of potential radially only and write

$$\nabla^2 V = \frac{1}{r}\frac{\partial}{\partial r}\left(r\frac{\partial V}{\partial r}\right) = 0$$

Integrating twice we obtain $V = K \ln r + \text{constant}$. We choose $V = 0$ at $r = r_0$ and write

$$V = K \ln \frac{r}{r_0} \tag{9.57}$$

Recall from Chapter 3 that this is the potential for an infinite line charge. The potential is plotted in Figure 9.18. Notice that the potential is zero at $r = r_0$, and for $r < r_0$ the potential is negative. We now construct a physical model – an elastic membrane – as shown in Figure 9.19(a). The membrane is attached to the xy plane. Also note that the membrane is stretched up by a rod of radius a.

If the potential in equation (9.57) is sketched so that $V = K \ln r/r_0 > 0$

Figure 9.18. A plot of potential versus the natural log.

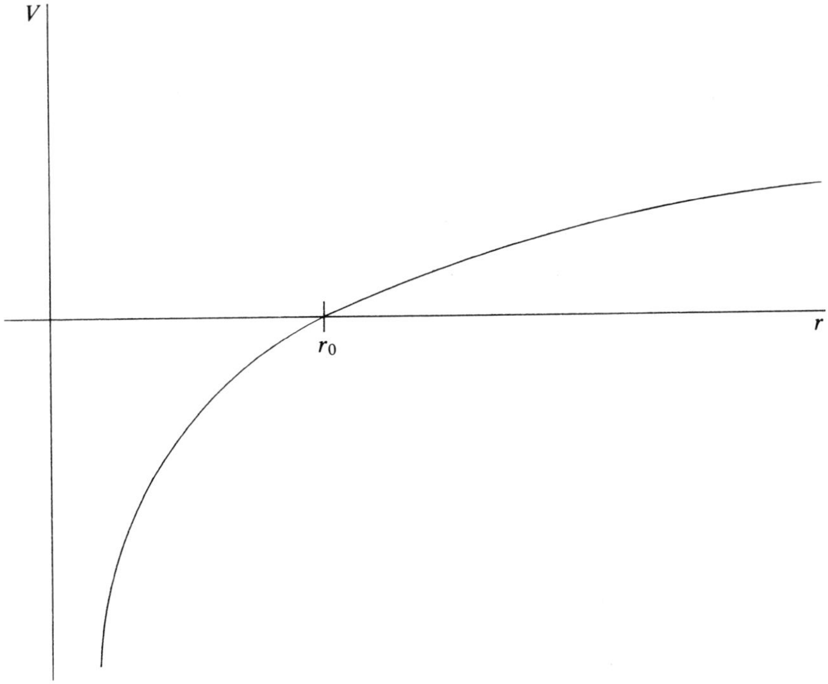

we model the potential, which is positive for $r > r_0$, with a membrane shaped like a soap bubble. This is shown in Figure 9.19(*b*).

9.11 Heat transfer

From Chapter 7 we learned that heat transfer occurs by conduction, convection and radiation. In this section we model heat transfer by convection. We are most familiar with convection as we experience heating and cooling of buildings and automobiles. In particular, a radiator is familiar. As the air (fluid) flows past the tubes, a heat transfer takes place. Either the heat within the tubes is transferred to the air flow – heating – or the heat in the air flow is transferred to the fluid within the tubes – cooling. Both are convective heat transfer.

In the case of an automobile radiator cooler air is forced through several rows of tubes. The tubes are usually made of thin copper and have fins mounted. Within the tubes is a mixture of an ethylene compound and water. This mixture carries heat from the engine block. The cooling happens when the fan or car movement forces air past the rows of tubes in the radiator. This is shown in Figure 9.20.

We can use potential measurements on Teledeltos paper to model fluid flow. By painting the correct boundaries with conductive paint and finding the equipotential lines we can model air flow past the radiator in Figure 9.20. From Figure 9.21 note that in Figure 9.21(*a*) the potential lines are the streamlines and in Figure 9.21(*b*) the potential lines are pressure lines and are orthogonal to the fluid flow. The Teledeltos measurements are relatively accurate in predicting the streamlines for

Figure 9.19. An elastic membrane model of $K \ln r$. (*a*) membrane, (*b*) soap bubble.

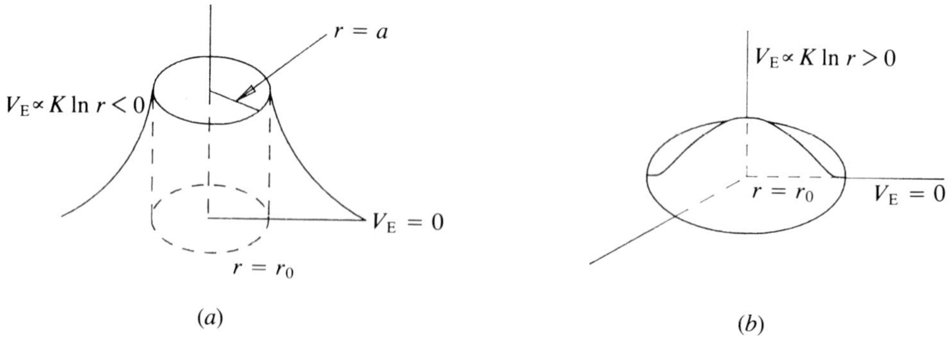

convective heat transfer. The heat loss by conduction in the tubes is neglected.

9.12 Permeability

Fluid flowing through a porous substance gives definition to an important engineering parameter – permeability. Oil or water moving through a permeable medium is of current interest. There is ongoing research in finding new underground oil sources. In addition, water seepage from wells and around dams needs to be understood better. Fluid flow through a permeable medium is governed by Darcy's law.

$$\mathbf{u} = -k\,\nabla p \tag{9.58}$$

Figure 9.20. Air flow past a typical car radiator.

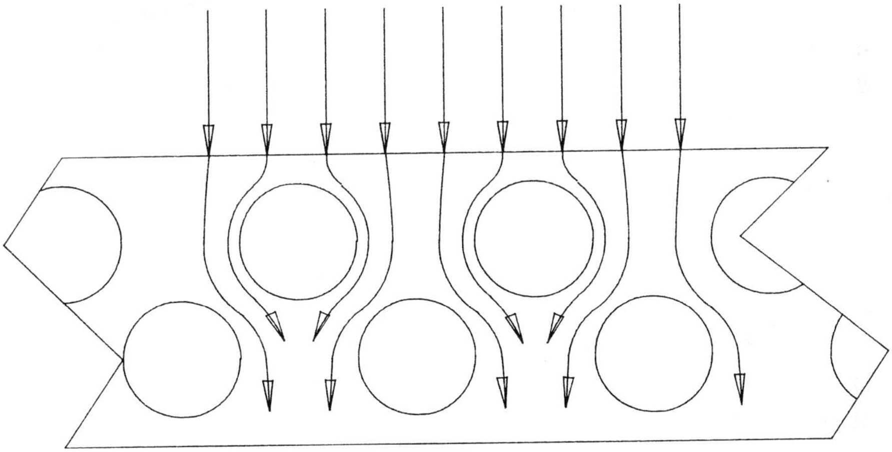

Figure 9.21. Field lines showing (*a*) streamlines, (*b*) pressure lines.

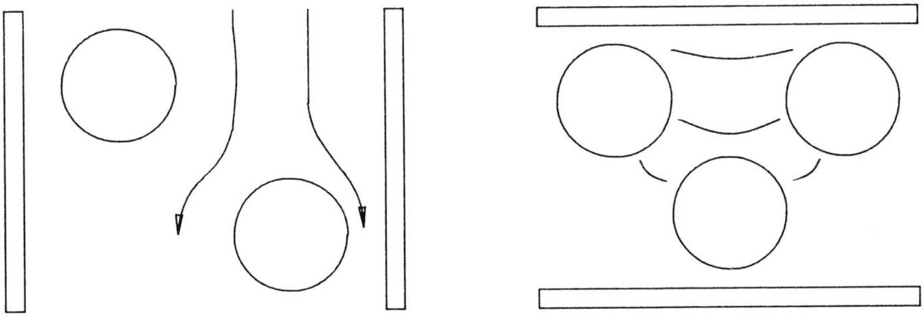

(*a*) (*b*)

where k is permeability of the porous medium in $m^3\,s/kg$, $\mathbf{u}$ is velocity, and p is pressure. If the fluid movement is vertical then gravity is a factor and equation (9.58) becomes

$$\mathbf{u} = -k(\nabla p + \rho g\,\nabla h) \tag{9.59}$$

where ρ is the volume density of the fluid and h is the height of displacement. The flow through a permeable medium is analogous to current flow through a resistance. That is, fluid flow is analogous to current flow; Darcy's law is analogous to Ohm's law.

Example 9.9 Consider an automobile engine which has a filter to remove foreign particles from the oil as shown in Figure 9.22. The oil pump applies a pressure (potential) which moves the oil (current) through the filter (resistance). The point at which the oil enters the filter can be modeled by a point source. The streamlines for a point source are well known. Refer to Chapter 3.

Example 9.10 Consider a deep water well in sandy soil as shown in Figure 9.23(a). The well can be modeled by a line source with radial flow as shown in Figure 9.23(b). Using cylindrical coordinates we see that the fluid flows in the radial direction. Constant pressure surfaces are concentric cylinders about the z axis.

Figure 9.22. An engine with oil filter.

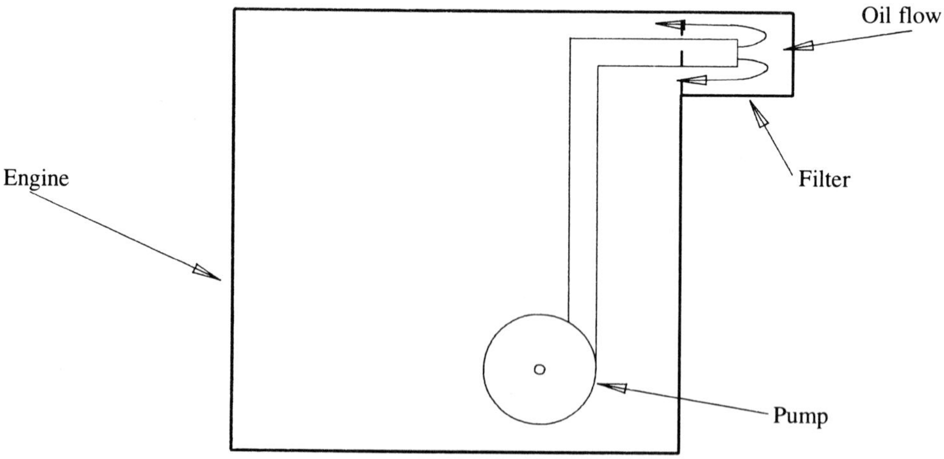

Example 9.11 A third example of fluid flow through a permeable medium is water seepage under a dam as shown in Figure 9.24(a). The constant pressure fronts are shown as broken lines; the streamlines are shown as solid lines. Figure 9.24(b) models water seepage under a dam with a line source/sink combination. However, the roles are reversed. The streamlines for the source/sink combination are the constant pressure lines in Figure 9.24(a), whereas the constant pressure lines in Figure 9.24(b) are the streamlines in Figure 9.24(a).

Figure 9.23. A deep water well (a) and a model (b).

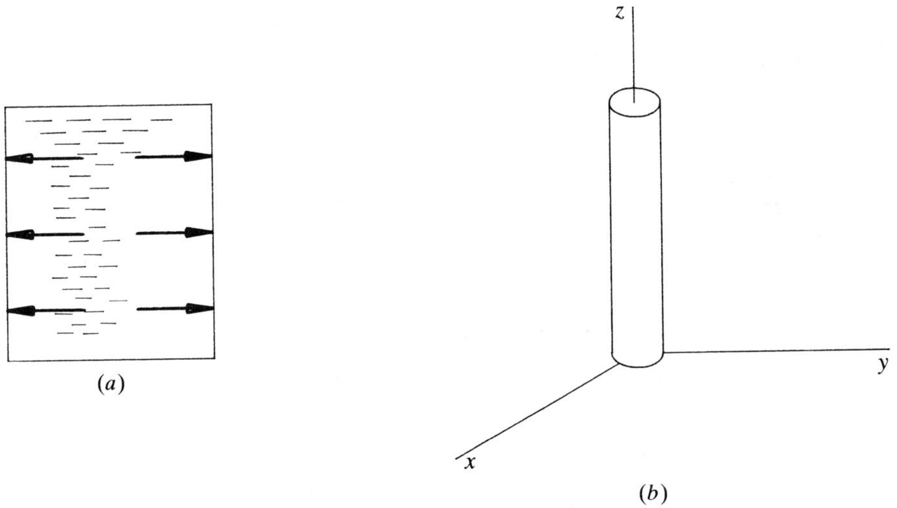

Figure 9.24. (a) Water seepage under a dam and (b) its model.

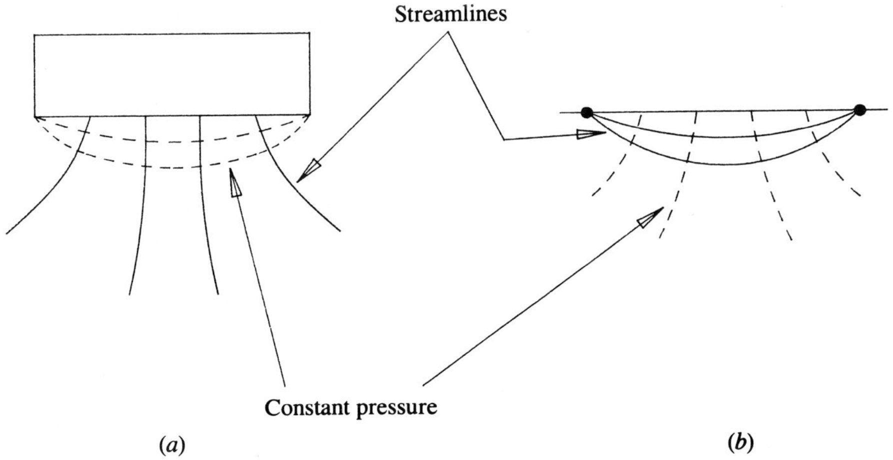

9.13 Fluid flow modeling

There is considerable interest in understanding fluid flow in the vicinity of stationary objects. For example, the study of air flow across a wing section of an aircraft is very important in designing airplanes. In underwater applications the shape of a submarine determines speed limitations and detectability. (Certain shapes produce more turbulence than others; turbulence can be detected.) Automobile designers attempt to reduce wind resistance for better fuel economy. Thus we see that modeling fluid flow is important in many engineering fields.

How can we model these and other engineering problems?

Some combinations of line sources and sinks and uniform flow have served as good mathematical models of engineering problems. Some of these are shown below. Figure 9.25 is the combination of a line source or sink and uniform flow. The only difference between a line source and uniform and a line sink and uniform is the direction of the stream lines. Figure 9.25(*a*) is the fluid flow around an object and Figure 9.25(*b*) is the math model. Point 'a' is the stagnation point where there is no fluid movement.

Figure 9.26 is the combination of a line source and a line sink and uniform flow. The shape of the oval is determined by the strength and separation of the source/sink combination and the strength of the uniform flow. Increasing the strength of the source/sink makes the oval more circular while increasing the uniform makes the oval more elliptical or flat. Figure 9.27 is the combination of a line source and more than one line sink and uniform flow. Notice that the model can represent the air flow around an aircraft wing. As the number of sinks is increased the wing becomes elongated. As the number of sinks is

Figure 9.25. The combination of a line source and uniform flow. (*a*) Flow around object, (*b*) model.

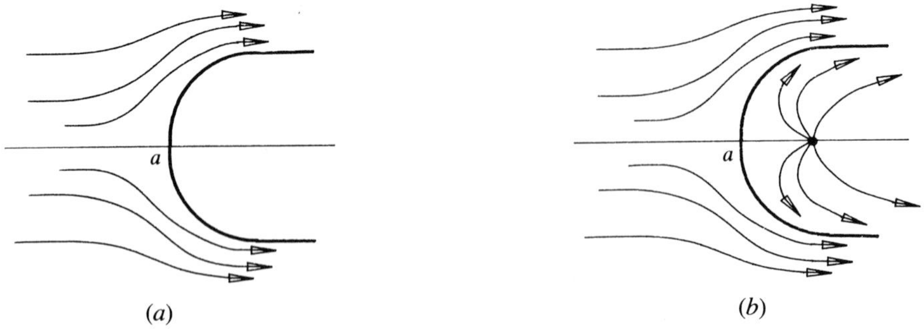

decreased the wing becomes more stubby. This is a convenient method for modeling engineering problems. The computation is simplified by using the computer.

9.14 The method of moments

The method of moments is a numerical method of solving integral equations. An integral equation is one in which the unknown appears inside the integral. Consider equations of the form,

$$V = \int \frac{\rho\, dv}{4\pi\epsilon_0 R}, \quad V = \int \frac{\rho_s\, dA}{4\pi\epsilon_0 R}, \quad V = \int \frac{\rho_l\, dl}{4\pi\epsilon_0 R}$$

Figure 9.26. The combination of a line source, line sink and uniform flow.

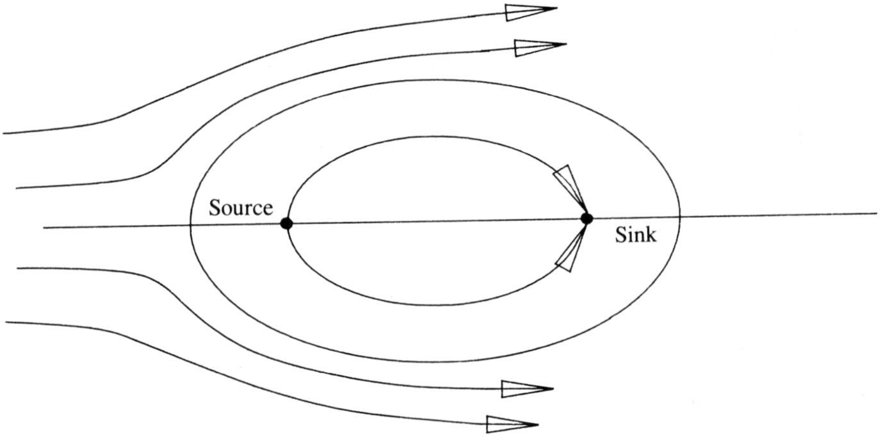

Figure 9.27. The combination of a line source, three line sinks and uniform flow.

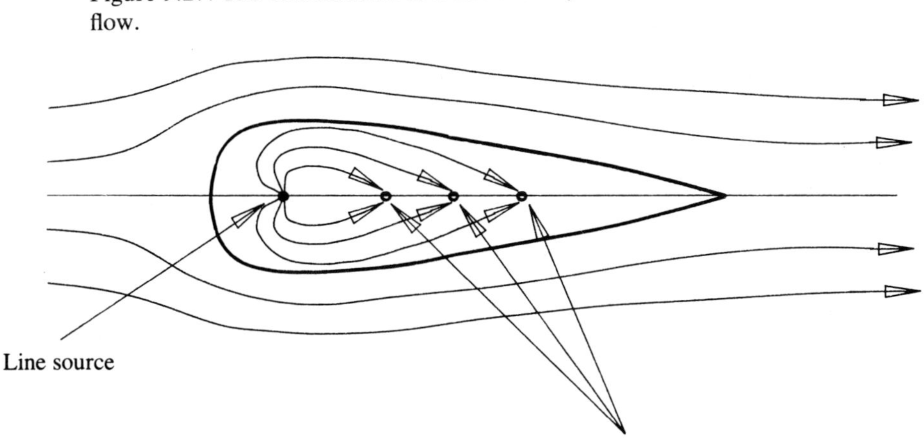

which we obtained from Coulomb's law in Chapter 3. In some cases the charge distribution, ρ, ρ_s or ρ_l, is not known. In fact the charge distribution is *assumed* in most cases in order to learn how to solve typical electromagnetic problems. In addition, it is true that most antenna problems are solved by first assuming a current (charge) distribution. Therefore we assume the charge is unknown and apply the method of moments.

In this section we apply the method of moments technique to a finite wire to find the charge distribution on the wire. Consider a wire of length l and radius a along the z-axis as shown in Figure 9.28. For $l \gg a$ consider the voltage constant and we write

$$V = 1 = \int_0^l \frac{\rho_l \, dl}{4\pi\epsilon_0 R} \tag{9.60}$$

$$4\pi\epsilon_0 = \int \frac{\rho_l \, dz}{R} \tag{9.61}$$

where

$$R = \sqrt{(z_n - z)^2} = |z_n - z|$$

Figure 9.28. A thin wire at constant potential.

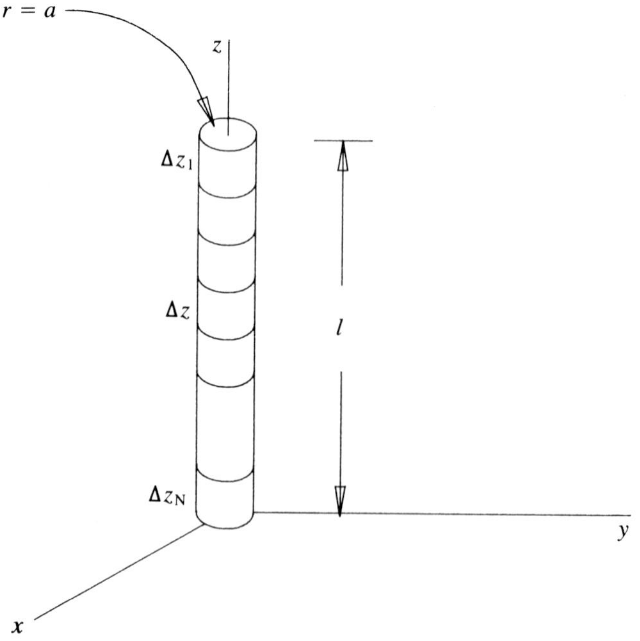

Recall that in integration we are actually summing areas, that is,

$$\int f(z)\,\mathrm{d}z \approx f(z_1)\,\Delta z + f(z_2)\,\Delta z + \cdots + f(z_N)\,\Delta z$$

Therefore we can write equation (9.61) as

$$4\pi\epsilon_0 = \sum_{n=1}^{N} \frac{\rho(z_n)\,\Delta z}{|z_n - z|} \tag{9.62}$$

where $\Delta z = l/N$ and N is the number of segments along the wire. Assume the charge on each segment is constant (it is not) and we write

$$\begin{aligned} 4\pi\epsilon_0 &= \frac{\rho_1\,\Delta z}{|z_1 - z_1|} + \frac{\rho_2\,\Delta z}{|z_1 - z_2|} + \cdots + \frac{\rho_N\,\Delta z}{|z_1 - z_N|} \\ &\cdots \\ 4\pi\epsilon_0 &= \frac{\rho_1\,\Delta z}{|z_N - z_1|} + \frac{\rho_2\,\Delta z}{|z_N - z_2|} + \cdots + \frac{\rho_N\,\Delta z}{|z_N - z_N|} \end{aligned} \tag{9.63}$$

Equation (9.63) can be solved using matrix techniques which are available on computer. However, the denominators of the diagonal terms go to zero. These must be dealt with more carefully. Over a small length Δz, the charge density may be assumed constant. Since the wire has a radius of a the charge distributes itself over the surface of the wire so at the center of each segment the singular terms may be expressed by

$$\begin{aligned} 4\pi\epsilon_0 &= \int_0^{2\pi}\int_{-\Delta z/2}^{\Delta z/2} \frac{\rho_s a\,\mathrm{d}\phi\,\mathrm{d}z}{[a^2 + z^2]^{1/2}} \\ &= 2\pi a\rho_s \ln\left\{\frac{\dfrac{\Delta z}{2} + \left[\left(\dfrac{\Delta z}{2}\right)^2 + a^2\right]^{1/2}}{-\dfrac{\Delta z}{2} + \left[\left(\dfrac{\Delta z}{2}\right)^2 + a^2\right]^{1/2}}\right\} \end{aligned}$$

which for $a \ll \Delta z$ reduces to

$$\approx 2\rho_l \ln\frac{\Delta z}{a}$$

where $\rho_l = 2\pi a\rho_s$.

We now have the values of the diagonal terms. In matrix form we can write

$$\begin{bmatrix} 4\pi\epsilon_0 \\ \vdots \\ 4\pi\epsilon_0 \end{bmatrix} = \begin{bmatrix} 2\ln\dfrac{\Delta z}{a} & \dfrac{\Delta z}{z_1 - z_2} & \cdots & \dfrac{\Delta z}{z_1 - z_N} \\ \dfrac{\Delta z}{z_2 - z_1} & 2\ln\dfrac{\Delta z}{a} & & \\ \vdots & & & \\ \dfrac{\Delta z}{z_N - z_1} & & \cdots & 2\ln\dfrac{\Delta z}{a} \end{bmatrix} \begin{bmatrix} \rho_1 \\ \vdots \\ \rho_N \end{bmatrix} \tag{9.64}$$

Equation (9.64) has the form

$$[C] = [L][\rho] \tag{9.65}$$

Multiplying equation (9.65) by $[L]^{-1}$ we obtain the charge distribution

$$[\rho] = [L]^{-1}[C]$$

where

$$[C] = \begin{bmatrix} 4\pi\epsilon_0 \\ \vdots \\ 4\pi\epsilon_0 \end{bmatrix}$$

The method of moments technique has been used to find the charge distribution on a thin wire at constant potential.

9.15 The finite element method (13)

Finite element analysis (FEM) is a numerical method for solving differential equations. It is similar to the finite difference method but is more versatile. That is, the finite element method is better able to handle irregularly shaped boundaries. There are other differences. The finite difference uses an array of grid points while the finite element uses (fewer) elements. Also the finite difference method is effective for two-dimensional problems; the FEM can solve one-, two-, and three-dimensional problems. Finite elements was used initially in structural problems. Today it is used in electromagnetics as well as other fields. While the FEM is relatively complex, there are a number of computer programs such as FINEL, ANSYS, NASTRAN, MacPoisson and MacElastic available for the engineer.

There are three steps in the FEM: (1) dividing the region, (2) writing equations for each region, and (3) solving the system of equations. This will be illustrated using a one-dimensional problem – the potential distribution in a Poissonian region between two plates.

Consider, for example, two infinite parallel conducting plates as shown in Figure 9.29. There exists a volume charge density, ρ, of uniformly distributed particles between the plates. Let the two plates be at a potential of 0 and 1.0 V and let the plates be separated by 1.0 m. Since the two plates are long, the problem is a one-dimensional problem. Assume the potential obeys Poisson's equation; in one dimension this becomes

$$\frac{d^2V}{dy^2} = -\frac{\rho}{\epsilon}$$

with the boundary conditions

$$y = 0 \qquad V = 0$$

and

$$y = 1.0\,\text{m} \qquad V = 1.0\,\text{V}$$

We can normalize Poisson's equation and write

$$\frac{d^2V}{dy^2} = -1$$

The basis for the finite element method is variational calculus. In this method one defines a *functional*

$$I \triangleq \frac{1}{2} \iiint [\epsilon\, |\nabla V|^2 - 2\rho V]\, dv \tag{9.66}$$

Figure 9.29. Two infinite parallel conducting plates.

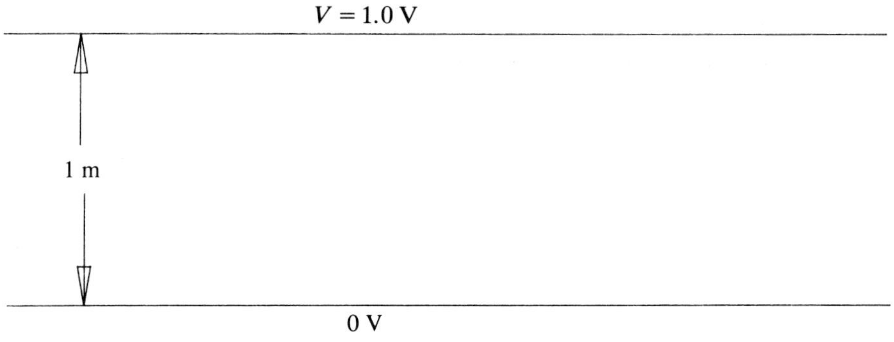

where the value of I depends on the *function* V. In our example equation (9.66) becomes

$$I = \frac{1}{2}\int \left[\frac{dV}{dy}\right]^2 dy - \int V\, dy \tag{9.67}$$

The problem of minimizing I is the subject of the *calculus of variations*.

At this point we choose trial functions $V_1(y)$, $V_2(y) \cdots V_n(y)$ such that the boundary conditions are satisfied, that is, $V = 0$ V at $y = 0$ and $V = 1.0$ V at $y = 1.0$ m. We now substitute the trial function into equation (9.67) and minimize I, that is, set $\partial I/\partial V_i = 0$. The result is a system of simultaneous linear equations.

For our example our trial function is

$$V = V_i + \frac{V_j - V_i}{y_j - y_i}(y - y_i)$$

and

$$\frac{dV}{dy} = \frac{V_j - V_i}{y_j - y_i}$$

Now substitute these into equation (9.67) and obtain

$$\begin{aligned} I &= \frac{1}{2}\int_{y_i}^{y_j} \left[\frac{V_j - V_i}{y_j - y_i}\right]^2 dy - \int_{y_i}^{y_j} \left[V_i + \frac{V_j - V_i}{y_j - y_i}(y - y_i)\right] dy \\ &= \frac{1}{2}\frac{(V_j - V_i)^2}{y_j - y_i} - \frac{1}{2}(V_j + V_i)(y_j - y_i) \end{aligned}$$

We choose, say, four elements (implying five nodes) and write

$$\begin{aligned} I &= I_1 + I_2 + I_3 + I_4 \\ &= \frac{1}{2}\frac{(V_2 - V_1)^2}{y_2 - y_1} - \tfrac{1}{2}(V_2 + V_1)(y_2 - y_1) + \frac{1}{2}\frac{(V_3 - V_2)^2}{y_3 - y_2} \\ &\quad - \tfrac{1}{2}(V_3 + V_2)(y_3 - y_2) + \frac{1}{2}\frac{(V_4 - V_3)^2}{y_4 - y_3} - \tfrac{1}{2}(V_4 + V_3)(y_4 - y_3) \\ &\quad + \frac{1}{2}\frac{(V_5 - V_4)^2}{y_5 - y_4} - \tfrac{1}{2}(V_5 + V_4)(y_5 - y_4) \end{aligned}$$

The minimum of I is the unique solution to

$$\frac{\partial I}{\partial V_i} = 0$$

so we write

$$\frac{\partial I}{\partial V_2} = 0, \qquad \frac{\partial I}{\partial V_3} = 0 \quad \text{and} \quad \frac{\partial I}{\partial V_4} = 0$$

These become

$$\begin{aligned}
\frac{\partial I}{\partial V_2} &= \frac{V_2 - V_1}{y_2 - y_1} - \tfrac{1}{2}(y_2 - y_1) - \frac{V_3 - V_2}{y_3 - y_2} - \tfrac{1}{2}(y_3 - y_2) = 0 \\
\frac{\partial I}{\partial V_3} &= \frac{V_3 - V_2}{y_3 - y_2} - \tfrac{1}{2}(y_3 - y_2) - \frac{V_4 - V_3}{y_4 - y_3} - \tfrac{1}{2}(y_4 - y_3) = 0 \\
\frac{\partial I}{\partial V_4} &= \frac{V_4 - V_3}{y_4 - y_3} - \tfrac{1}{2}(y_4 - y_3) - \frac{V_5 - V_4}{y_5 - y_4} - \tfrac{1}{2}(y_5 - y_4) = 0
\end{aligned} \tag{9.68}$$

If we choose equal length elements (not a requirement of FEM) we obtain $y_1 = 0$, $y_2 = 0.25$, $y_3 = 0.5$, $y_4 = 0.75$, and $y_5 = 1.0$. At these nodes the variational parameters are $V = V_2$, V_3 and V_4 at $y = y_2$, y_3, and y_4 respectively. We have set $V_1 = 0$ and $V_5 = 1.0$ V.

Solving equation (9.68) we obtain

$$\frac{V_2 - 0}{0.25} - \tfrac{1}{2}(0.25) - \frac{V_3 - V_2}{0.25} - \tfrac{1}{2}(0.25) = 0$$

$$\frac{V_3 - V_2}{0.25} - \tfrac{1}{2}(0.25) - \frac{V_4 - V_3}{0.25} - \tfrac{1}{2}(0.25) = 0$$

$$\frac{V_4 - V_3}{0.25} - \tfrac{1}{2}(0.25) - \frac{V_5 - V_4}{0.25} - \tfrac{1}{2}(0.25) = 0$$

The solutions are

$$V_2 = 0.343\,75 \text{ V}, \qquad V_3 = 0.625 \text{ V} \quad \text{and} \quad V_4 = 0.843\,75 \text{ V}$$

Figure 9.30 shows a plot of potential versus separation of the plates.

Now return to Poisson's normalized equation

$$\frac{d^2V}{dy^2} = -1$$

Integrating twice we obtain

$$V = -\frac{y^2}{2} + Ay + B$$

From the boundary conditions

$$V = -\frac{y^2}{2} + \frac{3y}{2}$$

This is the exact solution to Poisson's equation. Using this we obtain at $y = 0.25$, $V_2 = 0.343\,75$ V; at $y = 0.5$, $V_3 = 0.625$ V; and at $y = 0.75$ m, $V_4 = 0.843\,75$ V. Both results are shown in Figure 9.30. Note that the FEM and exact method solutions closely agree, but this is not true, in general.

Finite element analysis was first used in structure analysis. It is very useful for complex shapes. Figure 9.31 shows a simple structural problem. It also shows the structure divided into three elements A, B, and C, with five nodes chosen. This two-dimensional problem would be solved in the same manner as the one-dimensional. That is, after defining the elements, we would write equations for each element then assemble the coefficients in a *global* coefficient matrix. It should be clear that the complexity increases quickly as we go from one to two to three dimensions.

This exposure to the FEM should whet the student's appetite. The finite element method is powerful and is getting widespread usage. It is hoped that this brief discussion has given the student confidence in utilizing the FEM computer programs available.

Figure 9.30. A plot of potential versus separation.

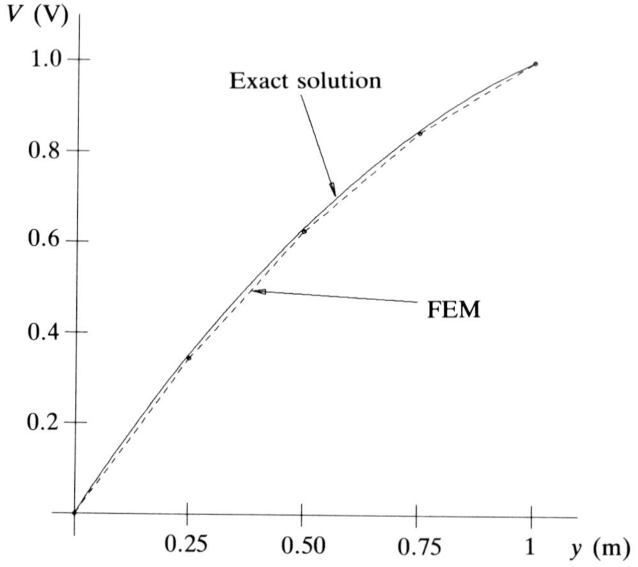

ASSIGNMENTS

Assignment 1

Design a rectangular Laplacian region using at least three boundary potentials or temperatures. Analyse the region using two methods – iteration and the matrix method. Divide the region into a minimum of a 4×4 grid. Iterate to 0.1% accuracy. Plot several equipotentials or isotherms and compare both methods. Label boundaries and equipotentials or isotherms.

Assignment 2

Design a four boundary, three potential or temperature Laplacian region. Analyse the region using the relaxation method at a minimum of 20 interior points. Calculate to 0.1% accuracy. Sketch several isotherms. Label boundaries and equipotentials or isotherms.

Assignment 3

Design a rectangular, three potential or temperature Laplacian region. Analyse using separation of variables and superposition. Sketch several

Figure 9.31. A vertically loaded structure and its finite elements.

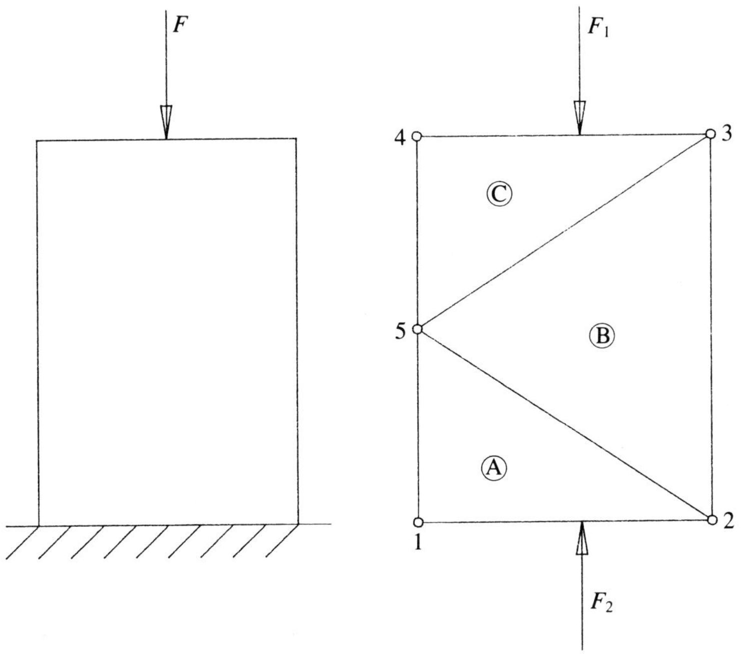

equipotentials. Label boundaries and equipotentials or isotherms. Clearly show equations.

Assignment 4

Design a two sided, two potential Laplacian region. Divide the region into a minimum of 20 sections. Calculate the DC resistance of each section in both directions. Let $\sigma = 1$. Assume a depth of one meter. Model the same Laplacian region in the laboratory with resistances, a circuit board and a DC power supply. Measure the potentials with a digital multimeter (DMM) at the intersections. Model the Laplacian circuit using software such as PSPICE, MICROCAP or similar programs. Sketch the equipotentials and label.

Assignment 5

Design at least a three boundary, three potential or temperature Laplacian region. Analyse the region using a conductive salt solution. Use number 16 or larger uncoated copper wire or other metals on the boundaries. Bend the conductors to conform to your boundaries. Use a 1 kHz signal source. Design a resistive voltage divider to provide the boundary potentials. Provide a centimeter grid for the bottom of the pan holding the solution. Measure the potentials at the grid points with a DMM or an oscilloscope. The probes can provide the electric field lines as well as the potentials. The **E** fields are normal to the equipotential lines. By rotating the probes to a maximum voltage we can define an **E** field line (or a streamline). Also from $\mathbf{E} = \nabla V$ we can calculate the electric field by dividing the voltage measured by the probe spacing. In addition when the probes are rotated to a minimum (zero) we define an equipotential line. The depth of the salt solution should be constant over the bottom of the pan. Care should be taken at the gaps. If the gaps are very large there will be much fringing. If the gaps are very small there will be high currents. Make three large sketches consisting of 1. the flux or streamlines, 2. the equipotentials and 3. a combined plot.

Assignment 6

Design a two sided, two potential non-symmetric Laplacian region. Analyse the region by painting the boundaries on Teledeltos paper with conductive paint. Apply a low DC voltage to the boundaries and measure the potentials with a DMM. Use a heavy pad under the

Teledeltos paper to protect it from tearing or puncture. Measure the DC resistance across the boundaries. Sketch and label the equipotentials on a large piece of paper.

Assignment 7

Design a two sided, two potential Laplacian region to model orthogonality of streamlines and isotherms. Paint the boundaries on Teledeltos paper and apply a low DC voltage. Measure the equipotentials. On another piece of Teledeltos paper the same size as the first paint boundaries which are orthogonal to the first set. Measure as before. Sketch the equipotentials from both on a separate piece of paper. You will note the lines from both parts tend to be orthogonal. There should be a duality between the two sketches. Measure and compare the resistance from both Teledeltos papers.

Assignment 8

Design a two boundary, two potential Laplacian region with an obstruction along the boundary. Paint the boundaries on Teledeltos paper. Apply DC voltage and measure potentials. Sketch the equipotentials. Using the method of sketching field maps above, sketch the potentials and streamlines. Compare with the equipotentials from the Teledeltos measurements. Analyse the region with available software. Estimate the capacitance and resistance of the region using equations (9.48) and (9.50) respectively.

Assignment 9

The paths of magnetic flux lines are important to the designers of audio and video record/playback devices. Figure 9.32 shows a typical record/playback. Determine and sketch the magnetic vector potential lines in the vicinity of the tape using equation (9.53). Neglect the gap from the magnetic vector potential lines and sketch the flux or streamlines. The high reluctance underlay material 'keeps' flux in the magnetic tape.

Assignment 10

Design an elastic membrane arrangement to model the problem of two long parallel cylindrical conductors with potentials $+V$ and $-V$. The conductors are surrounded by a conducting shield which is grounded. The height distribution of the membrane should correspond to the potential distribution of the conductors/shield combination. Figure 9.33

Figure 9.32. A typical record/playback head.

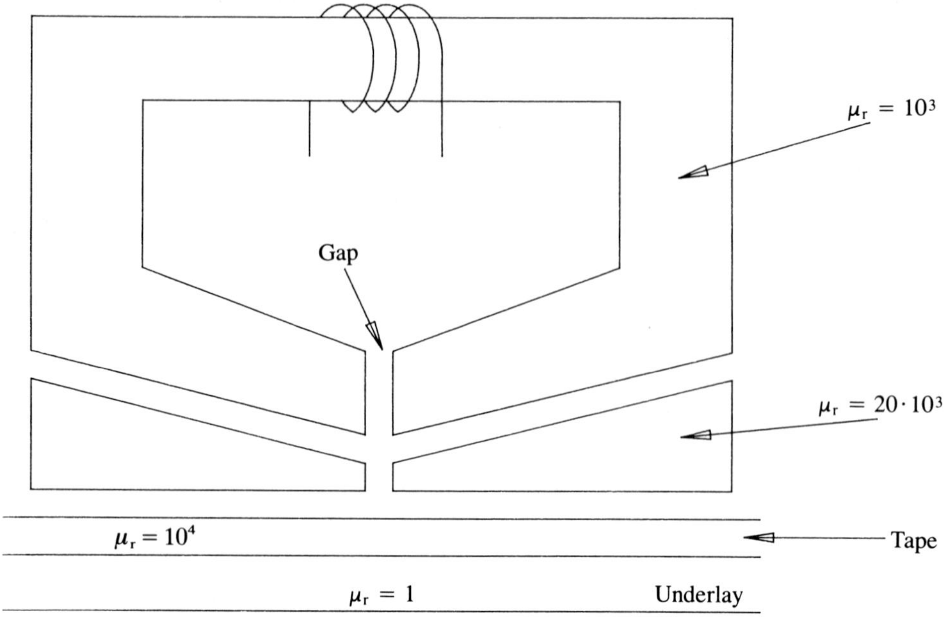

Figure 9.33. Two parallel conductors inside a shielded enclosure.

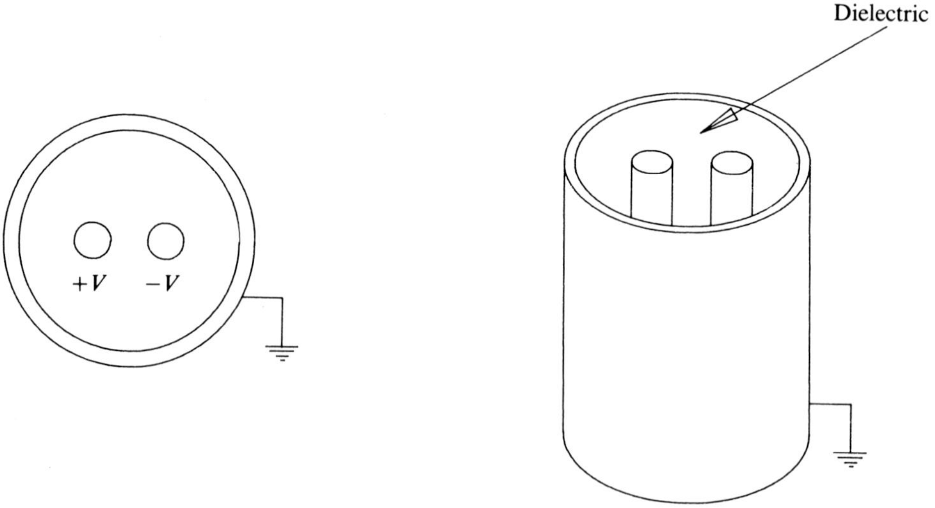

shows the shielded pair of conductors. Determine the potential distribution and plot it by comparing with the elevations at the corresponding locations for the rubber membrane.

Assignment 11

Design and analyse a heat exchanger for an automobile's cooling system. The size of the radiator is 0.05 × 0.5 m. Assume a depth of one meter. See Figure 9.34. Assume copper tubing. Assume the cooling fluid is air at 15 °C. The tubes contain water at 90 °C. From Chapter 7 the heat conduction equation is

$$Q_h = -kA\frac{\Delta T}{\Delta l}$$

and the heat convection equation is

$$Q_h = hA(T_2 - T_1)$$

The thermal conduction coefficient for copper is $k = 386$ W/m K and the surface convection coefficients are $h_{water} = 400$ W/m^2 K, $h_{forced\ air} =$

Figure 9.34. The heat exchanger of assignment 11.

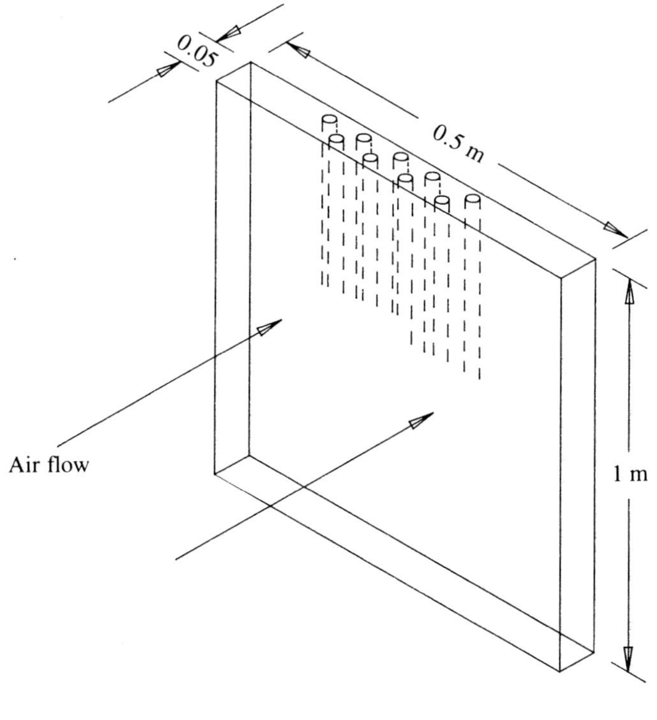

50 W/m^{2}K. Model your design in two dimensions using Teledeltos paper and conductive paint. Measure the potentials. On a separate piece of paper sketch the equipotentials and the streamlines. The streamlines will be orthogonal to the equipotentials so sketch the equipotentials first.

In addition evaluate your design on the basis of economics and safety. That is consider the trade off between thin tubes and safety and thick tubes and cost.

Assignment 12

Design a test to determine the permeability of a soil sample. Recall from Chapter 8 the fluid flow velocity in a soil is dependent on the pressure and soil permeability

$$\mathbf{u} = -k\,\nabla p$$

We obtain a volume rate by multiplying both sides by the cross-sectional area

$$\mathbf{u}(\text{Area}) \equiv \Phi = k\,|\nabla p|\,(\text{Area})$$

and we obtain a (fluid) volume by multiplying both sides by time

$$\Phi t \equiv \text{Vol} = k\,|\nabla p|\,(\text{Area})t$$

Finally we solve for the permeability; that is,

$$k = \frac{\text{Vol}}{(\text{Area})t\rho g\dfrac{\Delta h}{\Delta L}}$$

This is illustrated in Figure 9.35. Include a test procedure and apparatus setup in your design. Test two samples and compare with typical values given in the table. Show data.

soil	k m^{3}s/kg
gravel	10^{-1}–10^{-2}
sand/gravel	10^{-2}–10^{-5}
sand/silt/clay	10^{-6}–10^{-9}
clay	$<10^{-9}$

Assignment 13

With the combination of uniform and line sources and sinks we can model the streamlines of flow around objects in fluids. Sketch the streamlines, both analytically and graphically, for the following combinations:

(a) uniform and line source,
(b) uniform and one line source and one line sink,
(c) uniform and one line source and two line sinks.

Sketch the streamlines of the three combinations above using Teledeltos paper with painted boundaries. Do *not* paint a boundary around the 'object' in the fluid but cut away the Teledeltos paper to simulate the object. Compare the results.

Assignment 14

Calculate and plot using the method of moments the charge distribution for a long, thin wire at a constant potential $V = 1\ V$. Assume

(a) $l = 1.5$ m, $a = 2 \cdot 10^{-3}$ m and $N = 10$,
(b) $l = 1.5$ m, $a = 2 \cdot 10^{-2}$ m and $N = 10$,
(c) $l = 1.5$ m, $a = 2 \cdot 10^{-2}$ m and $N = 6$. See Figure 9.28.

Figure 9.35. A test setup for determining soil permeability.

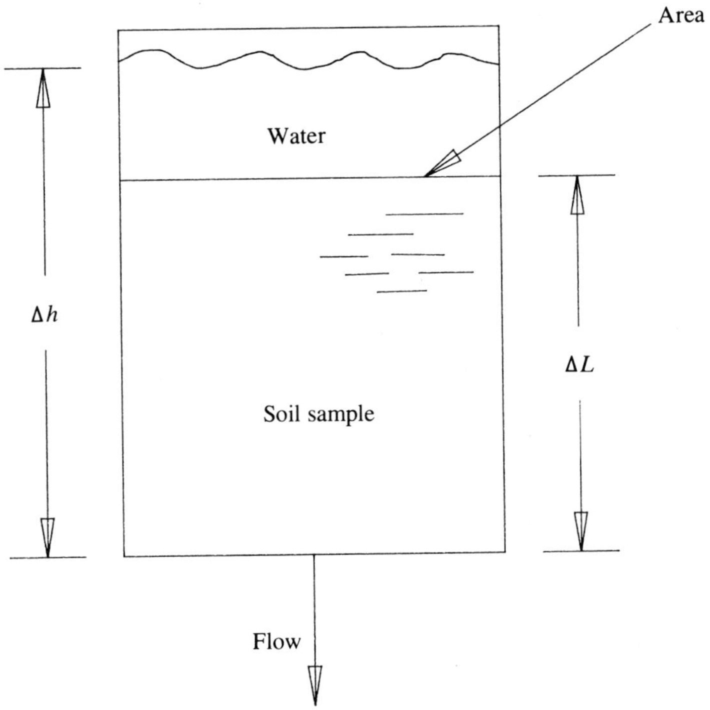

Assignment 15

Complete the following: Repeat the FEM example in section 9.15 but use unequal length elements. Plot your results and compare with the exact results.

APPENDIX A

Vector operations

Divergence $\nabla \cdot \mathbf{F}$

Cartesian
$$\frac{\partial F_x}{\partial x} + \frac{\partial F_y}{\partial y} + \frac{\partial F_z}{\partial z}$$

Cylindrical
$$\frac{1}{r}\frac{\partial}{\partial r}(rF_r) + \frac{1}{r}\frac{\partial F_\phi}{\partial \phi} + \frac{\partial F_z}{\partial z}$$

Spherical
$$\frac{1}{r^2}\frac{\partial}{\partial r}(r^2 F_r) + \frac{1}{r \sin\theta}\frac{\partial}{\partial \theta}(F_\theta \sin\theta) + \frac{1}{r \sin\theta}\frac{\partial F_\phi}{\partial \phi}$$

Laplacian $\nabla^2 f$

Cartesian
$$\frac{\partial^2 f}{\partial x^2} + \frac{\partial^2 f}{\partial y^2} + \frac{\partial^2 f}{\partial z^2}$$

Cylindrical
$$\frac{1}{r}\frac{\partial}{\partial r}\left(r\frac{\partial f}{\partial r}\right) + \frac{1}{r^2}\frac{\partial^2 f}{\partial \phi^2} + \frac{\partial^2 f}{\partial z^2}$$

Spherical
$$\frac{1}{r^2}\frac{\partial}{\partial r}\left(r^2\frac{\partial f}{\partial r}\right) + \frac{1}{r^2 \sin\theta}\frac{\partial}{\partial \theta}\left(\sin\theta\frac{\partial f}{\partial \theta}\right) + \frac{1}{r^2 \sin^2\theta}\frac{\partial^2 f}{\partial \phi^2}$$

Gradient ∇f

Cartesian
$$\frac{\partial f}{\partial x}\hat{\mathbf{a}}_x + \frac{\partial f}{\partial y}\hat{\mathbf{a}}_y + \frac{\partial f}{\partial z}\hat{\mathbf{a}}_z$$

Cylindrical
$$\frac{\partial f}{\partial r}\hat{\mathbf{a}}_r + \frac{1}{r}\frac{\partial f}{\partial \phi}\hat{\mathbf{a}}_\phi + \frac{\partial f}{\partial z}\hat{\mathbf{a}}_z$$

Spherical
$$\frac{\partial f}{\partial r}\hat{\mathbf{a}}_r + \frac{1}{r}\frac{\partial f}{\partial \theta}\hat{\mathbf{a}}_\theta + \frac{1}{r \sin\theta}\frac{\partial f}{\partial \phi}\hat{\mathbf{a}}_\phi$$

Curl $\nabla \times \mathbf{F}$

Cartesian $$\left(\frac{\partial F_z}{\partial y} - \frac{\partial F_y}{\partial z}\right)\hat{\mathbf{a}}_x + \left(\frac{\partial F_x}{\partial z} - \frac{\partial F_z}{\partial x}\right)\hat{\mathbf{a}}_y + \left(\frac{\partial F_y}{\partial x} - \frac{\partial F_x}{\partial y}\right)\hat{\mathbf{a}}_z$$

Cylindrical $$\left(\frac{1}{r}\frac{\partial F_z}{\partial \phi} - \frac{\partial F_\phi}{\partial z}\right)\hat{\mathbf{a}}_r + \left(\frac{\partial F_r}{\partial z} - \frac{\partial F_z}{\partial r}\right)\hat{\mathbf{a}}_\phi + \frac{1}{r}\left(\frac{\partial (rF_\phi)}{\partial r} - \frac{\partial F_r}{\partial \phi}\right)\hat{\mathbf{a}}_z$$

Spherical $$\frac{1}{r \sin\theta}\left[\frac{\partial (F_\phi \sin\theta)}{\partial \theta} - \frac{\partial F_\theta}{\partial \phi}\right]\hat{\mathbf{a}}_r + \frac{1}{r}\left[\frac{1}{r\sin\theta}\frac{\partial F_r}{\partial \phi} - \frac{\partial (rF_\phi)}{\partial r}\right]\hat{\mathbf{a}}_\theta + \frac{1}{r}\left[\frac{\partial (rF_\theta)}{\partial r} - \frac{\partial F_r}{\partial \theta}\right]\hat{\mathbf{a}}_\phi$$

Dot product $\mathbf{A} \cdot \mathbf{B} = |\mathbf{A}|\,|\mathbf{B}| \cos\alpha$

Cross product $\mathbf{A} \times \mathbf{B} = |\mathbf{A}|\,|\mathbf{B}| \sin\alpha\, \hat{\mathbf{a}}_N$

APPENDIX B

Analogous quantities

Engineering field	Gradient	Gauss' law
Electrostatic	$\mathbf{E} = -\nabla V$	$Q = \oint \mathbf{D} \cdot \mathrm{d}\mathbf{A}$
Magnetostatic	$\mathbf{H} = -\nabla V_{\mathrm{m}}$	$0 = \oint \mathbf{B} \cdot \mathrm{d}\mathbf{A}$
Gravitational	$\mathbf{g} = -\nabla V_{\mathrm{g}}$	$m = \oint \mathbf{D}_{\mathrm{g}} \cdot \mathrm{d}\mathbf{A}$
Fluid	$\mathbf{U} = -\nabla V_{\mathrm{f}}$	$\Phi = \oint \mathbf{U} \cdot \mathrm{d}\mathbf{A}$
Heat transfer	$\mathbf{q} = -k\, \nabla T$	$\Phi = \oint \mathbf{D}_{\mathrm{h}} \cdot \mathrm{d}\mathbf{A}$
Conductive	$\mathbf{J} = -\sigma\, \nabla V$	$I = \oint \mathbf{J} \cdot \mathrm{d}\mathbf{A}$
Permeable media	$\mathbf{U} = -k\, \nabla p$	$Q_{\mathrm{f}} = \oint \mathbf{U} \cdot \mathrm{d}\mathbf{A}$
Diffusive	$\mathbf{D}_d = -k_{\mathrm{d}}\, \nabla N$	$Q_{\mathrm{d}} = \oint \mathbf{D}_{\mathrm{d}} \cdot \mathrm{d}\mathbf{A}$
Acoustic	$\mathbf{U} = -\dfrac{1}{\mathrm{j}\omega\rho_{\mathrm{c}}} \nabla p$	—

APPENDIX C

Dielectric constants (approximate)

Material	Dielectric constant (ϵ_r)
Air	1
Glass	5
Mica	5.5
Nylon	8
Polypropylene	2.25
Polystyrene	2.55
Sodium chloride	6
Styrofoam	1.03
Teflon	2.1
Water	80

APPENDIX D

Relative permeability (approximate)

Material	Permeability (μ_r)
Aluminum	1
Cast iron	60
Ferrite	1000
Nickel	50
Pure iron	4000
Steel	300
Supermalloy	100 000
Transformer iron	3000

APPENDIX E

Physical constants

Density of water (20 °C)	10^3 kg/m^3
Acceleration of gravity	9.81 m/s^2
Mean radius of earth	6.37×10^6 m
Mean earth–sun distance	1.49×10^{11} m
Mean earth–moon distance	3.8×10^8 m
Gravitational constant	6.67×10^{-11} N m^2/kg^2
Mean earth density	5522 kg/m^3
Mass of earth	5.98×10^{24} kg
Mean orbital velocity of earth	29 770 m/s
Mean angular velocity of earth	7.29×10^{-5} rad/s
Earth's magnetic flux density	5.7×10^{-5} T
Earth's magnetic dipole moment	6.4×10^{21} A-m^2

APPENDIX F

Thermal properties

Thermal conductivities at 30 °C

Material	Thermal conductivity
Air	0.026
Aluminum	175–200
Asphalt	0.75
Brick	0.69
Cardboard	0.064
Copper	20–400
Concrete (1–2–4 mix)	1.37
Glass	0.78
Iron	30–70
Lead	35
Nickel	10–90
Oil	0.15
Plaster	0.48
Rock wool	0.04
Silver	400
Steel	10–70
Stone	1.33–2.94
Tungsten	160
Wood (across grain)	
Fir	0.11
Oak	0.166
Pine	0.112–0.147

APPENDIX G

Convective heat transfer coefficients

Medium	Heat transfer coefficient
Natural	
air	5–30
water	200–600
Forced	
air	10–500
water	100–210
oil	60–210
Boiling water	210–510

APPENDIX H

Absorptivity (300 K)

Material	α
Clay-white	0.9
Asbestos	0.9
Cork	0.85
Wood	0.85
Porcelain	0.95
Concrete	0.89
Graphite	0.6
Aluminum	0.07
Roof shingles	0.88

APPENDIX I

Soil permeability

Soil	Permeability (k m^3s/kg)
Gravel	10^{-1}–10^{-2}
Sand/gravel	10^{-2}–10^{-5}
Sand/silt/clay	10^{-6}–10^{-9}
Clay	$<10^{-9}$

BIBLIOGRAPHY

1 Artley, J. (1965). *Fields and Configurations*. New York: Holt, Rinehart & Winston.
2 Beranek, Leo L. (1954). *Acoustics*. New York: McGraw-Hill.
3 Burr, A. H. (1981). *Mechanical Analysis and Design*. New York: Elsevier.
4 Fuller, Baden A. J. (1973). *Engineering Field Theory*. Oxford: Pergamon Press.
5 Hayt, Jr, William H. (1981). *Engineering Electromagnetics*, 4th edn. New York: McGraw-Hill.
6 Hoole, S. Ratnajeevan (1989). *Computer Aided Analysis and Design of Electromagnetic Devices*. New York: Elsevier Science Publishing.
7 John, J. E. A. and Haberman, W. L. (1988). *Introduction to Fluid Mechanics*, 3rd edn. Englewood Cliffs, New Jersey: Prentice Hall.
8 Kraus, J. D. (1992). *Electromagnetics*, 4th edn. New York: McGraw-Hill.
9 McCarthy, D. F. (1988). *Essentials of Soil Mechanics and Foundations*. Englewood Cliffs, New Jersey: Prentice Hall.
10 Moon, P. and Spencer, D. E. (1988). *Field Theory Handbook*, 2nd edn. Berlin: Springer-Verlag.
11 Mundi, B. B. and McNabb, J. W. (1980). *Engineering Mechanics of Materials*, New York: Macmillan Publishing.
12 Neff, Jr, Herbert, P. (1987). *Basic Electromagnetic Fields*, 2nd edn. New York: Harper Row.
13 Pao, Y. C. (1986). A First Course in Finite Element Analysis. Newton, Mass.: Allyn & Bacon.
14 Paul, C. R. and Nasar, S. A. (1987). *Introduction to EM Fields*, 2nd edn. New York: McGraw-Hill.
15 Popovic, Branko D. (1971). *Introductory Engineering Electromagnetics*. Reading, Mass: Addison-Wesley Publishing.
16 Sadiku, Matthew N. O. (1989). *Elements of Electromagnetics*. New York: Holt, Rinehart & Winston.
17 Schwab, Adolf J. (1988). *Field Theory Concepts*. Berlin: Springer-Verlag.
18 Seely, S. and Poularikas, A. D. (1979). *Electromagnetics*. New York: Dekker.
19 Shames, Irving H. (1975). *Introduction to Solid Mechanics*. Englewood Cliffs, New Jersey: Prentice Hall.
20 Silvester, P. (1968). *Modern Electromagnetic Fields*. Englewood Cliffs, New Jersey: Prentice Hall.
21 Spiegel, Murray R. (1971). *Advanced Mathematics for Engineers and Scientists*. New York: McGraw-Hill.
22 Thomas, L. C. (1980). *Fundamentals of Heat Transfer*. Englewood Cliffs, New Jersey: Prentice Hall.
23 Weber, Ernst (1950). *Electromagnetic Fields*, Vol. 1. New York: Wiley.

INDEX